Telegeoinformatics

Location-Based Computing and Services

EDITED BY

Hassan A. Karimi

and

Amin Hammad

CRC Press
Taylor & Francis Group
Boca Raton London New York

CRC Press is an imprint of the
Taylor & Francis Group, an **informa** business

CRC Press
Taylor & Francis Group
6000 Broken Sound Parkway NW, Suite 300
Boca Raton, FL 33487-2742

First issued in paperback 2020

ISBN 13: 978-0-367-57840-4 (pbk)
ISBN 13: 978-0-415-36976-3 (hbk)

Visit the Taylor & Francis Web site at
http://www.taylorandfrancis.com

and the CRC Press Web site at
http://www.crcpress.com

Library of Congress Cataloging-in-Publication Data

Telegeoinformatics : location-based computing and services / edited by Hassan A. Karimi, Amin Hammand
p. cm.
Includes bibliographical references and index.
ISBN 0-415-36976-2 (alk. paper)
1. Mobile commuting. 2. Telecommunications systems. 3. Geographic information systems. I. Karimi, Hassan A. II. Hammand, Amin.

QA76.59T46 2004
004.6—dc22 2003064024

Library of Congress Card Number 2003064024

Contents

Part Three: Applications

Chapter Nine: Emergency Response Systems

Chapter Ten: Location-Based Computing for Infrastructure Field Tasks

Preface

We coined the term "Telegeoinformatics" based on the components it is comprised of and the type of processing it supports. The three main components of Telegeoinformatics are telecommunication for mobile computing, thus "tele," geospatial data and processing, thus "geo," and information processing, thus "informatics." The type of information Telegeoinformatics processes in addition to what is currently handled in distributed mobile computing environments is "location based." With these characteristics, we define Telegeoinfomatics as a new discipline emerging as a result of integrating mobile computing, wired and wireless communications, and geoinformatics including Geographic Information System (GIS), the Global Positioning System (GPS), and remote sensing techniques and technologies. In short, we define Telegeoinformatics as a distributed mobile computing environment where location-aware mobile clients are interconnected to other mobile clients, stationary clients, and servers via wired and wireless communication networks and where the bulk of computations is location based. This definition is used throughout the book to discuss the concepts and issues of Telegeoinformatics. With this in mind, Telegeoinformatics is a discipline that encompasses Location-Based Computing (LBC), providing the underpinning computing and communication infrastructure, and Location-Based Services (LBSs), supporting the set of technologies and data to process application-specific requests.

We envision that users of Telegeoinformatics will come from diverse backgrounds and interests wanting to solve a range of simple to complex problems emphasizing location information. Users of Telegeoinformatics range from individuals interested in obtaining such information as nearby locations of objects (e.g., restaurants) while driving, using in-car navigation systems or subscribing to LBSs, to dispatchers interested in obtaining such information as locations of the fleet of vehicles within an area in real time for planning and scheduling (e.g., delivery truck services), to engineers interested in obtaining such information as most up-to-date imagery and GIS data to repair damage on infrastructures (e.g., bridges). Like users, applications of Telegeoinformatics, both existing and emerging, are diverse and widespread. Example applications include transportation, transit, utilities, environmental studies, public health, navigation, planning, and emergency response.

Our goal was to publish a book with potential readers from different backgrounds and expertise to realize the fundamental intricacies of Telegeoinformatics. To achieve this goal, we divided the book into three major parts: Theories and Technologies, Integrated Data and Technologies, and Applications. Each part contains several chapters, each covering an important topic in Telegeoinformatics. We invited authors who are expert in one or more aspects of Telegeoinformatics to contribute. The book is intended to be used by researchers, professionals, and students from different disciplines including computer science, information technology, and engineering.

Although the later parts of the book assume some familiarity with the information introduced in the earlier parts, the chapters can be read in any order depending on the interest of the reader, and cross-referencing among chapters can be used as necessary. If the book is used as a resource in an undergraduate course, we suggest selected chapters in the Integrated Data and Technologies and Applications parts. If the book is used as a resource in a graduate course, we suggest selected chapters in all three parts depending on the background of the students and the specific focus of the course. Below is a brief description of each chapter.

Part 1 contains Chapters 1 to 4 covering the theories and technologies related to GISs and geoprocessing, remote sensing, the GPS, and wireless communication systems.

Chapter 1 provides an introduction to Telegeoinformatics. It discusses geoprocessing and the current trends in stand-alone and Internet-based GISs which may be customized and enhanced to meet the requirements of Telegeoinformatics. The chapter also provides an insight into some new features and capabilities, such as predictive computing and adaptation, which will make Telegeoinformatics easy to use and automatic.

Chapter 2 discusses remote sensing techniques and technologies. It gives an overview of satellite imaging, different satellite remote sensing currently available, and characteristics of different satellite imaging approaches. The chapter also discusses the process of extracting data from satellite images and aerial photos which are then used in GISs and Telegeoinformatics.

Most of Chapter 3 is devoted to the GPS as it is expected to be the predominant positioning technology used in Telegeoinformatics. The chapter describes GPS technology, operation, accuracy, and other characteristics and issues of importance to Telegeoinformatics. In addition to the GPS, the chapter overviews other positioning technologies that may be considered in place of the GPS or for GPS augmentation.

Chapter 4 overviews various wireless communication systems for Telegeoinformatics. Specific characteristics of wireless communication systems are described, and current trends and plans for the future are discussed. The potential use of wireless communication systems in Telegeoinformatics is alsodiscussed.

Part 2 contains Chapters 5 to 8 covering some of the main issues related to integrated data and technologies in LBC, LBSs, mediated reality, and mobile augmented reality systems.

Chapter 5 discusses the concept and issues of LBC. Several topics including LBC architecture, interoperability, especially with respect to geospatial data, and data distribution and processing using advanced new technologies, such as the Geographic Markup Language (GML), are described.

Chapter 6 is devoted to LBSs.The latest developments with respect to LBSs along with emerging applications are discussed. The chapter overviews different technologies supporting LBSs including location-aware devices and spatial database systems. LBS providers and vendors are also referenced.

Chapter 7 introduces the concepts of mediated reality, personal imaging, and humanistic intelligence. The chapter reviews some of the existing wearable systems. An innovative vision-based head tracking technology, called *VideoOrbits*, and how it can mediate the reality of its user by blocking spam, etc. is discussed.

Chapter 8 gives an extensive review of Mobile Augmented Reality Systems (MARS) as a powerful user interface to context-aware computing environments. This chapter introduces MARS and reviews important issues, such as wearable displays, computing hardware, tracking, registration, user interaction, collaboration, heterogeneous user interfaces, and user interface management.

Part 3 contains Chapters 9 to 12 covering several applications of Telegeoinformatics including Emergency Response Systems (ERSs), mobile inspection data collection systems, Intelligent Transportation Systems (ITSs) and a final chapter predicting the impact and penetration of LBSs.

Chapter 9 provides an overview of ERSs as an application area of Telegeoinformatics. The chapter also discusses a mobile location-aware communication support system for information exchange in cases of earthquake hazards and other disasters.

Chapter 10 discusses how Telegeoinformatics can be applied in a mobile data collection system for civil engineering infrastructure field tasks. The requirements and architectures of a framework for this application are described. The potential and limitations of the framework are demonstrated through developing a prototype system and testing it in a case study.

Chapter 11 focuses on the applications of Telegeoinformatics in ITSs. The architecture and development of ITSs, the role of positioning and communication systems in ITSs, and some Telegeoinformatics applications, such as Driver Assistance, Passenger Information and Vehicle Management are reviewed.

Chapter 12 makes general predictions of the impact of LBSs based on statistical data and a simple model of LBS market penetration. It concludes with scenarios for future growth using an analogy between the ways GISs have developed and the diffusion and penetration of mobile devices.

Part One: Theories and Technologies

CHAPTER ONE

Telegeoinformatics: Current Trends and Future Direction

Hassan A. Karimi

1.1 INTRODUCTION

Telegeoinformatics is enabled through advances in such fields as geopositioning, mobile computing, and wireless networking. There are different architectures possible for Telegeoinformatics, but one that is expected to be widely used is based on a distributed mobile computing environment where clients (stationary or mobile) are location aware, that is capable of determining their location in real time, and interconnected to intermediary servers via wired or wireless networks. One of the most distinctive features of Telegeoinformatics is that most computing- and application-related decisions are made based on locations of objects and the relationships among them. These could be locations of fixed objects of interest, stationary clients, or mobile clients in the physical environment. Managing and processing location-related information of such objects require geoprocessing. For this, the core activities in Telegeoinformatics are centered around geoprocessing. However, current geoprocessing techniques supported by today's Geographic Information Systems (GISs), stand-alone or Internet-based, are not all applicable to and efficient for Telegeoinformatics. This is because Telegeoinformatics deals with new challenges and issues not currently addressed in distributed real-time mobile computing environments. In this chapter, first we discuss geoprocessing, in particular with respect to stand-alone and Internet-based GISs that are applicable to Telegeoinformatics, and then we describe in detail new key features and capabilities for Telegeoinformatics that require new research and development activities.

Telegeoinformatics is still in its infancy but advances in several fields, especially geopositioning, mobile computing, and wireless networking are paving the way for its fast emergence. Telegeoinformatics can be based on different architectures depending on the objectives and requirements of applications. A prevalent architecture is envisaged to be a distributed computing environment where both mobile and stationary clients and servers are interconnected via wired and wireless networks. We focus our attention on such an architecture as it comprises all possible components of Telegeoinformatics and deals with much more complex issues and problems than other architectures. Therefore, throughout this chapter Telegeoinformatics refers to a distributed mobile computing environment where the clients are location aware and the core activities center around geoprocessing.

Compared with other distributed mobile computing environments, Telegeoinformatics has the following unique features: (1) mobile devices are equipped with one or more geopositioning technologies, e.g., the Global Positioning System (GPS); i.e., they are location aware, (2) much of the data needs and processing is on spatial objects, (3) managing and processing distributed spatial databases require new techniques, and (4) location-based techniques are needed to make computing- and application-related decisions. These features are possible through geoprocessing, which is the underpinning of many activities in Telegeoinformatics. Geoprocessing provides location-based solutions through spatial queries and computations and is performed in all Telegeoinformatics computing platforms including mobile devices, desktop machines, and high-end servers. Telegeoinformatics require new geoprocessing solutions not currently available in conventional geospatial technologies, predominantly stand-alone GISs and Internet-based GISs. To that end, new methodologies, mathematical and statistical models, algorithms, and computing strategies are needed for efficient and effective Telegeoinformatics implementation. In particular, geoprocessing in Telegeoinformatics must support such new functionality as location-based computing (optimized computation, networking and protocols), Intelligent Query Analyzer (IQA), and predictive computing and services based on different approaches such as on-the-fly data mining.

Telegeoinformatics components are hardware, software, data, and people. Figure 1.1 depicts components of Telegeoinformatics. The hardware component includes mobile devices—sensors, personal digital assistants (PDAs), laptops, etc. –desktops, and servers. However, in special applications where computation-intensive tasks are involved, high-performance computing platforms such as supercomputers (massively parallel processors) may also be used. This hardware equipment is interconnected via Wide-Area Networks (WANs), Local-Area Networks (LANs), and wireless networks.

The software component comprises new geoprocessing capabilities in addition to those supported by conventional GISs and must optimally perform distributed spatial data processing and functionality in real time. The distributed geoprocessing in Telegeoinformatics must handle heterogeneous data, functionality, and computing platforms and be interoperable. Scalability is another feature of the software as the number of clients in typical Telegeoinformatics applications ranges from tens to thousands to millions. In short, Telegeoinformatics software requires new geoprocessing solutions that can handle distributed heterogeneous spatial databases and functionality, performs in real time, is interoperable with respect to data and functions, and is scalable.

Spatial data in Telegeoinformatics comes from a variety of sources. Common spatial data sources in Telegeoinformatics include GPS, remote sensing, existing databases, digitized maps, and scanned maps. Spatial data obtained from these or other sources are either in vector or raster format. Typical Telegeoinformatics applications are expected to include very large spatial databases requiring special techniques for their handling. Furthermore, spatial database systems for Telegeoinformatics must support means for incorporating real-time data into spatial databases.

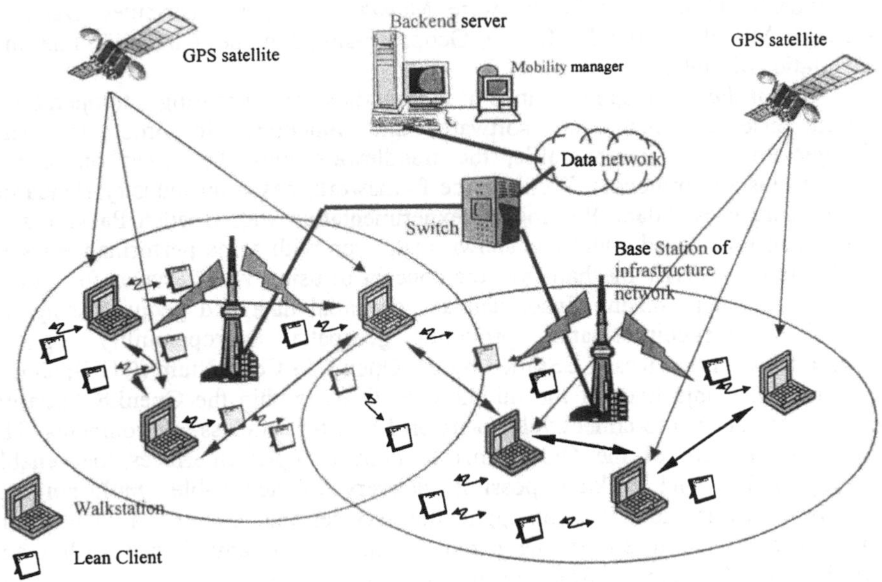

Figure 1.1 Telegeoinformatics Components

The people component includes end users, both traditional GIS users and mobile users, and developers. End users require geoprocessing on mobile devices (walking or driving), desktops, servers, or field data collectors using a range of sensors in the field, or they may be operators specialized in dispatch operations and activities. Developers design and develop Telegeoinformatics applications.

With these components, Telegeoinformatics is a distributed real-time mobile computing environment which is interoperable and scalable. In this chapter, the current trends in Internet-based GISs as they relate to Telegeoinformatics are overviewed but much of the attention is devoted to the future direction. The structure of the chapter is as follows. Section 2 discusses a common Telegeoinformatics architecture. An overview of Internet-based GISs is given in Section 3 followed by an overview of spatial databases in Section 4. In Section 5 the IQA is discussed followed by a discussion of predictive computing in Section 6 and a discussion of adaptation in Section 7. In Section 8 final remarks are given.

1.2 ARCHITECTURE

Telegeoinformatics can be based on different architectures to meet different requirements of applications. Of these, the three-tiered architecture where clients and servers are connected via wired and wireless networks is expected to be the most common one. In such an architecture, Tier 1 is the clients, Tier 2 is the middleware, and Tier 3 is the database engine. The middleware is used for performing many computations and activities and linking the different components. An ideal middleware for Telegeoinformatics comprises the following components (Figure 1.2): Middleware Manager Module, Distributed Database Manager Module, Network Module, Geoprocessing Module, Client Module, and Adaptation Module.

One of the key responsibilities of the middleware is ensuring interoperability among heterogeneous data, software, and functions. In order to make Telegeoinformatics interoperable, the middleware must be based on special mechanisms and protocols. Web service framework, based on industry standards, is an example candidate. Preliminary experimentation with Hewlett Packard's E-Speak and Java-based client has shown that, even with some performance issues related to the security mechanisms, the concept of using Web services for mobile environment is promising. Insomuch as geospatial data and geoprocessing are central to Telegeoinformatics, providing geospatial interoperability, e.g., in Location-Based Services (LBSs) led by the Open GIS Consortium (OGC), should be one of the objectives of the middleware. OGC, within the OpenLS Initiative project, has chosen to collectively focus on LBSs for wireless environments. The OpenLS Initiative of the OGC aims to deliver "open interfaces that enable interoperability and making possible delivery of actionable, multi-purpose, distributed, value-added location application services and content to a wide variety of service points, wherever the person might be, on any device within any application" (OpenLS, 2001). The ultimate aim of the OpenLS Initiative is to provide the consumer market with implemented technology products built with the agreed upon specifications.

OGC has defined the key words of their vision to assure that their focus remains true to the aims and goals of the OpenLS Initiative with the following: interoperability; actionable (users immediately benefit and complexity is hidden); multi-purpose (supports a variety of applications); distributed value-added (enhancing service by exploiting location, location content and location context while maintaining a consumer interest); sService point (the point of location the service is delivered); and platform independent.

OGC is currently pursuing an OpenLS testbed that is expected in its initial version to result in at least nine interfaces that will test and establish the fundamental end-to-end device. This testbed will draw on the resources and knowledge of the OGC sponsors and participants. The interfaces are expected to take into account the following in their design: directory services (query and display); route determination and display; map/feature interact and display; device-location service; content-transcoder service; location content access services (i.e., Web Map Service, Web Feature Service); geocode, geoparse and gazetteer services; coordinate transformation services; multiple location information services (weather-, traffic-, event-based information).

With respect to semantics in GIS databases, research in ontologies in GISs has continually increased in the past few years. Grossmann (1992) defined ontology in the general context which was extended to ontology in geospatial domain building the foundation for semantic interoperability. Semantic interoperability can be defined as the ability of a user to access, consistently and coherently, similar (though autonomously defined and managed) classes of digital objects and services distributed across heterogeneous repositories, with federating or mediating software compensating for site-by-site variations (Ng, 1998).

Although there is currently no solution for semantic interoperability in the domain of geoprocessing, there are works underway on general semantic interoperability and ontologies in the field of GIS; for example, refer to Goodchild (1999) for interoperability in GIS, to Egenhofer (2002) for a general discussion of semantic geospatial Web, and to Fonseca (2002) for a discussion of ontologies for integrated GIS.

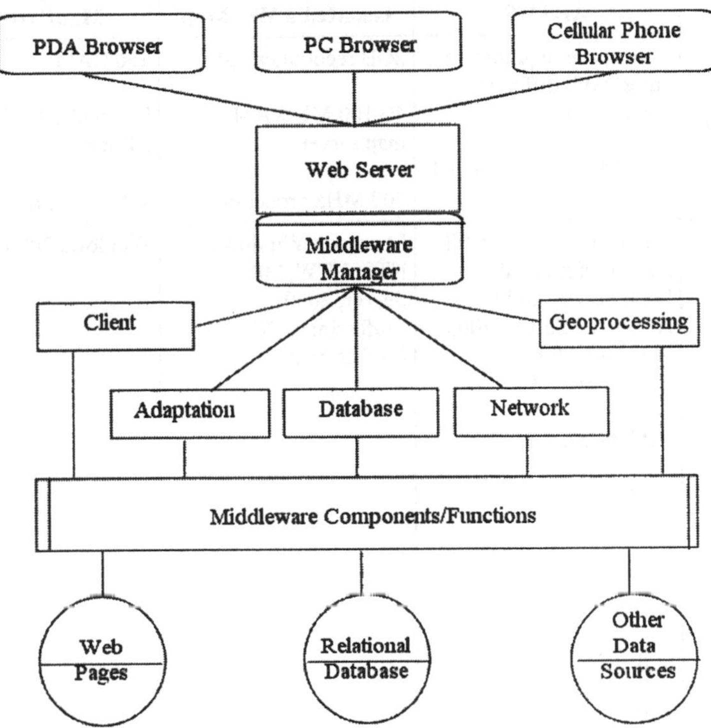

Figure 1.2 A Telegeoinformatics Architecture

1.3 INTERNET-BASED GIS

Internet-based GISs have been growing rapidly for the past several years. These systems can provide simple to complex spatial solutions to a wide range of problems available through the Internet. While early developments were mainly for users to access and process mapping functions through Web browsers on desktop machines, latest developments are allowing mobile users, in addition to desktop Web users, to perform mapping and GIS functions using mobile devices. Wireless networks are the fundamental enabling technology for emerging Internet-based GISs providing access and solutions to both desktop and mobile users. Current Internet-based GISs that support mobile location-aware devices, interconnected with servers and other mobile devices via wired and wireless networks, can be customized, enhanced and extended (new features and capabilities) to meet the requirements of Telegeoinformatics.

Table 1.1 Server Requirements

	ESRI ArcIMS	**Intergraph GeoMedia WebMap**	**MapInfo MapExtreme**
Hardware	Disk space requirement varies with different components 256 RAM recommended	2GB recommended 40-100 MB RAM per map server 800 MHz processor	256 RAM 9-200 MB for Java edition 400 MHz processor
Operating System	Microsoft Windows NT, 2000 Professional, 2000 Server, 2000 Advanced Server, 2003 Standard, 2003, Enterprise , XP HP HP-UX IBM AIX Linux-Intel Red Hat, Red Hat Advanced Server SGI IRIX SUN Solaris	Microsoft Windows NTS, NTW, 2000 Server, 2000 Professional, XP Professional	Windows 2000, NT
Web Server	Microsoft IIS Apache Web Server IBM HTTP Web Server SUN ONE iPlanet Oracle Application Server WebLogic	Microsoft IIS	Microsoft IIS

Today there are many Internet-based mapping and GIS software packages available in the market. For readers unfamiliar with these systems and their potential use in Telegeoinformatics, three off-the-shelf systems are overviewed in this section. These systems are ArcGIS by ESRI, GeoMedia WebMap by Intergraph, and MapXtreme by MapInfo. Each of these systems has its own proprietary functionality, compatible data formats, and skill level. Tables 1.1-1.4 summarize the requirements, features, and capabilities of these three Internet-based GISs. The data in tables 1.1-1.4 are compiled from the information available on the Web sites of the vendors. Empty boxes in the tables indicate that the data for those specific items were unavailable on the Web sites at the time of writing of this chapter.

Table 1.2 Client Requirements

	ESRI ArcIMS	**Intergraph GeoMedia WebMap**	**MapInfo MapExtreme**
Operating System		Microsoft Windows 2000 Microsoft Windows NTW 4.0 (Service Pack 5 recommended) Microsoft 98	
Viewer	Microsoft IE 5.0 or later Netscape Communicator 4.61 or later	Microsoft IE 5.0 or later Netscape Communicator 4.0 or later	Microsoft IE Netscape Communicator (6.0 and 6.1 only)
Plug-ins	JRE required to view ArcIMS Java Viewer based Web sites	InterCAP ActiveX control for ActiveCGM version 7.0sp02 or higher (Vector map for IE) InterCAP ActiveCGM Plugin version 7.0sp02 or higher (Vector map for Netscape)	

Table 1.3 Capabilities

	ESRI ArcIMS	**Intergraph GeoMedia WebMap**	**MapInfo MapExtreme**
Data Format	Arcview Shapefiles ArcInfo Coverages ArcSDE Features ADRG, BIL, BIP, BSQ, BMP, CIB, ERDAS, GeoTIFF, GIF, GRID, IMAGINE, IMPELL, JPG, MrSID, NITF, SUN, TIFF, USRP	ArcView, ArcInfo, MapInfo, GeoMedia Smartstore, AutoCAD, Microstation, MGE, MGSM, MGDM, FRAMME, Access, Oracle Object Model, Oracle Relational Model, SQL Server, ODBC Tabular, GML, WFS	MapInfo TAB files and .gst files Oracle Spatial SpatialWare for SQL Server SpatialWare for Informix Spatial Ware for DB2 MrSID, ECW, Vertical Mapper Continousus Grids, Vertical Mapper Classified Grids, ADRG, CADRG, CIB, ASRP, NITF
Vector Map Posting	Shapefile, ArcSDE layers	CGM graphics file	
Raster Map Posting	JPEG, GIF, PNG	Active CGM, JPEG, PNG	JPG, GIF, TIF, PNG, WMF, BMP, PSD
Sample Functions	Scale-based display of features	Scale-based display of features	Scale-based display of features
	Programmable actions using developing tools	Programmable actions using developing tools	Programmable actions using developing tools
	Annotations using different fonts, symbols, and labels. Support multi-language font	Annotations using different fonts, symbols, and labels. Support multi-language font	Annotations using different fonts, symbols, and labels
		Visualize data with thematic mapping	Visualize data with thematic mapping
	Combine data from different sources	Combine different map projection and coordinate systems into one	

	Geocoding in street address, Zip Code-level		Geocoding in Zip Code-level
	Scalable architecture		Scalable architecture
	Secure access to map services		
	Integration with ESRI's ArcGIS Desktop products		
		Create, combine, buffer, or erase features	Create, combine, buffer, intersect or erase features
			Write, insert, update data from *MapXtreme* to Oracle8*i* Spatial and MapInfo *SpatialWare*
			Full support of coordinate systems and map projection
			Publish web site remotely via FTP

Table 1.4 Client Functionality

		ESRI ArcIMS		Intergraph GeoMedia WebMap	MapInfo MapExtreme
		All Viewers	**Java Viewer**		
Mapping	**Zoom to Full Extent**	√	√	√	
	Zoom To Active Layer	√	√		
	Zoom In	√	√	√	√
	Zoom Out	√	√	√	√
	Zoom To Particular Width				√
	Previous Extent	√	√	√	
	Next Extent		√		
	Pan	√	√	√	√
	Pan One Direction	√	√	√	
Query	**List Attribute Feature**	√	√	√	
	Feature Identification	√	√	√	√
	Measure Distance	√	√	√	√
	Measure Cumulative Distance			√	
	Measure Area			√	
	Set Map Unit	√	√	√	
	Set Map Projection			√	
	Build Query	√	√	√	√
	Pinpoint Address	√	√	√	√

	Click and Drag To Select a Group Of Features	√	√		
	Find Map Features From Text	√	√	√	√
	Insert Buffer Zone	√	√	√	√
	Add Layers		√		
Project	**Open Project**		√		
	Close Project		√		
	Save Project		√		
	Copy Map Image		√		
	Print	√	√		√

Users of Internet-based GISs can benefit from lower costs since many systems are designed and offered as services for a set of target users (Sumrada, 2002). This is possible since those who host the server(s) tend to absorb the cost of the maintenance and implementation while the users are able to utilize the system targeted for their use for a much lesser cost. In addition, the distribution of the application can in turn reduce processing load for the system administrator. If the client has ample computing resources, it could take responsibility for performing some of the tasks the system must implement. Internet-based GISs are not only capable of reaching a wider audience, but also are capable of accommodating users of various skill levels. This is to say that while stand-alone GISs are typically designed for expert users, Internet-based GISs are for both novice and expert users.

Internet-based GISs offer other benefits in addition to be expected by only making GIS functionality available over the Internet (Limp, 1999). Also as many benefits become available with Internet-based GISs, so do many new issues. One major issue is related to the particular design upon which the Internet-based GIS is based. The two common design options for an Internet-based GIS are *client-side* and *server-side*, both based on a client/server architecture (Gifford, 1999).

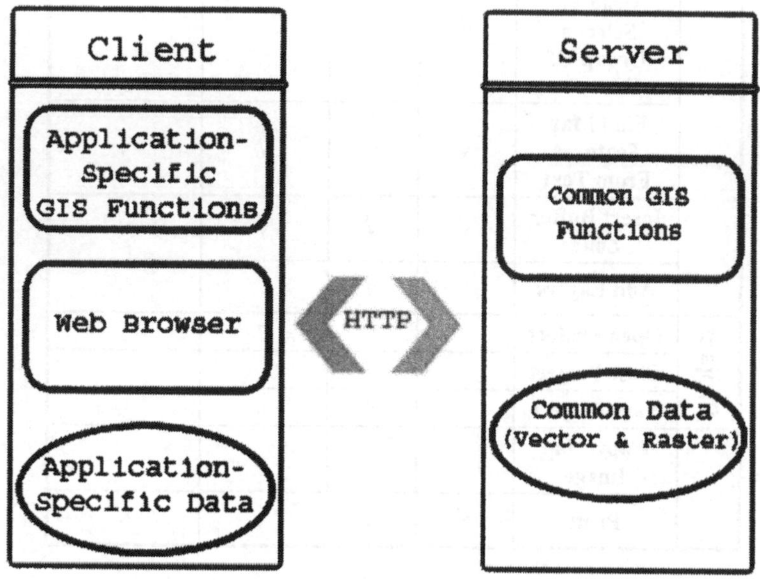

Figure 1.3 The Client-Side Design

In the *client-side* design (see Figure 1.3), the responsibility of the server is limited primarily to data storage and retrieval operations. The client is enabled to support GIS functionality, which allows the generation of spatial requests and presentation of the response. In this design, the client takes the role of centralized GIS application and the server takes the role of centralized GIS data repository. The client performs most of the computations, including data rendering, provided that it has the adequate power and resources. In doing so, a process or application, which is implemented by enhancing the Web browser with a Java applet, ActveX, or plug-ins, is run on the client. In case of a Java applet, the code, which provides a Graphical User Interface (GUI) for the GIS application, is transferred to the Web browser. Upon requesting a GIS operation, the required data sets (which mostly are in the vector format) to implement the operation are retrieved from the server and transferred to the client for geoprocessing.

In the *server-side* design (see Figure 1.4), most of the computations are performed on the server, relieving the processing burden from the client but can also become a bottleneck. In addition to storage and retrieval operations, the server is responsible for computations and rendering data. The *server-side* design is suitable for mapping applications (major Internet portals support *server-side* Internet-based mapping). In this design, the client sends a request to the server where it is processed and receives the results as an embedded image in a HTML page via standard HTTP. Data transmitted to a client are in the standard HTML format viewable by using any Web browsers. By having all the major software and all the spatial and attribute data and processing on the server, the design would allow for supporting the highest possible performance given the available resources.

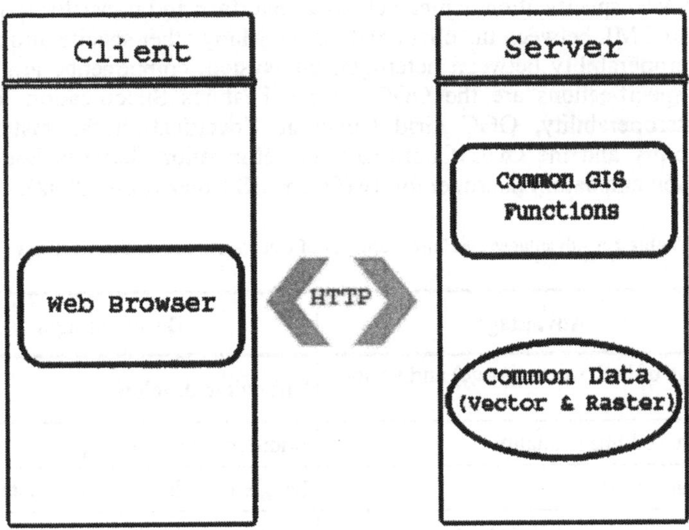

Figure 1.4 The Server-Side Design

There are advantages and disadvantages to both the *client-side* and *server-side* designs. Table 1.5 summarizes the advantages and disadvantages of each design. The prevailing concern in either case is the static processing in most systems, which is fixed. Thus, the server can become congested in the *server-side* design or the client may not have the required processing power in the *client-side* design. There is also the issue of concurrency control, in both designs, where several users try to update one data set simultaneously. Users spread across the Internet are not the ones who would need to ensure that the database stays in a consistent state if the same data set is concurrently updated. Because data storage and retrieval operations are the responsibility of the server in either design, the database management system on the server must be equipped with techniques and tools to handle such issues as update concurrency.

Interoperability plays a major role in Internet-based GISs, which is the ability of existing and emerging systems to work with one another. While each of these systems can function without difficulty when using their own proprietary data formats and functionality, they may not be able to use another data format or function from another system. For example, the ArcXML data format proprietary of ArcIMS (a component of ArcGIS) applications may not work inherently with other software such as Intergraph's WebMap. However, WebMap can use messages from an ArcIMS map server with the aid of specifications described by the OGC. The emergence of the Geographic Markup Language (GML) as a standard for encoding vector data is expected to promote interoperability among

Internet-based GISs. For example, through an ArcIMS' adaptor, data from ArcXML can be converted to GML which can be read by another application supporting OGC's GML specification. The generation and recognition of GML by heterogeneous map servers and clients is made possible by the Web Mapping Service (WMS) specification, which acts as an interface and translates requests and responses to GML between the client and server. Many other specifications exist to ensure interoperability between heterogeneous system components, among those approved specifications are the OGC Simple Features Specification for vector dataset interoperability, OGC Grid Coverage Specification for raster dataset interoperability and the OGC Coordinate Transformation Services Specification for projection and datum interoperability (Open GIS Consortium, 2002).

Table 1.5 Advantages and Disadvantages of Client- and Server-Side Designs

	Advantages	**Disadvantages**
Client-side	Usage of both raster (image) and vector data	Difficult to develop
	Enhanced image quality	Additional software required
	Enhanced GUI	Longer download times for data
	Can cache datasets for reuse	Performance hindrance for large datasets
		Possibility of lack of adherence to standards
		Processing dependent on client machine's capabilities
		Requires more skill from users
		Platform/browser incompatibility
Server-side	Simple to develop	Low-performance GUI
	Easy to display	Low image quality
	Easy to maintain	Requires each request to be processed by server
	Adheres to standards	Performance depends heavily on network conditions
	User of standard web browsers	Ignorant of client machine's capabilities
	Low bandwidth required	

1.4 SPATIAL DATABASES

Most current GIS databases comprise two components: a Spatial Database Engine (SDE) component and a Database Management System (DBMS) component. The SDE component is for storing, managing, and processing spatial data and the DBMS component for storing, managing, and processing attribute data. GIS software packages use proprietary data structures in their SDEs. Many GIS software packages use or provide interface to standard DBMSs, such as Oracle, IBM, Informix, and MS Access. With the separation between the two database components, a module is required to link the data in the databases when answering a query requires data from both databases. For example, to answer geometric queries only the SDE component is required, while to answer queries such as the name of a house at a given location requires that the data in the SDE component be linked to the data in the DBMS component.

Spatial database systems are increasingly gaining popularity, especially because of the emergence of LBSs. Spatial database systems are at the heart of LBSs and perform many tasks that otherwise may be performed inefficiently and ineffectively with conventional GIS software packages. Contrary to GIS databases where there are two database systems, SDEs and DBMSs, the trend in spatial database systems is to have both spatial and attribute data in one single database. Such a single-database approach has several benefits including seamless integration of spatial and attribute data, better spatial data management by reducing complexity of system management, maintaining data integrity like any other database system, and extending geospatial functionality.

Database technology vendors have begun to realize the importance and benefits of spatial databases; some vendors have been offering spatial databases as their core products. Table 1.6 compares four spatial database systems by four database technology vendors.

Spatial database systems in Telegeoinformatics establish the basis of geoprocessing. However, their efficient and effective use requires that they be able to handle both vector and raster data. More specifically, with respect to vector data processing they must support such common functions as on-the-fly geocoding, mapping, and routing and with respect to raster data processing they must support such common functions as georeferencing, edge detection and extraction, classification, and change detection. These functions must perform for real-time response and automatically. Furthermore, certain Telegeoinformatics applications rely on 3D modeling and analysis requiring spatial database systems that support 3D data structures and 3D functions. An example is a Telegeoinformatics application utilizing an ad hoc network where the dynamic location of mobiles at all times is used in optimal network routing or hand-offs; the database system must contain terrain data (e.g., cliffs) and 3D objects (e.g., buildings) on the terrain and be capable of specific 3D computations such as line-of-sight analysis.

Table 1.6 Comparing Four Spatial Database Systems (adapted from Lemmen, 2002)

Company		**Computer Associates**	**IBM/ Informix**	**Oracle**	**Sybase**
Product Name		Ingress II	Informix Dynamic Serer	Spatial	Adaptive Server enterprise 12.5 with Spatial Query Server Option
Data Model	**Vector**	Yes	Limited	All spatial data types	Yes
	Raster	Limited through Binary Large Object Feature	Limited	.jpg, .gif, .bmp, .geotiff	No
Data Processing	**Vector**	Yes	No	Yes	Yes
	Raster	No	No	Yes	No
Spatial Indexing		R-Tree	R-Tree & B Tree	Quad Tree & R-Tree	Quad Tree
Topology Processing		Limited	Limited	Limited	Minimal
3D Processing		Yes	No	Yes	Yes
Temporal Processing		No	No	Yes	Yes
Internet Access		Yes	No	Yes	Yes

1.5 INTELLIGENT QUERY ANALYZER (IQA)

With current GISs, stand-alone or Internet-based, a user submitting a query from a client device, stationary or mobile, requires to know in advance the various data sets needed to implement the query. However, users are often unsure of exactly what data sets are required and where they can be found. This results in time-consuming trial and error and in outcomes that may not be optimal. In Telegeoinformatics this problem is exacerbated as there are different interfaces at different clients and processing is required in real time. The concept of the IQA in Telegeoinformatics is for overcoming this problem.

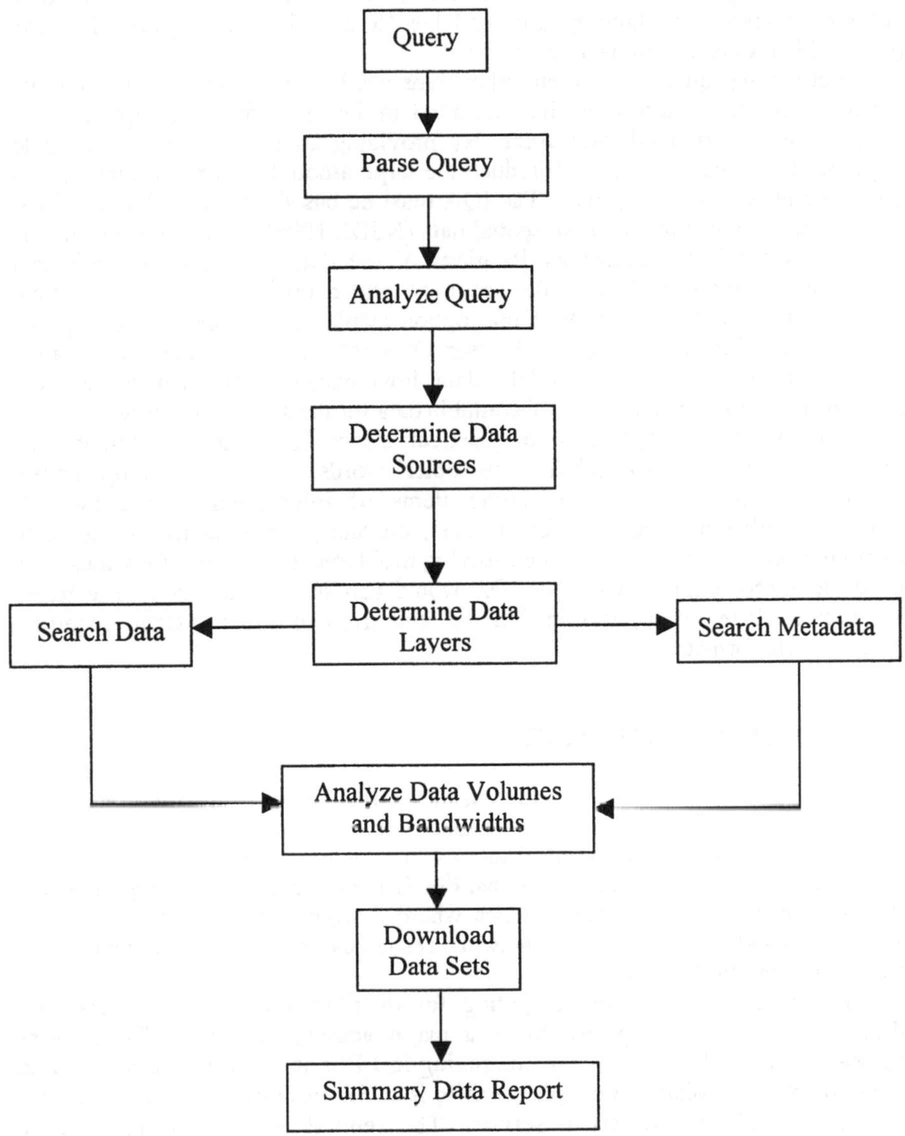

Figure 1.6 Steps of the IQA

Upon submission of a query through one of the devices, the IQA will analyze the query for the data sets required, analyze the networking options to determine the fastest access to those data sets at the current time, and consult the cache and other sources for any previously derived result sets that may apply. The steps of the IQA are depicted in Figure 1.6 and are: (1) parse and analyze the query to determine data sources required, (2) determine the data sets required, (3) search metadata and spatial data clearinghouses for the data sets, (4) analyze the data volumes and the networking options available, (5) download the required data, and (6) provide a summary of data downloaded.

For parsing queries, different algorithms need to be developed, as there are various ways that queries can be submitted in Telegeoinformatics applications. These include text, GUI, and voice. By providing users with data required to implement queries, the IQA will reduce the large amount of time currently spent on searching and gathering data. The IQA must be based on acceptable standards in metadata and data transfers for spatial data (NSDI, 1998). Through metadata, the IQA will identify the repository locations of the data needed along with data quality information (such as scale, accuracy, and date), and choose an optimal communication link, if more than one option available, to transmit the required data to the machine determined by the user. Once the required data are downloaded to the local client, a summary of the data downloaded is provided to the client machine so that there is a record of available data for the submitted query.

Clearly, for the IQA to be of practical use in Telegeoinformatics, it must support semantic interoperability. In other words, semantic interoperability facilitates transparent cross-correlating items of information across multiple sources to solve problems. Different users, domains, and systems use different terminologies and conventions when solving problems in Telegeoinformatics. The semantic interoperability aspect of IQA would handle queries posed by different users, from different application domains and from different systems requiring different data formats.

1.6 PREDICTIVE COMPUTING

Current stand-alone and Internet based GISs are primarily interactive environments, with some custom designed for real-time applications. Current systems are capable of providing solutions to simple and complex problems but lack capabilities to predict computations. Predictive computing is an important and attractive feature of Telegeoinformatics where it would be used for a variety of purposes including computer and network resources optimization, optimal query processing, and fault avoidance.

Research on predictive computing for distributed mobile computing has already begun and is expected to be a major activity for LBSs. The primary purpose of research on predictive computing in LBSs is to overcome networking and computing problems. A simple scenario is as follows (see Figure 1.7). At time t_0 a mobile client requests a service. The request along with its location information (which is determined by GPS or other positioning sensors) is sent via the wireless network to the server. Upon receiving this information, the server will process the request and will send the results back to the user, again via the wireless network. While this sequence of activities is what is expected to be performed by today's LBSs, future LBSs are expected to predict upcoming computations facilitating means for users to access information at any location and any time by overcoming computing and networking problems. To that end, the basis of predictive computing, e.g., in LBSs, is to predict future locations where the mobile client will visit. For this, location prediction has become the focal point of the current research in predictive computing. However, much of the current research is based on approximations using grid or cell structures. These structures do not always result in accurate and reliable solutions because they can only predict location of the mobile within the extent of a grid or cell.

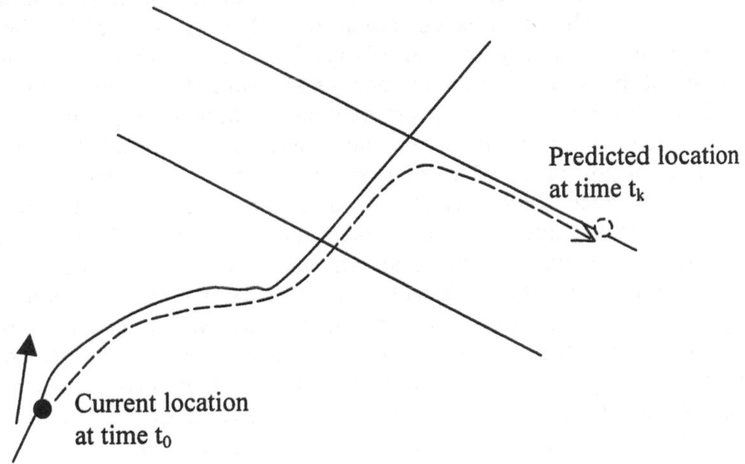

Figure 1.7 Location Prediction in Telegeoinformatics

We envision a new approach for predicting location in LBSs which results in more accurate and reliable solutions. This new approach is based on new models that take into account such information as current location of the mobile, map database, especially geometric and topological information, travel patterns, and auxiliary information obtained from other sources.

1.7 ADAPTATION

As mentioned earlier, hardware, software, data, and people constitute the components of Telegeoinformatics where each is subject to a large number of variations resulting in a large number of combinations. When the components and their respective variations are considered together, compatibility and optimization become a challenging problem. To overcome this challenge, adaptation is considered in Telegeoinformatics. Adaptation in Telegeoinformatics is a set of methodologies and algorithms that assist the heterogeneous components achieve compatibility and optimization. Without adaptation, all Telegeoinformatics components would be required to adhere to strict standards to achieve compatibility and optimization, which is a daunting task.

One adaptation strategy in Telegeoinformatics is to adapt to the client machine the user would use to access the system. There are several client variations: a PDA, a desktop PC, a laptop computer, a cell phone. Differences include storage capacity, processing power, and user interface. Storage capacity can vary both in terms of permanent storage and temporary storage. By having knowledge about the client's storage capacity and the requirement of a specific query, the system can provide the user with alternatives such as storage compression or use of another client with adequate storage capacity. In a similar way, the system can adapt to different response times accommodating the different processing powers of different clients. With respect to user interface, both input and output interfaces should be taken into account. For instance, if the client is a cell phone equipped with only a numeric keypad, Telegeoinformatics should have knowledge about this and adjust its interactions according to possible options with a numerical keypad. Similarly, the client may have a limited output interface. For example, the client with a cell phone may only have a two-color display capable of displaying a small portion of what the user requires to view. Telegeoinformatics should have knowledge about this client's limitation with respect to the output interface and adjust its output information presentation to the capabilities available.

Heineman (1999) has analyzed and evaluated various adaptation techniques for software components. One such technique is active interfaces. This technique acts on port requests between software components, which is where method requests are received. These requests may be augmented, replaced or denied either before or after the request execution. It is through these ports that decisions can be made to adapt the behavior of the component. Another technique is Automatic Path Creation (APC) which is a data format and routing technique that allows multi data format adaptation and adapts to current network conditions (Zao and Katz, 2002). An APC service instance establishes a data path from the requesting end of the system to the response end and contains the necessary operators to convert data from a format by the responding component to a format understandable by the requesting component. These services also contain support for client mobility and maintenance of user perceived Quality of Service (QoS).

Much research is focused on LBSs and the ability of the network to route resources to the location of the user's client. The goal is to supply the client with the necessary bandwidth and resources in a WAN (Kubach and Rothermel, 2001). Movement Prediction with Information Hoarding is a technique that places islands of high bandwidth with info-stations throughout a low bandwidth WAN. The information requested by the user is hoarded (retrieved) near the islands of high bandwidth as much as possible while the user moves throughout the network until the next high bandwidth island is encountered. The resources are directed towards the island that is determined statistically the user will most likely to visit.

Layered Information Retrieval is a technique that goes hand-in-hand with Movement Prediction with Information Hoarding but deserves individual attention (Ye *et al.*, 1998). Whenever the user approaches the info-stations in the high bandwidth islands, the client's speed is taken into account for determining the level of detail in maps the user has requested. If the user is moving quickly, the system will return less detailed information in the form of basic layers. Upon slowing down, the system adapts by responding with a greater number of detailed layers underneath the basic layers. The assumption used in the system is that a user moving quickly will need basic information such as boundaries and roads, while a user moving slow will need more detailed information.

Heineman and Ohlenbusch (1999) have contrasted adaptation with evolution and customization. Adaptation is not the modification of components by system designers; this is considered component evolution. Adaptation should be accomplished automatically by the system with little user intervention (Stephanidis, 2001). Any manual modifications by the user would again be considered customization. Instead adaptation should be considered modifications made by the application or application designer utilizing the component.

In Telegeoinformatics, like many other information systems, most adaptation strategies center around the user component. For this, the need of the broadest end user population should be considered in designing adaptation strategies. It is important to note that adaptation should not be a single system design for these end users, since a single design would violate the premise of adaptation. Adaptation should be considered a design capable of adapting to users with respect to the user's needs (Stephanidis, 2001). These user needs can be interpreted as a barometer, which the adaptation module of Telegeoinformatics should meet.

With a special emphasis on the user component in developing adaptation strategies in Telegeoinformatics, personal profiles for each user to store preferences and behaviors are needed. Users who commonly use a system expect a certain QoS, while new or infrequent users who do not know what to expect from the system may accept a poor system performance. User preferences stored in a user profile can allow the system to adapt to their preferences without requiring the user or the system to commit repeated actions. One possible adaptation technique in Telegeoinformatics is a design known as the Universal Inbox which allows the user to set a preferred contact method in the system, through which it can adapt to the incoming calls for a user to their specified contact method (Ramen *et al.*, 2000).

To accommodate user-oriented adaptation in Telegeoinformatics, the user's interactions with the system should be taken into account. These interactions can be collected as either explicit preferences expressed by the user or implicit behaviors expressed by the user. There are differences between preferences by different users. Telegeoinformatics must be able to adapt to different users rather than support one design for all users. User preferences would be explicitly set by the user first time he/she uses the system. Common user preferences include default language, desired screen name, desired access levels and subscribed services. User behaviors, on the other hand, would be gathered by the system through interactions with the user and without the user's explicit actions (or awareness). Typical user aggregated behaviors in Telegeoinformatics applications include commonly visited locations or a common moving speed. With such an adaptation for the user component, Telegeoinformatics would adapt to the user instead of the user attempt to adapt to the system.

There are also variations on the network the client is equipped with. For example, there are two types of networks to establish communications, a packet switched network and a circuit switched network. The network component within the system can also vary based on available bandwidths and disconnections. Telegeoinformatics should be able to detect a client's disconnection from the network and reconnect it immediately. The amount of bandwidth fluctuates in proportion to the number of user requests and the returning responses, which require bandwidth to move information throughout the network. There should be provision for adaptation allowing the required bandwidth to be available according to the request/response traffic generated by the users.

The server in Telegeoinformatics is where the middleware is housed which is responsible for performing most computations and activities of the system including communication between the different components. An example of adaptation in the servers is to route user requests to another server capable of meeting the user's expected QoS. The middleware should handle the decisions of adaptation between the client and server components. The middleware will make all decisions regarding the processing of user requests and system responses. One of these decisions includes the best location (client or server) to process the user requests. For instance, if the user has a high-end personal computer and the server is presently responding to several other clients, the server could send most of the raw data processing to be processed on the client. Conversely, if the client machine was a low-end PDA with no hard disk and 64 MB of memory, the server could allocate the resources to itself to process this request. The result is optimal utilization of system resources.

The information component can vary in terms of format and size. Due to the nature of data, that is spatial data, required in Telegeoinformatics there are many possible data formats and structures. Each client in Telegeoinformatics may support only a small subset of all possible data formats. For example, a client in Telegeoinformatics may be equipped with a geoprocessing software which supports only the shapefile data format to import and export data. Telegeoinformatics should have information about this limitation and take it into account when sending data to and receiving it from this client.

1.8 FINAL REMARKS

Telegeoinformatics introduces new challenges and issues not addressed in other distributed mobile computing and applications. Although certain existing techniques and technologies of stand-alone GISs, Internet-based GISs, and distributed mobile computing applications can be applied to Telegeoinformatics, customizations and new developments are needed in order to meet the specific requirements of Telegeoinformatics. In this chapter, some of the unique features of Telegeoinformatics that call for new research and development were discussed. These features include the IQA, predictive computing, and adaptation.

REFERENCES

Egenhofer, M, 2002, Towards the Semantic Geospatial Web, in *Proceedings of the 10 th ACM International Symposium on Advances in Geographic Information Systems*, November 8-9, McLean, VA.

Fonseca, F., Egenhofer, M., Agouris, P. and Camara, G., 2002, Using ontologies for integrated geographic information systems, *Transactions in GIS*, 6(3), pp. 231-257.

Gifford, F., 1999, Client v Server—Which Side is Right for You? Online. Available HTTP: <http://www.geoplace.com/ma/1999/0899/899foc.asp> (accessed 4 October, 2002).

Goodchild, M., Egenhofer, M., Fegeas, R. and Kottman, C. (Eds.), 1999, Interoperating Geographic Information Systems, (Boston: Kluwer).

Grossmann, R., 1992, *The Existence of the World: An Introduction to Ontology*, (New York: Routledge).

Heineman, G.T., 1999, An Evaluation of Component Adaptation Techniques, in *2 nd ICSE Workshop on Component-Based Software Engineering*, Orlando, FL.

Heineman, G.T. and Ohlenbusch, H.M., 1999, *An Evaluation of Component Adaptation Techniques*, Technical Report WPI-CS-TR-98-20, Department of Computer Science, Worcester Polytechnic Institute, Worcester, MA.

Kubach, U. and Rothermel, K., 2001, Exploiting Location Information for Infostation-Based Hoarding, in *Proceedings of the Seventh ACM SIGMOBILE Annual International Conference on Mobile Computing and Networking (MobiCom 2001)*, Rome, Italy, pp. 15-27.

Lemmen, C., 2002, Product Survey on RDBMS, *GIM International*, 2(16), pp. 46-47.

Limp, F.W., 1999, Web Mapping. Online. Available HTTP: <http:// www.geoplace.com/gw/1999/1199/1199lmp.asp> (accessed 17 October, 2002).

Ng, D., 1998, Semantic Interoperability for Geographic Information Systems. Online. Available HTTP: <http://ai.bpa.arizona.edu/tng/pub/DLI98_Berkeley/> (accessed 21 October, 2002).

NSDI (National Spatial Data Infrastructure), 1998, Content Standard for Digital Geospatial Metadata Workbook. Online. Available HTTP: <http:// www.fgdc.gov/standards/documents/standards/remote_sensing/MetadataRemote SensingExtens.pdf> (accessed 4 January, 2003).

Open GIS Consortium, Inc. 2002, Online. Available HTTP: <http://www.opengis.org/> (accessed 16 December, 2002).

OpenLS (Open Location Services)., 2001, Initiative. Online. Available HTTP: <http://www.openls.org/index.htm> (accessed 16 December, 2002).

Ramen, B., Katz, R.H. and Joseph, A.D., 2000, Universal Inbox: Providing Extensible Personal Mobility and Service Mobility in an Integrated Communication Network, in *Proceedings of the Third IEEE Workshop on Mobile Computing Systems and Applications*, December 7-8, Los Alamitos, CA, pp. 95-106.

Sumrada, R., 2002, Towards distributed application of GIS technology, *GIM International*, 16, pp. 40-43.

Stephanidis, C., 2001, Adaptive Techniques for Universal Access, *User Modeling and User-Adapted Interaction*, 11(1-2), pp. 159-197.

Ye, T., Jacobsen, H.A. and Katz, R., 1998, Mobile Awareness in a Wide Area Wireless Network of Info-Stations, in *Proceedings of the Fourth Annual ACM/IEEE International Conference on Mobile Computing and Networking*, Oct 25-30, Dallas, TX.

Zao, M. and Katz, R., 2002, Achieving service portability using self-adaptive data paths, *IEEE Communications Magazine*, 40(1), pp. 108-114.

CHAPTER TWO

Remote Sensing

Jonathan Li and Michael A. Chapman

2.1 INTRODUCTORY CONCEPTS

Remote sensing has arguably emerged as the most important discipline in Telegeoinformatics employed in the collection of spatially related information for use in geospatial databases. This chapter introduces the fundamental principles of remote sensing from a Telegeoinformatics perspective. It reviews the nature of electromagnetic radiation and how the reflected or emitted energy in the visible, near-infrared, middle-infrared, thermal infrared, and microwave portions of the electromagnetic spectrum can be collected by a variety of sensor systems. Emphasis in this chapter is placed on the extraction of thematic information using digital image interpretation techniques and the extraction of metric information using digital photogrammetric techniques from remotely sensed data. This chapter shows non-experts what remote sensing is and what imaging software does and provides them with sufficient expertise to use it. It also gives specialists an overview of these totally digital processes from A to Z. Along with other Telegeoinformatics technologies, remote sensing represents the primary means of generating data for geospatial databases.

The success of any Telegeoinformatics application depends on the quality of the geospatial data used. Collecting high-quality geospatial data for input to geospatial databases is therefore an important activity. Traditionally, geospatial data can be collected directly in the field using in-situ (ground surveys) methods. This type of data collection normally makes use of an instrument that measures a phenomenon directly in contact with the ground. In-situ data collection can be expensive because it is labor-intensive and time-consuming. Today, remote sensing is a preferred method to use if data covering a large area are required for a Telegeoinformatics application.

2.1.1 What is Remote Sensing?

Remote Sensing is the science and technology of gathering and processing reliable information about objects and the environment without direct physical contact. Remote sensing data can be analog or digital in form as well as small or large in scale, according to the type of sensor and platform used for acquiring the data. The digital form of remote sensing data is preferred because a large amount of data can be directly processed in a computerized environment. It should be noted that remote sensing is not only a data-collection process, but also a data analysis process for extracting meaningful spatial information from the remote sensing data for direct input to the geospatial databases to support various Telegeoinformatics applications.

The advantage of remote sensing is that it provides a bird's-eye view, so that data covering a large area of the earth can be captured instantaneously and can then be processed to generate map-like products. Another advantage of remote sensing is that it provides multispectral and multiscale data for the geospatial databases. However, the adoption of remote sensing does not eliminate in-situ data collection, which is still needed to provide ground truth to verify the accuracy of remote sensing. Also, as the earth and its environment constantly change, spaceborne remote sensing provides the easiest means to keep the geospatial databases up to date.

Table 2.1 Major Milestones in the Development of Remote Sensing

Year	Innovative Developments
1960	The term "remote sensing" was first introduced by the personnel from U.S. Office of Naval Research.
1960s	Side-Looking Airborne Radar (SLAR) and higher resolution Synthetic Aperture Radar (SAR) were declassified and began to be used for civilian mapping applications.
1972	NASA launched Landsat-1 carrying a Multispectral Scanner (MSS). Remote sensing has been extensively investigated and applied since then.
1965	International Society for Photogrmmetry and Remote Sensing (ISPRS) started to publish *ISPRS Journal of Photogrammetry & Remote Sensing*.
1982	NASA launched 2nd generation of Landsat sensor – Landsat 4 carrying Thematic Mapper (TM) and MSS.
1986	France launched first SPOT satellite carrying a twin High Resolution Visible (HRV) sensor. Since then, Landsat MSS and TM, and SPOT HRV have been the major sensors for data collection for large areas all over the world.
1980-1990	Earth resources satellites were launched from other countries such as India, Japan, Russia, and China were launched. A new type of sensors called an imaging spectrometer has been developed.
1990 -	Earth Observation Systems (EOS) aiming at providing data for global change monitoring and various sensors. Several SAR imaging satellites (e.g., Japanese JERS-1, ERS-1, and Canada's RADARSAT-1) have been proposed.
1995	Canada launched RADARSAT-1 SAR imaging satellite. India launched IRS-1C (5 m resolution).
1998	ASPRS published *Manual of Remote Sensing – Radar*.
1999	NASA launched Landsat-7 carrying an Enhanced Thematic Mapper Plus (ETM+). Space Imaging launched the world's first high-resolution commercial imaging satellite – IKONOS providing 1 m panchromatic and 4 multispectral imagery. American Society for Photogrammetry and Remote Sensing (ASPRS) published *Manual of Remote Sensing – Geoscience*.
2001	DigitalGlobe launched QuickBird commercial imaging satellite providing the highest spatial resolution of 60 cm in today's world.
2002	French Space Agency (CNES) launched SPOT 5 providing up to 2.5 m spatial resolution.

2.1.2 The Evolution of Remote Sensing

Table 2.1 gives a brief list of the times when innovative developments of remote sensing were documented. More details may be found in Lillesand and Kiefer (2000).

2.1.3 Electromagnetic Radiation Principles in Remote Sensing

The first requirement for remote sensing is to have an energy source to illuminate the earth. Remote sensing systems then record data of reflection and/or emission of electromagnetic energy from the earth's surface (Figure 2.1). The major source of the electromagnetic energy is the sun, although the earth itself can emit geothermal and man-made energy.

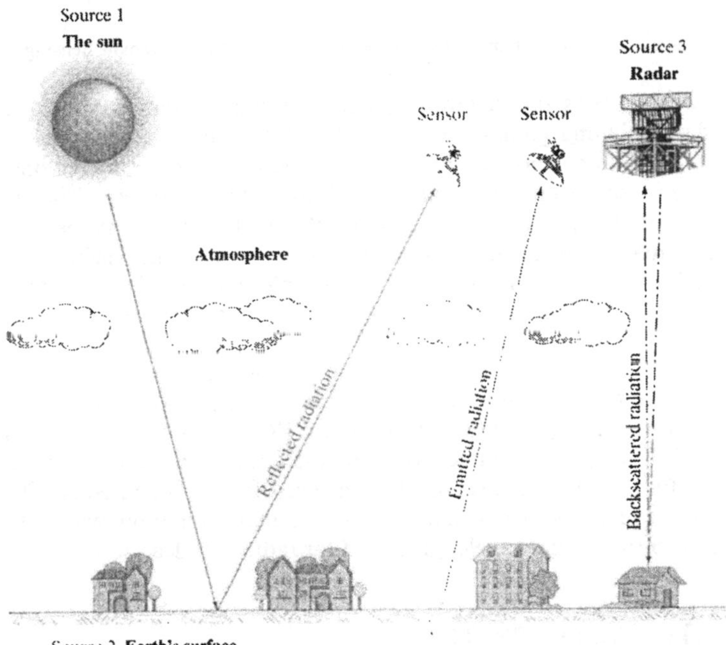

Figure 2.1 Sources of Electromagnetic Energy in Remote Sensing (Source: Lo and Yeung, 2002)

Electromagnetic radiation is a form of energy derived from oscillating magnetic and electrostatic fields. Two characteristics of electromagnetic radiation are particularly important for understanding remote sensing. These are the wavelength and frequency, which are related by the formula: $c = \lambda f$, where, λ is wavelength (m), f is frequency (cycles per second, Hz), and c is the speed of light

(3×10^{-8} m/s). Electromagnetic radiation occurs as a continuum of wavelengths and frequencies from short-wavelength, high-frequency cosmic waves to long-wavelength, low-frequency radio waves (Figure 2.2).

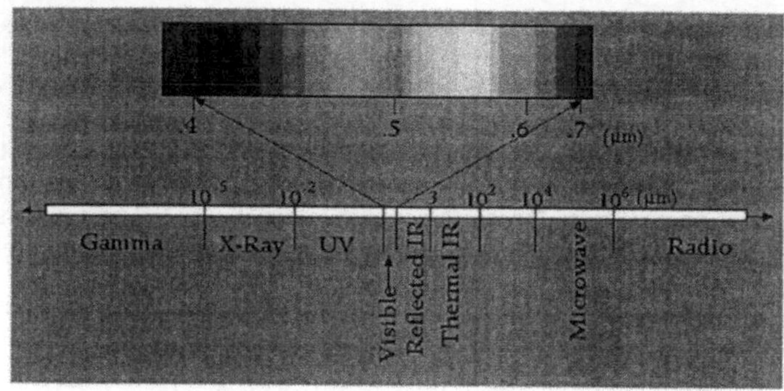

Figure 2.2 Useful Portions of the Electromagnetic Spectrum in Remote Sensing.

This is known as the electromagnetic spectrum. It ranges from the shorter wavelengths (including gamma and x-rays) to the longer wavelengths (including microwaves and broadcast radio waves). There are several regions of the electromagnetic spectrum which are useful for remote sensing. The ultraviolet (UV) portion of the spectrum has the shortest wavelengths. Some earth's surface materials, primarily rocks and minerals, fluoresce or emit visible light when illuminated by UV radiation. Although the visible portion of the electromagnetic spectrum (0.4-0.7 μm) is very narrow, it is important in remote sensing because aerial photography directly records the electromagnetic radiation in this region. The infrared (IR) region covers the wavelength range from approximately 0.7 μm to 100 μm, more than 100 times as wide as the visible portion. Two categories are based on their radiation properties, the reflected IR (0.7-3.0 μm) and the emitted or thermal IR (3.0-100 μm). The microwave region covers the wavelength range from about 1 mm to 1 m, the longest wavelengths used for remote sensing. The shorter wavelengths have properties similar to the thermal IR region while the longer wavelengths approach the wavelengths used for radio broadcasts.

2.2 REMOTE SENSING SYSTEMS

Basically, remote sensing systems can be classified into two types: passive and active. Passive remote sensing systems sample emitted and reflected radiation from the earth's surface when the sun illuminates the earth. The sun's energy is either reflected, as it is for visible wavelengths, or absorbed and then re-emitted, as it is for thermal IR wavelengths. Good examples are camera and thermal IR detectors. Active remote sensing systems, on the other hand, provide their own energy source at a specified wavelength for illumination and sample the portion reflected back to the detecting devices. A good example is imaging radar. Both passive and active remote sensing can be further subdivided into analog and digital types (Table 2.2). There is also another way to classify remote sensing system depending on the type of imaging platform used, mainly airborne and spaceborne.

Table 2.2 Classification of Remote Sensing Systems

	Analog systems	**Digital systems**
Passive systems	Aerial camera, video recorder	Multispectral scanner, linear and area array scanner, digital camera, spectroradiometer
Active systems	Side-Looking Airborne Radar (SLAR)	Synthetic Aperture Radar (SAR), laser scanner or Light Detection and Ranging (LIDAR)

An increasing number of satellites and sensors are used to monitor different aspects of the environment. Most satellite imaging systems are electromechanical scanners, linear array devices, or imaging spectrometers. Characteristics of selected satellites appear in Table 2.3.

2.3 IMAGING CHARACTERISTICS OF REMOTE SENSING SYSTEMS

The success of remote sensing data collection requires an understanding of the basic characteristics of remote sensing systems. There are four major resolution characteristics determining the type of geospatial data that can be detected by remote sensing systems: (1) spatial resolution, (2) spectral resolution, (3) radiometric resolution, and (4) temporal resolution (Lillesand and Kiefer, 2000).

2.3.1 Spatial Resolution

Spatial resolution determines the ability of a remote sensing system in recording spatial details and refers to a measure of a smallest object that can be identified by the sensor, or the area on the ground represented by each pixel. The finer the resolution, the lower the number. Spatial resolution is by far the most important (but also conceptually complex) characteristic of a remote sensing system. Figure 2.3 shows a 60 cm natural color image of the Pentagon in Washington, D.C. area that was collected by QuickBird on August 2, 2002.

The ratio of distance on an image or map, to actual ground distance is referred to as scale. Also the term “large-scale” and “small scale” often refers to spatial resolution. Large-scale in remote sensing refers to imagery in which each pixel represents a small area on the ground, while in small-scale imagery each pixel represents a large area on the ground. Images or maps with small "map-to-ground ratios" are referred to as small scale (e.g., 1:100,000), and those with larger ratios (e.g., 1:2,000) are called large scale. Typically, different spatial resolutions are required for particular applications (Table 2.4). Other considerations include accuracy and scene size.

Table 2.3 Characteristics of Selected Satellites

Satellite (Country)	Launch status	Spatial Resolution (m)	Spectral Resolution	Temporal Resolution (days)	Footprint (km)	Stereo
RADARSAT 1 (Canada)	1995	8-50	C-band (5.6 cm) SAR, HH	24	50-500	IF
RADARSAT 2 (Canada)	2004	3-100	C-band (5.6 cm) SAR, HV	3	20-500	IF
TerraSAR-1 (ESA)	2005	1 (X-band) 3-15 (L band)	X-band SAR L-band SAR	2	10 x 10 40 x 60	N/A
SPOT 4 (France)	1997	10 (Pan) 20 (XS)	4 (XS)	26	60×60	CT
SPOT 5 (France)	2002	2.5 (super-mode) 5 (Pan) 10 (XS)	1 (Pan) 3 (XS)	1-4	60×60	CT
MOMS-02/D2 (Germany)	1993	4.5 (HR pan) 13.5 (VNIR)	4 (MSS)	N/A	37×37 78×78	AT
IRS-1C/D (India)	1995/97	5.8 (Pan) 23.5 (VNIR) 70.5 (SWIR)	1 (Pan)	5 (P) 24 (M)	70×70 (P) 150×150 (M)	CT
JERS-1 (Japan)	1992	18	L-band SAR	44	75×75	AT
ADEOS 9Japan)	1996	8 (Pan) 16 (VNIR)	4 (MSS)	41	80×80	CT
KOMPSAT-1 (Korea)	1999	6.6 (Pan) MSS	1 (Pan) (MSS)	N/A	17 x 17	AT
KOMPSAT-2 (Korea)	2004	1 (Pan) 4 (MSS)	1 (Pan) 4 (MSS)	N/A	17 x 17	AT
SPIN-2 (Russia)	Periodic	2 (Pan) 10 (Pan)	N/A	8	180×180 200×200	N/A
Landsat 7 ETM+ (USA)	1999	15 (Pan) 30 (VNIR, SWIR) 60 (TIR)	1 (Pan) 7 (TM)	16	185×185	N/A
IKONOS (USA)	1999	1 (Pan) 4 (MSS)	1 (Pan) 3 (MSS)	3.5-5	11×11	AT and CT
EROS A1 (USA)	2000	1.8 (Pan)	1 (Pan)	15	12.5×12.5	AT and CT
Terra's ASTER (USA)	2000	15 (VNIR) 30 (SWIR) 90 (TIR)	14 (MSS)	16	Variable	AT
QuickBird (USA)	2001	0.6 (Pan) 2.4 (MSS)	1 (Pan) 4 (MSS)	1.5-4	16.5×16.5	AT and CT
OrbView-3 (USA)	2001	1 (Pan) 4 (MSS)	1 (Pan) 4 (MSS)	< 3	8×8 (Pan, MSS)	N/A

N/A = not applicable
AT = along track
CT = cross track
IF = interferometry
Pan = parametric
MSS = multispectral
VNIR = visible and near-infrared
SWIN = short-wave infrared

ASTER = Advanced Spaceborne Thermal Emission and Reflection Radiometer
EROS = Earth Remote Observation Satellite
KOMPSAT = Korea Multi-Purpose Satellite
ADEOS = Advanced Earth Observing Satellite
MOMS = Modular Optoelectronic Multispectral Stereo Scanner

Figure 2.3 QuickBird 60 cm Resolution Image of the Pentagon in Washington, D.C. Area (Source: DigitalGlobe, 2002)

Table 2.4 Spatial Resolution and Application

Satellite Imagery	Spatial Resolution (m)	Map Scale	Applications
QuickBird IKONOS	0.6 (Pan) 1 (Pan)	1:2,000	Land parcels, buildings Sidewalks, vehicles, utilities
SPOT 4 SPOT 5	10 (Pan) 5 (Pan)	1:25,000	Road or street mapping
SPOT 4/5 Landsat 7	20 (XS)/10(XS) 30 (TM)	1:62,500	Land surface properties, land-cover and land-use mapping

2.3.2 Spectral Resolution

Spectral resolution refers to the electromagnetic radiation wavelengths to which a remote sensing system is sensitive. It consists of two components: the number of wavelength bands (or channels) and the width of each band. A large number of bands and a narrower bandwidth will give rise to a higher spectral resolution.

Many remote sensing systems record energy over several separate wavelength ranges at various spectral resolutions. These are referred to as multi-spectral sensors and will be described in some detail in following sections. Advanced multi-spectral sensors called hyperspectral sensors, detect hundreds of very narrow spectral bands throughout the visible, near-infrared, and mid-infrared portions of the electromagnetic spectrum. Their very high spectral resolution facilitates fine discrimination between different targets based on their spectral response in each of the narrow bands. Table 2.5 compares the spectral resolutions of the Landsat-7 ETM+, SPOT-5, and IKONOS satellite images for different applications.

Table 2.5 Spectral Wavelengths and Applications

Color/Spectrum	Spectral Wavelength (μm)			Applications
	Landsat 7	SPOT 5	IKONOS	
Blue	TM1: 0.45-0.52	N/A	Band 1 0.45-0.52	Map shallow water, discriminate soil vs. vegetation, urban feature identification
Green	TM2: 0.52-0.60	XS1: 0.50-0.59	Band 2 0.52-0.60	Vegetation health, urban feature identification
Red	TM3: 0.63-0.69	XS2: 0.61-0.68	Band 3 0.63-0.69	Vegetation species, urban feature identification
Near IR	TM4: 0.79-0.90	XS3: 0.79-0.89	Band 4 0.76-0.90	Vegetation health/stress, vegetation classification
Short-wave IR	TM5: 1.55-1.75	XS4: 1.58-1.75	N/A	Land/water boundary, moisture in soil and vegetation; discriminating snow and cloud-covered areas
Thermal IR	TM6: 10.4-12.5	N/A	N/A	Heat sources
Short-wave IR	TM7: 2.08-2.35	N/A	N/A	Discriminating mineral and rock types; sensitive to vegetation moisture content

2.3.3 Radiometric Resolution

Radiometric resolution is the smallest difference in radiant energy that can be detected by a sensor. The finer the radiometric resolution of a sensor, the more sensitive it is to detecting small differences in reflected or emitted energy. Imagery is represented by positive digital numbers that vary from 0 to (one less than) a selected power of 2. This range corresponds to the number of bits used for coding numbers in binary format. The maximum number of brightness levels available depends on the number of bits used in representing the energy recorded. Thus, if a sensor used 8 bits to record the data, there would be $2^8 = 256$ digital values available, ranging from 0 to 255. However, if only 4 bits were used, then only

$2^4=16$ values ranging from 0 to 15 would be available. Thus, the radiometric resolution would be much less.

2.3.4 Temporal Resolution

Temporal resolution is the frequency of remote sensing data collection and refers to the length of time it takes for a satellite to complete one entire orbit cycle. The revisit period of a satellite sensor is usually several days. Therefore, the absolute temporal resolution of a remote sensing system to image the exact same area at the same viewing angle a second time is equal to this period. However, because of some degree of overlap in the imaging swaths of adjacent orbits for most satellites and the increase in this overlap with increasing latitude, some areas of the earth tend to be re-imaged more frequently. Also, some satellite systems are able to point their sensors to image the same area between different satellite passes separated by periods from one to five days. Thus, the actual temporal resolution of a sensor depends on a variety of factors, including the satellite/sensor capabilities, the swath overlap, and latitude.

An important application of remote sensing is change detection, which is possible only with high temporal resolution. By imaging on a continuing basis at different times we are able to monitor the changes that take place on the earth's surface, whether they are naturally occurring (such as changes in natural vegetation cover or flooding) or induced by humans (such as urban development or deforestation).

2.4 ACTIVE MICROWAVE REMOTE SENSING

2.4.1 What is Radar and IFSAR?

Radar stands for *radio detection and ranging* and is an active remote sensing technology. Radar sends a signal to the object and then receives the reflectance of the signal from the object. Radar operates at the microwave region of the electromagnetic spectrum, which penetrates cloud cover and other atmospheric particles due to its relatively long wavelengths, giving radar a compelling advantage over optical sensors. Unique among all imaging technologies, radar is a cloud-penetrating, day-night remote sensing system, independent of the weather. While the electromagnetic radiation is characterized in the visible and infrared portions of the spectrum primarily by wavelength, frequencies are usually used together with wavelengths to designate the bands. The most commonly used bands for imaging radar are: X-band (2.4-3.8 cm in wavelengths, or 12.5-8 GHz in frequencies), C-band (3.8-7.5 cm, or 8-4 GHz), S-band (7.5-15 cm, or 4-2 GHz), and L-band (15-30 cm, or 2-1 GHz).

The last two decades has seen the development from initial concept to commercial systems of a mapping technology based on interferometric synthetic aperture radar (SAR). Conventional SAR has been used extensively since its inception in the late 1950's for fine resolution mapping and remote sensing applications (Elachi, 1988). Operating at microwave frequencies (3-40,000 MHz) these systems are used to generate imagery that provides a unique look at the electromagnetic and structural properties of the surface being imaged, at day or night and in nearly all weather conditions. Conventional SAR systems typically measure only two image coordinates. One coordinate is measured along an axis oriented parallel to the flight direction while the other coordinate is the range (or distance) from the SAR to the point being imaged. By augmenting a conventional SAR system to have two spatially separated antennas in the cross-track plane it is possible to measure the 3D location of imaged points to a high degree of accuracy. Measurement of the third coordinate is based on an interferometric technique developed in the radio astronomy community over several decades for fine angular measurements. Combining of SAR and interferometry techniques into a single system is called interferometric synthetic aperture radar (IFSAR or InSAR).

2.4.2 Introduction to SAR

Before proceeding directly to IFSAR concepts and systems a brief introduction to SAR systems and terminology is provided. SAR can be used to produce high-resolution imagery from either airborne or spaceborne platforms (Raney, 1999). Unlike optical sensors that form images from reflected solar radiation, SAR systems transmit their own radiation and record the signals reflected from the terrain. With optical systems images are generally formed instantaneously, whereas for SAR, data collected from multiple points along the flight path are required in order to achieve useful resolution in the along track, or azimuth, direction. Rather sophisticated image processing is required to form recognizable images from the raw data. The resolution and quality of the imagery depends on a number of system parameters as well as how the data are collected and processed. SAR takes advantage of the motion of the platform to synthesize a large antenna that may be many hundreds of meters in length to achieve fine along track resolution. Figure 2.4 shows the typical SAR imaging geometry with the SAR platform moving along in flight.

SAR produces a very long antenna synthetically by using the forward motion of the platform to carry a relatively short real antenna to successive positions along the flight line. These successive portions are treated as though each were an individual element of the same antenna. The synthetic antenna's length is directly proportional to range, as across-track distance increases, antenna length increases. This produces a synthetic beam with a constant width, regardless of range (Avery and Berlin, 1992). In operation, the relatively short SAR antenna transmits wide beams of microwave radiation in the across-track direction at regular intervals along the line of flight. Due to the wide beamwidth, features will enter the beam, move through the beam, and finally leave it; the time period that an object is illuminated increases with increasing range (see Figure 2.4). The SAR receiver measures and records the time delay between transmission and reception of each pulse that establishes range resolution. However, the discriminate different objects at the same distance from the ground track, the azimuths to all features must also be measured and recorded to establish azimuth resolution. Typical airborne SAR image resolutions are 0.5 to 10 m, while spaceborne SAR image resolutions are 5 to 100 m.

Because SAR is an active remote sensing system, the type of signal to be transmitted and received can be controlled through ***polarization***, which is the direction of vibration of the electrical field vector of electromagnetic radiation. There are two types of polarization in SAR: either horizontal (H) or vertical (V). It is possible for SAR to transmit the signals with horizontal polarization and receive the return signals with horizontal polarization (HH). Similarly, radar signals can be transmitted with vertical polarization and received with vertical polarization (VV). Both HH and VV are called *like polarization*. Alternatively, the radar signals can be transmitted with horizontal polarization and received with vertical polarization (HV), or just the reverse (VH). Both HV and VH are known as *cross polarization*. Radar return signals are usually strong in like polarization but weak in cross polarization. VV polarization is more sensitive to vertical objects such as tree trunks and plants because of their vertical orientation, while HH polarization is more sensitive to physical and cultural surfaces that exhibit a dominant horizontal configuration.

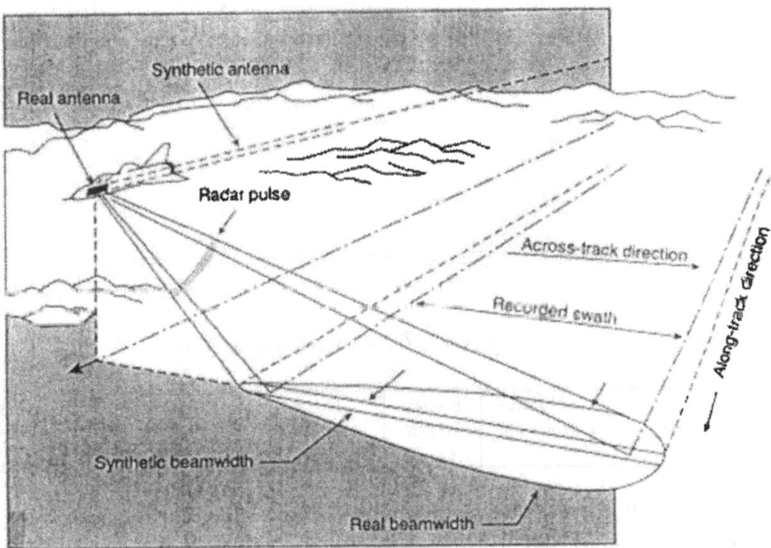

Figure 2.4 Operating Principle of a SAR System (Source: Avery and Berlin, 1992)

2.4.3 Interferometric Synthetic Aperture Radar (IFSAR)

The previous discussion of SAR systems was primarily concerned with collecting a single image of the terrain. It is possible to acquire multiple SAR images of the terrain from airborne or spaceborne platforms to extract valuable 3D information. *Imaging radar interferometry* is the process whereby radar images of the same location on the ground are recorded by antenna at (1) different locations, or (2) different times. Analysis of the resulting two interferograms allow very precise measurements of the range to any specific point found in each image of the interferometric pair. Topographic mapping based on SAR interferometry or IFSAR relies on acquiring data from two different look angles and assumes that the scene imaged did not move between data acquisitions. The two measurements can be from two radars placed on the same platform separated by a few methers. This is called *single-pass interferometry* (see Figure 2.5a). The first single-pass IFSAR is the NASA's Shuttle Radar Topography Mission (SRTM) flight occurred during February 11-22, 2000. Interferometry may also be conducted using a single radar that obtains two measurements on different orbital tracks that are closely spaced but a day or so longer apart. This is the methodology used for the Spaceborne Imaging Radar-C (SIR-C) and European Space Agency ERS-1 and 2 interferometry and is called multiple-pass or *repeat-pass interferometry* (see Figure 2.5b). Characteristics of the spaceborne IFSAR platforms and sensors are listed in Table 2.6.

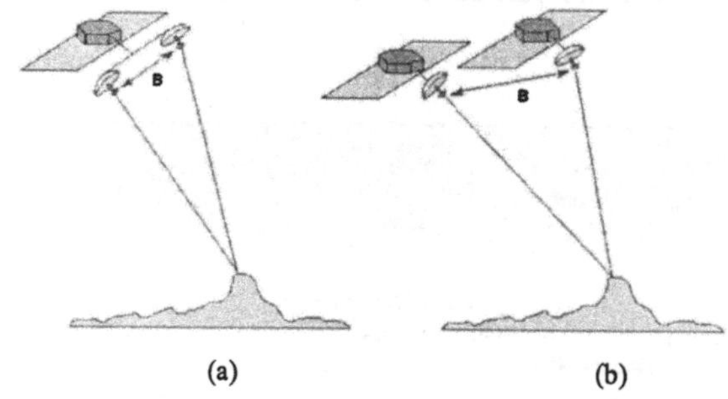

Figure 2.5 IFSAR Systems Using (Left) Single-Pass and (Right) Repeat-Pass Interferometry

Table 2.6 Repeat-Pass Spaceborne IFSAR Systems

Sensor	Band	Incidence angle (°)	Swath width (km)	Range resolution (m)	Azimuth resolution (m)	Exact repeat cycle (days)
Almaz-1 (1991 - 1993)	S (9.6 cm)	30-60	20-45	15-30	15 (4 looks)	N/A
ERS-1 (1991)	C (5.6 cm)	23	100	26	30 (6 looks)	3, 35, 176
ERS-2 (1995)	C (5.6 cm)	23	100	26	30 (6 looks)	35
JERS-1 (1992-1998)	L (23.5 cm)	39	75	18	18 (3 looks)	44
RADAR SAT-1 (1995)	C (5.6 cm)	20-59	45-500	10-100	10-100	24
RADAR SAT-2 (2004)	C (5.6 cm)	30-40	20-500	3-100	9-100	24

Typically, SAR signal data are recorded on high-density digital tapes. These tapes are either read directly by a SAR processor, or transcribed into raw signal data sets that are then processed by a SAR processor into a complex imagery. The next step is to interferometrically process a pair of SAR images, converting them to map products. Figure 2.6 gives the major steps for IFSAR processing software applications to perform these steps in one form or another.

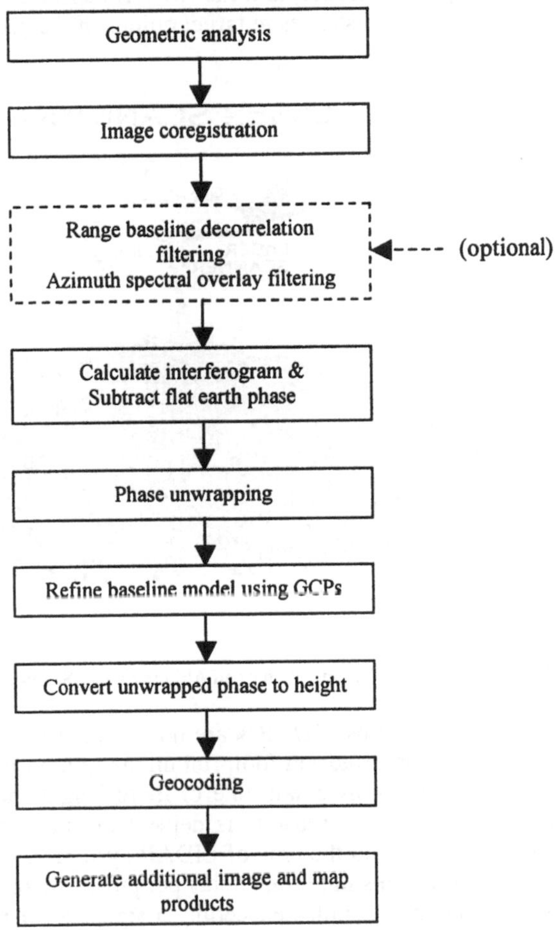

Figure 2.6 Major Steps of IFSAR Processing (filtering is not always necessary)

Spaceborne IFSAR has the ability to measure terrain elevation with height accuracies on the order of a few meters and horizontal resolution as small as 10 m. This 3D resolution capability combined with the ability to map large areas up to 100 km with a few pairs of images makes spaceborne IFSAR a cost-effective method for producing up-to-date and accurate topographic maps.

2.4.4 LIDAR

LIDAR stands for *light detection and ranging*. LIDAR, like IFSAR, is an active sensor providing its own illumination and similarly records the time delay between transmit and receipt of reflected signals from the surface. In an airborne laser scanning system (see Figure 2.7), airborne Global Positioning System (GPS) accurately measures the 3D coordinates of the LIDAR sensor in the air; the inertial navigation system (INS) or inertial measurement unit (IMU) measures the roll, pitch, and heading of the sensor; and the LIDAR sensor or laser scanner measures the scan angles and laser distances to target points on the ground.

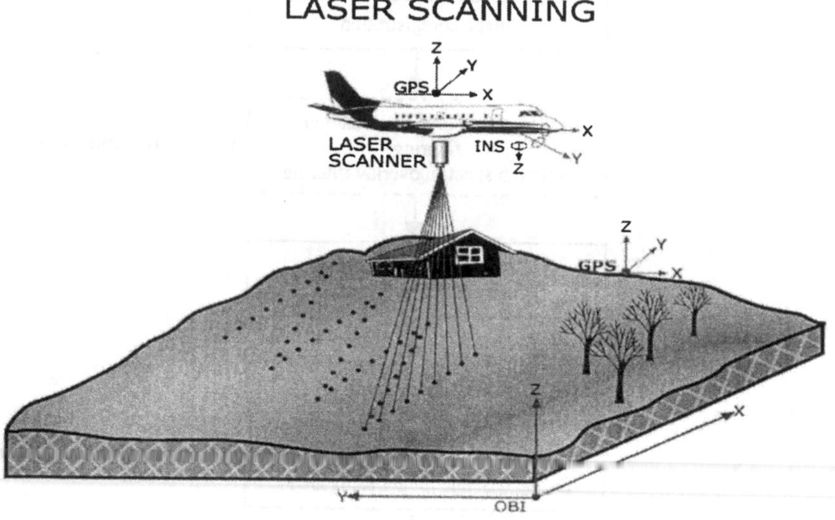

Figure 2.7 The Operation of an Airborne Laser Scanning System

Unlike IFSAR systems LIDAR's are not imaging sensors. Employing a very narrow beam so that the protected footprint on the ground is typically 10 m or less from space, LIDAR systems obtain one or more height measurements per pulse. The number of height measurements is dependent upon the vertical structure of objects within the beam and the type of LIDAR system. Some LIDAR systems are equipped to only record a single time delay per pulse whereas other systems record time delays for multiple samples exceeding a signal level threshold. By scanning cross track to either side of the nadir point of the aircraft and rapidly pulsing the laser reasonable mapping swaths are obtained. Operating at optical instead of microwave frequencies, LIDAR systems do not penetrate clouds and other atmospheric obscurants (Maune, 2001).

The launch date of the world's first spaceborne LIDAR, the *Vegetation Canopy LIDAR (VCL)* developed through NASA's Earth System Science Pathfinder (ESSP) program, currently has not been determined. The VCL will have five separate lasers operating with a wavelength of 1.064 μm. Each laser has a ground footprint with a diameter of 25 m.

2.5 EXTRACTION OF THEMATIC INFORMATION FROM REMOTELY SENSED IMAGERY

Information from imagery is derived by spatial, spectral, and semantic attribution of features identified by remote sensing specialists trained in interpretation of panchromatic, multispectral and other remotely sensed data and by photogrammetrists trained in the science of making reliable 3D measurements from images. The data are calibrated, spatially registered and classified from which selected datasets are then cartographically depicted for use in digital, graphical and orthophoto forms as maps, charts and overlays.

2.5.1 Visual Image Interpretation

The process of visual image interpretation can be divided into five stages: detection, identification, analysis and deduction, classification, and theorization-verification or falsification of hypotheses. In assisting the identification of objects, seven basic elements have to be used. These are, in order of complexity: tone/color, size, shape, texture, pattern, shadow, and association.

Tone/color refers to the relative brightness or color of objects in an image. Generally, tone is the fundamental element for distinguishing between different targets or features. Variations in tone also allow the elements of shape, texture, and pattern of objects to be distinguished. On panchromatic film, tone varies from black through gray to white, making use of the spectral and radiometric resolution of the image. In general, the more light reflected by the object, the lighter its tone on the image. **Color** is used for true color and color infrared image interpretation. Vegetation is green and water is blue in true color images, while vegetation is usually red in color infrared images.

Shape refers to the general form, structure, or outline of individual objects. Shape can be a very distinctive clue for interpretation. Straight edge shapes typically represent urban or agricultural (field) targets, while natural features, such as forest edges, are generally more irregular in shape, except where man has created a road or clear cuts. Farm or cropland irrigated by rotating sprinkler systems would appear as circular shapes.

Size of objects in an image is a function of scale. It is important to assess the size of a target relative to other objects in a scene, as well as the absolute size, to aid in the interpretation of that target. A quick approximation of target size can direct interpretation to an appropriate result more quickly. For example, if an interpreter had to distinguish zones of land use, and had identified an area with a number of buildings in it, large buildings such as factories or warehouses would suggest commercial property, whereas small buildings would indicate residential use.

Pattern refers to the spatial arrangement of objects and is also highly dependent on scale. Typically an orderly repetition of similar tones and textures will produce a distinctive and ultimately recognizable pattern. Examples include drainage patterns, field patterns, and settlement patterns.

Texture refers to the arrangement and frequency of tone change within an image that arises when a number of small features are viewed together. It gives the visual impression of the roughness or smoothness of an object. Texture is scale dependent and is affected by the spatial resolution of the image.

Shadow can provide some idea of the profile and relative height of an object if no stereoscopic reviewing is possible. Shadows tend to emphasize linear features and shapes. Shadow is also useful for enhancing or identifying topography and landforms, particularly in radar imagery. However, shadows obscure features.

Association is the spatial relationship of objects and phenomena. The identification of one object or phenomenon will point toward the occurrence of associated objects and phenomena. For example, aircrafts and runways are associated with an airport.

2.5.2 Digital Image Classification

The digital image analysis approach is used to extract thematic information (attributes) from digital images, such as those acquired by imaging satellites (e.g., Landsat, SPOT, IKONOS, and QuickBird). The same process used in visual interpretation with the seven basic elements is applicable to digital image classification. Tone, which is represented as a digital number in each pixel of the digital image, is obviously the most important element used in digital image classification. Size, shape, texture, and pattern represent spatial arrangement of tone and color. Texture is particularly useful in the analysis of cultural features. Texture is as important as tone in radar image classification. Association provides the basis for the development of contextual image classification to improve image classification accuracy.

Several commercially available software systems (e.g., PCI Geomatica, ERDAS Imagine, ER Mapper, ENVI) have been developed specifically for image classification and analysis. Most of the common image processing functions available in those image analysis systems can be categorized into the following three categories: (1) preprocessing (radiometric and geometric correction), (2) enhancement, and (3) classification.

Image Preprocessing functions involve those operations that are normally required prior to the main data analysis and extraction of information, and are generally grouped as radiometric or geometric corrections. ***Radiometric corrections*** include correcting (1) errors in the detectors of the remote sensing system, such as missing scan lines, striping or banding (caused by a detector out of adjustment), or line–start problem (scanning system not able to collect data at the beginning of a scan line); (2) atmospheric attenuation caused by scattering and absorption in the atmosphere; and (3) topographic effects caused by slopes and aspects of terrain (Jensen, 1996). Radiometric correction is necessary when multi-date images are used for change detection. Otherwise the same land cover types on images of different years will not give the same tone reflectance. Absolute radiometric correction requires knowledge of the atmospheric conditions at the time of imaging, which is often not easily available. Relative normalization is a more economical approach by which a reference image in a time sequence of images is chosen, and digital values of the other images within the full range from very dark to very light objects are then related to this reference image. There are two commonly used methods of relative radiometric normalization: (1) histogram matching and (2) image regression. The former matches the histogram of brightness values of one image to that of another image so that the apparent distribution of brightness values in the two images are made as close as possible, while by the latter the brightness value of each pixel of the subject image is related to that of the reference image band by band to produce a linear regression equation.

Geometric corrections is to correct the systematic and nonsystematic errors in the remote sensing system during the process of image acquisition. Systematic errors include scan skew caused by the forward motion of the platform during the time for each mirror sweep, irregular mirror-scan and platform velocities, while nonsystematic errors include variations in satellite orbital altitude and attitude (roll, pitch, and yaw) errors. Geometric correction involves a mathematical transformation of the coordinates (pixels fixed by columns and rows in a raster format). For geometric correction, the method of image-to-map rectification followed by attribute interpolation by resampling is used.

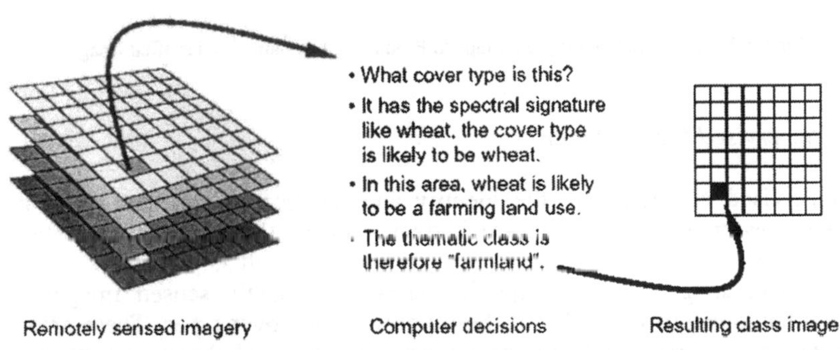
• What cover type is this?
• It has the spectral signature like wheat, the cover type is likely to be wheat.
• In this area, wheat is likely to be a farming land use.
• The thematic class is therefore "farmland".
Remotely sensed imagery
Computer decisions
Resulting class image

Figure 2.8 The Process of Making a Classed Thematic Map from a Digital Image

2.5.3 Image Classification Approaches

Image classification uses the spectral information represented by the digital numbers (DNs) in spectral band(s) and attempts to classify each individual pixel based on this spectral information. The objective is to assign all pixels in the image to particular classes or themes (e.g., water, vegetation, soil, etc.). The resulting classified image is essentially a thematic "map" comprising a mosaic of pixels, each of which belongs to a particular theme. Image classification is the process of creating a meaningful digital thematic map from an image (see Figure 2.8). The classes in the map are derived either from known cover types (water, soil) or by algorithms that search the data for similar pixels. Once data values are known for the distinct cover types in the image, a computer algorithm can be used to divide the image into regions that correspond to each cover type or *class*. The classified image can be converted to a land use map if the use of each area of land is known. The term *land use* refers to the purpose that people use the land for (e.g., city, national parks or roads), whereas *land cover* refers to the material that an area is made from (e.g., concrete, soil or vegetation).

Image classification is usually used to extract thematic information (attributes) from multispectral images. The major steps to produce a thematic or classified image are shown in Figure 2.9. There are three approaches to digital image classification: supervised, unsupervised, and hybrid.

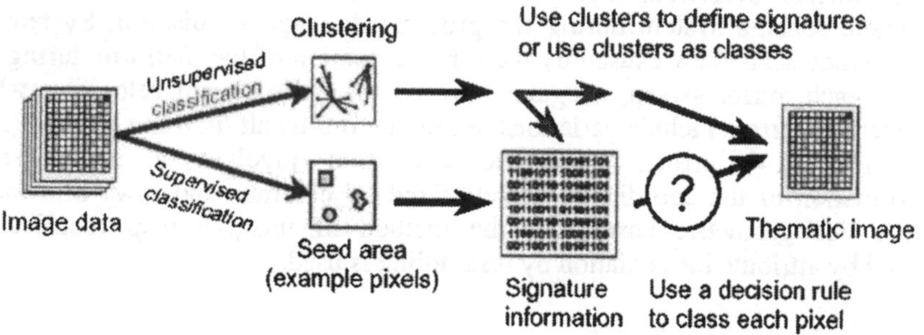

Figure 2.9 Diagram Showing the Steps to Produce a Thematic or Classified Image

2.5.3.1 Supervised Classification

In a supervised classification, the identity and location of the land-cover types (e.g., soil, water, or vegetation) are known a priori through a combination of fieldwork, analysis of aerial photographs and maps, as well as personal experience. The analyst attempts to locate specific sites in remotely sensed imagery that represent homogenous examples of these known land-cover types. These areas are commonly referred to as *training areas* because the spectral characteristics of these known areas are used to train the classification algorithm for eventual land-cover mapping of the reminder of the imagery. Multivariate statistical parameters (means, standard deviations, covariance matrices, correlation matrices, etc.) are calculated for each training area. Each pixel both within and outside these training areas is then evaluated and assigned to the class of which it has the highest likelihood of being a member. This is often referred to as a *hard classification* (Jensen, 1996). The selection of training areas is a very important step because it will affect the accuracy of the final classification.

There are three commonly used *classifiers*: *parallelepiped*, *minimum distance*, and *maximum likelihood* classifiers. They are all per-pixel classifiers, i.e., individual pixels are used as the basic unit for classification.

Parallelepiped classifier considers the range of values in each class's training set. This range may be defined by the highest and lowest DNs in each band and appears as a rectangular box in the 2D scattergram (Figure 2.10). When a pixel lies inside one of the boxes, then it is classified into the corresponding class (e.g., pixel 2 in Figure 2.10). If a pixel lies outside all regions, then it is classified as "unknown." Difficulties are encountered when class ranges overlap where a pixel has to be classified as "not sure" or be arbitrarily placed on one of the two overlapping classes.

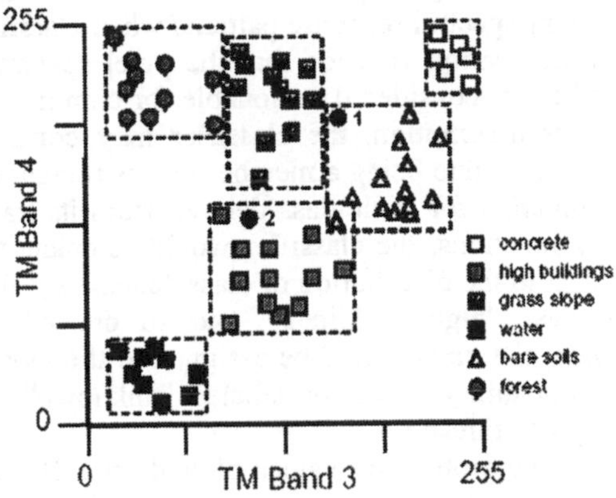

Figure 2.10 Parallelepiped Classification Strategy

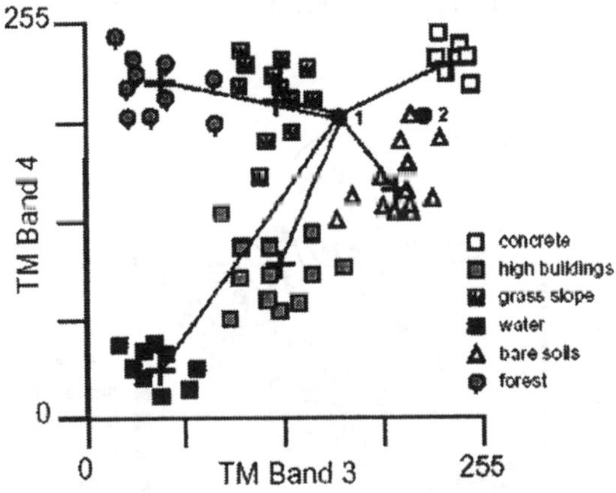

Figure 2.11 Minimum Distance Classification Strategy

Minimum distance classifier comprises three steps. First, the mean of the spectral value in each band for each class is computed (represented in Figure 2.11 by symbol "+"). Then, the distance between the spectral value of an unknown pixel and each of the category means can be computed. The pixel is then assigned to the "closest class." This classifier is mathematically simple and it overcomes the poor representation problem of rectangular decision region used by parallelepiped classifier. For example, pixel 1 shown in Figure 2.11 would be correctly classified as "grass slope." This strategy, however, has its limitations. It is insensitive to different degrees of variance in the spectral response data. In Figure 2.11, pixel 2 would be classified as "concrete" in spite of the fact that the pixel would probably more appropriate to be "bare soils" because of the class' greater variability.

Maximum likelihood classifier quantitatively evaluates both the variance and covariance of the category spectral response patterns when classifying an unknown pixel. To do this, an assumption is made that the pixel spectral cluster forms a normal distribution, which is considered reasonable for common spectral response distributions. Under this assumption, the classifier may compute the statistical probability of a given pixel value being a member of a particular class by applying a *probability density function* for each class derived from its training data. Using the probability density functions, the classifier would calculate the probability of the pixel value occurring in the distribution of class "concrete," then the likelihood of its occurring in class "high buildings," and so on. After evaluating the probability in each class, the pixel would be assigned to the most likely class that presents the highest probability value, or labeled "unknown" if the probability values are all below a given threshold.

Figure 2.12 shows the probability values plotted on a 2D scattergram where the contour lines are associated with the probability of a pixel value being a member of one of the classes. Basically the maximum likelihood classifier delineates ellipsoidal equal-probability contours, the shape of which shows the sensitivity of the classifier to both variance and covariance. For example, both pixel 1 and pixel 2 would be appropriately assigned to the class "grass slope" and "bare soils," respectively.

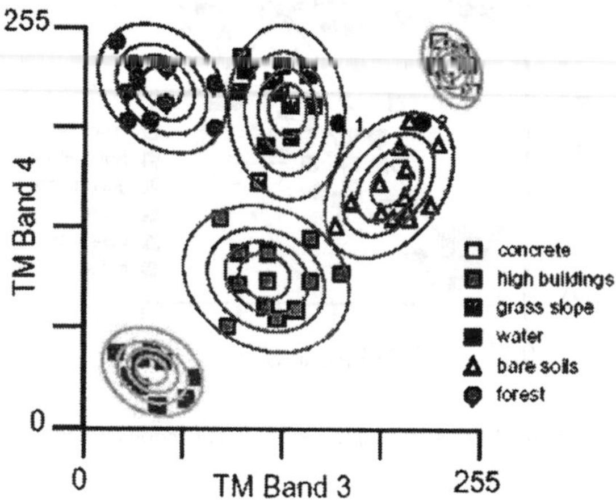

Figure 2.12 Equal-Probability Contours Defined by a Maximum Likelihood Classifier

The principal drawback of maximum likelihood classification is the extensive demand on computation to classify each pixel. When a large number of spectral bands are involved or a large number of classes must be differentiated, the maximum likelihood classifier would perform much slower than the other classifiers described. This drawback was one of the major limitations in the past, but is becoming much less critical today with rapid development of computer hardware.

2.5.3.2 *Unsupervised Classification*

In unsupervised classification, the identity of land-cover types to be specified as classes within a scene are not generally known a priori because ground reference information is lacking or surface features within the scene are not well defined. The computer is required to group pixels with similar spectral characteristics into unique *clusters* according to some statistically determined criteria. The analyst then combines and relabels the spectral clusters into land information classes through comparing classified data with *reference data* (Jensen, 1996). Clustering algorithms use predefined parameters to identify cluster locations in data space, and to determine whether individual pixels are in those clusters or not (see Figure 2.13).

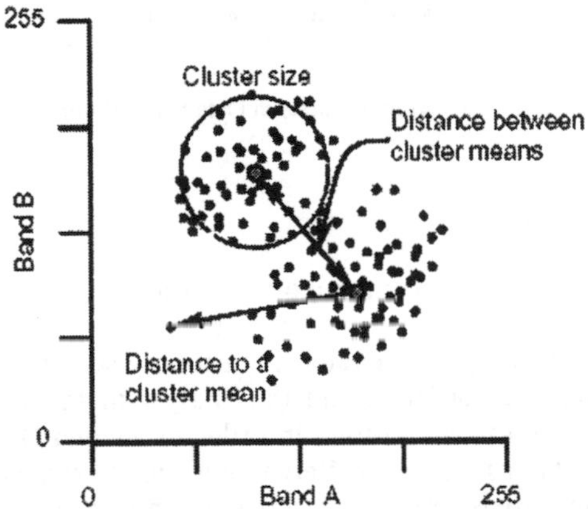

Figure 2.13 Measures that define a cluster include the size of a cluster and the distance between clusters (Source: ERDAS, 1999)

A typical multiple-pass, or iterative, clustering algorithm works as shown in Figure 2.14. Pass One: (A) Cluster centers are arbitrarily assigned. (B) Each pixel is assigned to the cluster center nearest to it in data space (spectral distance). (C) The cluster means are then calculated from the average of the cluster members (the middle cluster is shown with gray points) and the pixels are reassigned to the new cluster centers. Pass Two: (D) the process is repeated. The iteration stops when the cluster centers (or means) move by less than a pre-set amount during each iteration. With a number of iterations the location of clusters tend to stabilize as the location of cluster centers between each pass changes less and less.

Algorithms that pass through the image only once tend to be more affected by the initial conditions than iterative algorithms that repeatedly adjust the cluster means. After each pass through the data, cluster means can be calculated along with other measures such as standard deviation. In addition to simple straight-line distances, statistical measures of distance can be used where the distance to clusters is weighted by the size and importance of that cluster. The result from unsupervised classification may also need post-classification process as described above. In addition, because the real-world nature of spectral classes derived from the classification is largely unknown, considerable analysis and interpretation will be required. Often the resulting classes need to be merged into fewer classes to make the classified image more acceptable as a thematic map.

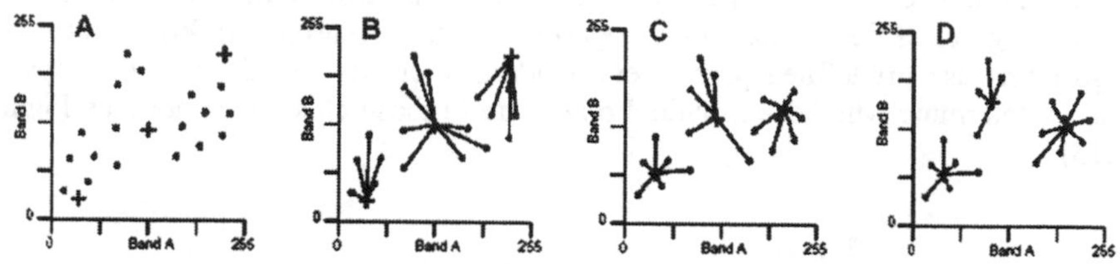

Figure 2.14 Iterative Clustering of Points in Data Space (Source: ERDAS, 1999)

2.5.3.3 Hybrid Classification

A *hybrid classification* approach involves aspects of both supervised and unsupervised classification and is aimed at improving the accuracy and efficiency of the classification process. For example, unsupervised training areas might be delineated in an image in order to aid the analyst in identifying the numerous spectral classes that need to be defined in order to adequately represent the land-cover information classes to be differentiated in a supervised classification. Unsupervised training areas are image subareas chosen intentionally to be quite different from supervised training areas. Training statistics are then developed for the combined classes and used to classify the entire scene (e.g., by a minimum distance to means or maximum likelihood algorithm). Hybrid classifiers are particularly valuable in analyses where there is complex variability in the spectral response patterns for individual land-cover types present.

In addition, *fuzzy classification*, which takes into account the heterogeneous and imprecise nature of the real world and attempts to handle the mixed-pixel problem by employing the fuzzy set theory in which a given entity may have partial membership in more than one category, may be used in conjunction with supervised and unsupervised classification algorithms (Jensen, 1996; Lillesand and Kiefer, 2001; Richard and Jia, 2000). Sometimes it is necessary to include nonspectral ancillary data when performing a supervised, unsupervised, and/or fuzzy classification to extract the desired information.

2.5.4 Accuracy Assessment

Classification accuracy analysis is one of the most active research fields in remote sensing. Meaningless and inconclusive assessment on the image classification results sometimes precludes the application of automated land cover classification techniques even when their cost is more favorable with more traditional means of data collection. A classification is not complete until its accuracy is assessed (Lillesand and Kiefer, 2000).

To correctly perform classification accuracy assessment, it is necessary to compare two sources of information: (1) classified information generated from remote sensing imagery and (2) reference information from ground truth data. The relationship between these two sets of information is commonly summarized in an *error matrix* (or confusion table) (see Table 2.7). An error matrix is a square array of numbers laid out in rows and columns that express the number of sample unites (i.e., pixels, clusters of pixels, or polygons) assigned to a particular category relative to the actual category as verified in the field. The columns normally represent the reference information, while the rows indicate the classified information. The accuracy of each category is clearly described in the error matrix, along with both the errors of inclusion (*commission errors*) and errors of exclusion (*omission errors*).

Table 2.7 Error Matrix Expressing Classification Accuracy (Source: Jenson, 1996)

	Reference information					
Classification	**Residential**	**Commercial**	**Wetland**	**Forest**	**Water**	**Row Total**
Residential	70	5	0	13	0	88
Commercial	3	55	0	0	0	58
Wetland	0	0	99	0	0	99
Forest	0	0	4	37	0	41
Water	0	0	0	0	121	121
Column Total	73	60	103	50	121	407

Overall Accuracy = 382/407 = 93.86%

Producer's Accuracy (measure of omission error)		**User's Accuracy (measure of commission error)**	
Residential =70/73= 96%	4% omission error	Residential =70/88= 80%	20% commission error
Commercial=55/60=92%	8% omission error	Commercial=55/58=95%	5% commission error
Wetland=99/103=96%	4% omission error	Wetland=99/99=100%	0% commission error
Forest=37/50=74%	26% omission error	Forest=37/41=90%	10% commission error
Water=121/121=100%	0% omission error	Water=121/121=100%	0% commission error

Computation of kappa coefficient

$N = 407$

$$\sum_{i=1}^{n} x_{ii} = 70 + 55 + 99 + 37 + 121 = 382$$

$$\sum_{i=1}^{n} (x_{i+} \times x_{+j}) = (88 \times 73) + (58 \times 60) + (99 \times 103) + (41 \times 50) + (121 \times 121) = 36,792$$

$\kappa = 92\%$

For the global assessment of classification accuracy, a discrete multivariate technique called *kappa analysis* is often employed. Kappa analysis yields a *kappa coefficient* (κ) or a kappa statistic (an estimate of kappa) that is a measure of agreement or accuracy (Congalton, 1991). The kappa coefficient is computed as

$$\kappa = \frac{N\sum_{i=1}^{n} x_{ii} - \sum_{i=1}^{n}(x_{i+} \times x_{+j})}{N^2 - \sum_{i=1}^{n}(x_{i+} \times x_{+j})} \tag{2.1}$$

where n is the number of rows in the matrix, x_{ii} is the number of observations in row i and column j, x_{i+} and x_{+j} are the marginal totals for row i and column j, respectively, and N is the total number of samples. The optimal κ score is 1.0 (perfect classification). In this case, $\kappa = 92.1\%$ (see Table 2.7).

2.5.5 Change Detection

In remote sensing, change detection is the process of identifying differences in the state of an object or phenomenon by observing it at different times. Digital change detection has been greatly facilitated by the use of geospatial information systems (GISs), particularly in the following two approaches (Lo and Yeung, 2002): (1) map-to-map comparison, and (2) image-to-image comparison.

Map-to-map comparison is also known as post-classification comparison (Jensen, 1996). If satellite images for two different years are used, the two images need to undergo geometric rectification and registration so that they can be matched up exactly. Each image will then be classified using one of the approaches explained in image classification. The overlay function of GISs is used to compare the two classification maps pixel by pixel. A change detection matrix, which is a cross-tabulation between the two maps, is produced. A change map may also be generated. The map-to-map comparison may also be performed on the basis of polygon types rather than pixels.

Image-to-image comparison is used for comparing the radiance value of each pixel of the two satellite images acquired at two different dates (Jensen, 1996). Because any change in radiance value is assumed to reflect a genuine change of the land cover, the two satellite images will have to be geometrically rectified and accurately registered. The two images are compared either by means of ***image differencing*** or by ***band ratioing***. In the image differencing case, the results can be negative or positive. Commonly, a constant value is added to convert all negative values into positive values. Green et al. (1994) used a constant of 100 in their application of multidate Landsat TM images for vegetation change detection. Thus, values of 100 in the resulting image difference file indicate "no change" in reflectance value between the two dates; areas of increased reflectance are indicated by values greater than 100; and areas of decreased reflectance are indicated by values less than 100. This method will allow the areas of gain or loss to be mapped and quantified. In band ratioing, the same spectral bands of the images for the two years are divided, with the result that changes will have values of either greater than or less than one, thus revealing the direction of land cover reflectance change for a pixel between the two years.

Because overlay functionality is involved in both map-to-map and image-to-image comparison approaches, the accuracy of the change map produced is adversely affected by the accuracy of each map that forms the overlay. The more maps to overlay, the less will be the accuracy of the final overlay.

2.6 EXTRACTION OF METRIC INFORMATION FROM REMOTELY SENSED IMAGERY

The concept of using images for the extraction of precise 3D geometric information is routinely employed most notably in the production of topographic maps. Traditionally, aerial photographs have been taken using film-based aerial cameras. These aerial cameras represent high precision optical systems that are capable of recording information at the micrometer level at image scale. A typical aerial camera has a format of 230 mm by 230 mm (9" by 9") and a focal length of 152 mm (6"). Other focal lengths are available such as 305 mm (12") which is typically used in areas of high relief. Registration marks such as fiducial and reseau marks are recorded on the image in order to recover the interior geometry of the photographs. Modern aerial cameras are now incorporating digital image sensors such as charged-coupled devices (CCDs). These sensors employ 1D or 2D arrays comprised of two to six thousand picture elements (pixels) in any one dimension.

In order to extract meaningful measurements from photographs (film-based) or digital images, it is necessary to understand the internal geometry of these sensors including the lens system, scanning mechanisms and orientation control. An overview of the mathematical relationships that have been developed to describe these imaging systems is presented in the following sections.

2.6.1 Fundamentals of Photogrammetry

A brief definition of related terminology is given for a basic understanding of the underlying principles of photogrammetry. An exhaustive treatment of this topic is beyond the scope of this chapter and the reader is referred to a more comprehensive text such as that of Wolf and Dewitt (2000). For the purposes of the first part of this discussion, it is assumed that frame cameras are being employed for image acquisition. A frame camera captures an entire frame at each exposure time. Sensors using linear arrays or scanning mechanisms build up images in pieces such as lines or sets of lines. Further treatment of such imaging systems will be given in a later part of this chapter.

Photographic or Image Scale defines the scale relationship between objects in reality and as recorded on images. Since an image is a 2D representation of 3D objects, it is apparent that some information is lost when the image is recorded. As such, every point on an image has its own photographic scale. For the sake of simplicity, an average photographic scale is used. This scale can be computed as the ratio of the distance between two points as computed in object space (the real world) and on the plane of the photograph. It is recommended that the two points that are chosen should be at an average distance as compared to all points recorded on the image. It should be noted that unless there is a distinction to be made, the terms photograph (film-based) and image (digital sensor based) will be used interchangeably.

Camera or Sensor Orientation relates to the position and attitude parameters of the camera corresponding to the time of exposure (image recording). For most cameras, it is sufficient to define its orientation in terms of six geometric degrees of freedom. These six parameters define the camera's position in object space (X^s, Y^s, and Z^s) and its attitude with respect to the three primary axes (X, Y and Z) of the object space coordinate system. The three angular parameters describing the camera's attitude are ω, ϕ, and κ and are about the X, Y and Z axes, respectively.

Interior Orientation is the process of re-establishing the interior geometry of the imaging sensor (i.e., camera) corresponding to a supplied calibration report or through a calibration procedure. The fiducial marks permit the recovery of a consistent image coordinate system to which corrections are subsequently applied to account for principal point offsets and lens distortion as well as atmospheric refraction and, potentially, earth curvature corrections (for details see Wolf and Dewitt, 2000).

Relative Orientation of Stereomodels involves the re-establishment of the relative geometry of a stereo pair of photographs using the interior orientation parameters and a minimum of five points yielding independent equations. Generally, six or more points are used to minimize the chance of accepting an erroneous observation. Upon completing the relative orientation process, the pair of photographs can be viewed in stereo such that a scaled version (usually smaller scale) of object space is viewable. The 3D "model" that can be viewed forms the so-called stereo model. The model, in its initial state, has no direct link to the coordinate system used in object space which explains the use of the term relative orientation.

Absolute Orientation represents the process after relative orientation whereby the stereomodel is transformed in the coordinate system of object space by means of control points that are visible in the pair of photographs, and which points have known object space coordinates permitting the transformation into an "absolute" coordinate system.

Exterior Orientation is the process that combines both relative and absolute orientation into a single operation. Most modern photogrammetric systems permit the use of either a two step (relative then absolute) orientation or one step orientation (exterior) in order to establish the relationship between measurements on a photograph and those in object space.

Restitution involves the collection of 3D information from the stereo models and assembling these data to produce a map (e.g., topographic) of the entire project area. Traditionally, this was done using photogrammetric stereoplotters or analytical plotters. Generally, all natural and man-made features are collected for the production of a digital map or the introduction into a GIS.

Digital Elevation Model (DEM) Generation occurs once the absolute or exterior orientation process is completed. This process involves the collection of a dense distribution of points (X, Y and Z) within each stereo model. These points represent a discrete portrayal of the earth's surface. DEM data from several stereo models can be combined to cover a larger area. Other terms such as digital terrain model (DTM) and digital surface model (DSM) are often associated with DEMs. The distinction between these terms is derived from the fact that a DEM represents the earth's surface, a DTM may also include natural features such as vegetation covering the earth's surface while a DSM may also include man-made features such as buildings.

Orthoimage Generation is the process that occurs after a DEM is acquired for the area imaged on a photograph. Since a photograph is a perspective projection of the object being imaged several distortions remain in the image which make it unusable for many mapping operations. Most significant are the effects of relief displacement and camera attitude. Through the use of the orientation parameters determined through the absolute or exterior orientations processes, effects of camera attitude can be effectively eliminated. The effects of relief can then be removed (or at least minimized) by introducing the DEM corresponding to the area recorded by the photograph.

2.6.2 Photogrammetric Processing of Multiple Photographs

In the case where project areas cover a larger area, multiple photographs can be employed to ensure complete coverage. The complete set of these photographs for a given project is termed a *block*. Mathematical models have been developed to simultaneously process blocks involving from one to several thousand photographs. An example of a project area covered by six photographs is given in Figure 2.15.

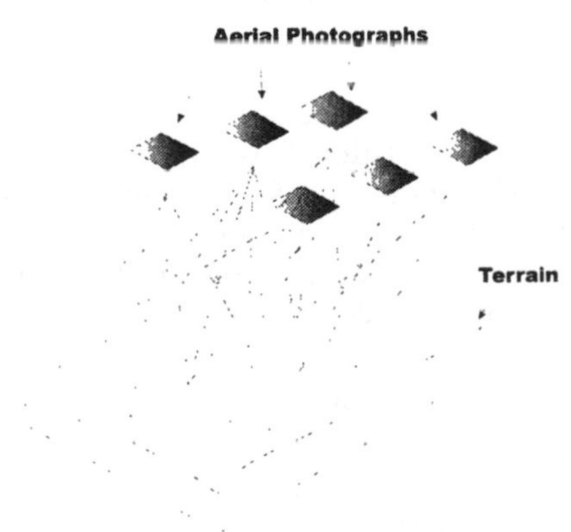

Figure 2.15 Block of Aerial Photographs

To fully appreciate the notion of using multiple photographs for measurement purposes, a brief overview of analytical photogrammetry is given. The following expressions represent the mathematical relationship between points on the ground (i.e., object space) and their images on the photographs (i.e., image space) (see Figure 2.16). The planes for both the original image negative and corresponding image positive are given since measurement work is frequently carried out in the positive image plane.

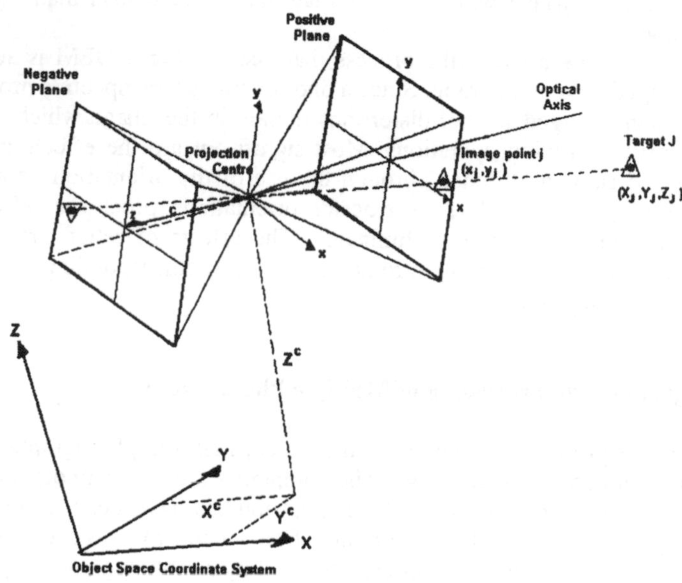

Figure 2.16 Imaging Geometry for a Single Image

Seven parameters define the mathematical transformation from object space to image space. These parameters include a scale, three rotations and three translations. The notations for rotational parameters about *X*, *Y* and *Z* axes are ω, ϕ and κ, respectively. The terms R_ω, R_ϕ, R_κ are the rotational matrices about *X*, *Y* and *Z* axes, respectively and S is a scale factor. The relationship between a point (*i*) in object space and an image (*s*) is developed for illustrative purposes. The following collinearity equations are derived as:

$$\begin{bmatrix} x \\ y \\ -f \end{bmatrix} = S * R_\kappa * R_\phi * R_\omega * \begin{bmatrix} X_i - X_s \\ Y_i - Y_s \\ Z_i - Z_s \end{bmatrix} \tag{2.2}$$

where: $R = R_\kappa * R_\phi * R_\omega = \begin{bmatrix} r_{11} & r_{12} & r_{13} \\ r_{21} & r_{22} & r_{23} \\ r_{31} & r_{32} & r_{33} \end{bmatrix}$

which can be rewritten as:

$$\begin{bmatrix} x \\ y \\ -f \end{bmatrix} = S * \begin{bmatrix} r_{11} & r_{12} & r_{13} \\ r_{21} & r_{22} & r_{23} \\ r_{31} & r_{32} & r_{33} \end{bmatrix} * \begin{bmatrix} X_i - X_s \\ Y_i - Y_s \\ Z_i - Z_s \end{bmatrix} \tag{2.3}$$

Furthermore, the equations can be expanded to:

$$x = \frac{-f * S * (r_{11} * (X_i - X_s) + r_{12} * (Y_i - Y_s) + r_{13} * (Z_i - Z_s)}{S * (r_{31} * (X_i - X_s) + r_{32} * (Y_i - Y_s) + r_{33} * (Z_i - Z_s)}$$

$$y = \frac{-f * S * (r_{21} * (X_i - X_s) + r_{22} * (Y_i - Y_s) + r_{23} * (Z_i - Z_s)}{S * (r_{31} * (X_i - X_s) + r_{32} * (Y_i - Y_s) + r_{33} * (Z_i - Z_s)} \tag{2.4}$$

These expressions correspond to the basic measurement equations relating image space coordinates to object space coordinates (see Figure 2.16).

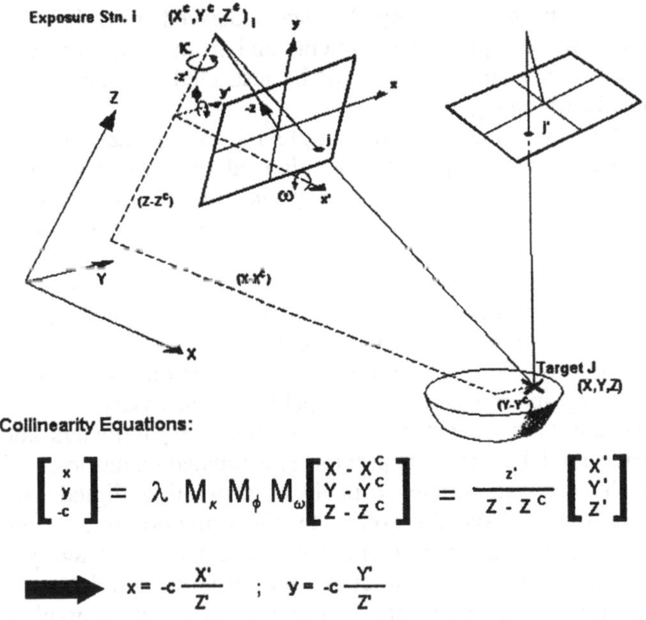

Figure 2.17 Geometric Relationships for Two Images and Object Space

When a collection of these equations are assembled for all photographs in the block, then a photogrammetric adjustment can occur that produces a homogeneous set of results for the entire block (Wolf and Dewitt, 2000). The two-dimensional nature of the image restricts its use since much of the explicit 3D information corresponding to the recorded object is suppressed. In order to reconstruct the 3D characteristics of the object, it is often necessary to employ two or more images of the object taken from different exposure stations. The concept of using two images for 3D reconstruction is illustrated in Figure 2.17.

2.6.3 Softcopy Photogrammetry

Traditional photogrammetry employed film-based cameras. With the advent of high-resolution digital cameras, we now have the possibility to collect data in a computer compatible form. The notion of digital or softcopy photogrammetry grew out of the realization that (1) conventional analytical instrumentation was expensive and required costly maintenance, (2) high-precision scanners were becoming available and there would be an eventual availability of high-resolution digital cameras, (3) computer technology had matured to a level that it would address the computational and storage needs of high-resolution digital images, and (4) image processing algorithms would permit the enhancement and improved extraction of quality geometric information. While there are several commercially available softcopy photogrammetry packages on the market, they all possess similar basic features. These features include:

- The ability to accept digital imagery derived from the scanning of film-based photography from cameras and from digital cameras themselves. Discussion in this section will be restricted to aspects of scanned film-based aerial images.
- A stereo viewing capability employing one of many possible techniques such as anaglyph, split-screen, polarized monitors and flicker mode.
- The ability to move a measuring and floating mark about the image or stereo model, respectively, in order to measure points observed within the imagery.
- With the use of various software modules, the ability to complete the tasks of interior, relative, absolute or exterior orientation as commonly done on analytical photogrammetric plotters.
- In order to effectively complete the orientation tasks, some form of error reporting and remeasurement capability is necessary.
- The ability to save the derived orientation parameters and to reset the stereo model using these previously estimated quantities.
- In the case of systems capable of generating digital orthoimages, the system must have the ability to either import or generate a DEM for applying subsequent corrections to the perspective imagery.
- For the purpose of restitution, the system should have the ability to incorporate existing vector data (e.g., dual screen or graphical overlay) so that map completion or updating can be completed.
- Finally, the system should be able to store and/or print the generated orthoimage and vector data either separately or as a composite image.

The mandated features of a softcopy photogrammetric system are very similar to those of many existing analytical photogrammetric systems. The degree to which each of these softcopy systems can achieve its objectives (i.e., meet specifications) will determine its appropriateness for a particular application.

2.6.3.1 Softcopy and Analytical Photogrammetry: a Comparison

Given that analytical photogrammetric procedures have been adopted as a mapping standard during the 1980s and 1990s, it is useful to contrast any new approach with this accepted norm. A rigorous comparison requires that each of the stages employed in each of the processes be compared and contrasted in a qualitative and quantitative manner. Table 2.8 represents the framework from which subsequent comparisons will be made.

Table 2.8 A Comparison between Analytical and Software Photogrammetry

Stage	Analytical	Softcopy
Image source	cut film negative or diapositive digitized	cut or roll film negative or diapositive
Measurement system	software controlled precision stages with zoom, variable measuring marks and illumination	software controlled screen display of images with zoom, image enhancement and variable measuring marks
Interior orientation	software processed manual measurements of fiducials (4 or 8) and reseau	software processed manual or automated measurement of (4 or 8) fiducials and reseau
Relative orientation	software processed manual measurements of unlimited number of points	software processed manual or automated measurements of an unlimited number of points
Absolute orientation	software processed manual measurements of unlimited number of points pass/tie points indicated via point marking (pugging)	software processed manual or automated measurements of unlimited number of points pass/tie points indicated via point marking (pugging) or by automated image chip matching
Exterior orientation	software processed manual measurements of unlimited number of points, pass/tie points indicated via point marking (pugging)	software processed manual or automated measurements of unlimited number of points, pass/tie points indicated via point marking (pugging) or by automated image chip matching
Restitution	manual collection in stereo, possible vector superimpostion and captured vector display	manual collection in mono with DEM or in stereo, possible vector superimpostion automated feature extraction envisaged
DEM generation	manual or software controlled scanning mode	manual or software controlled scanning mode with automated matching (auto-correlation)
Orthophoto generation	off-line processing with hardware or software system, image processing possible	processing completed with software module, image processing possible

2.6.3.2 Image Sources

In the case of softcopy systems, the choice between the use of scanned diapositives and negatives is generally a question of (1) the availability of the negative roll and (2) the ability of the scanner to accept rolls of film. The scanned negative can be tone-reversed using software to produce an equivalent diapositive. As such, a film image must be scanned at about 8-10 μm (as per the Nyquist theorem) in order to preserve the geometric information content of a typical high quality negative. The required scanning resolution for maintaining the geometric quality of diapositives is, therefore, somewhere between 10 μm and 20 μm. This also presupposes that the scanner has passed the calibration tests to be set out at a later date.

2.6.3.3 Measurement System

At this point it is of value to note the differences between various terms used to describe the geometric quality of a system. The use of the word "accuracy" expresses the closeness of a measurement to the true value. On the other hand, "precision" describes the degree of grouping of a set of measurements. The word "resolution" expresses the least count of a system. In the case of scanners, this corresponds to the pixel spacing. The word "acuity" is sometimes used to express the resolution of the elementary storage value. For example, if a pixel dimension is represented by 8 bits, then it would suggest a sub-pixel resolution (or acuity) of 1/256 or ~0.04 pixels. This does not imply such an accuracy is achievable but only representable. However, it does permit sub-pixel pointing which is essential in order for softcopy photogrammetry to be competitive with its analytical counterpart.

Based upon the ability of a softcopy system to make sub-pixel measurements on a target size of 1 pixel, then the probability of the target center being at a point other than the pixel center is ± d (d is the pixel width). Thus, the quantizing error, in terms of the expected value $E(x - \mu)$ is expressed as:

$$\sigma^2_q = E(x-\mu)^2 = \int_{0.5}^{+0.5} (x-\mu)^2 p(x)dx = 0.083 \quad \text{pixel units}^2 \tag{2.5}$$

where σ^2_q is the variance of quantization (q) expressed in pixel units^2, p(x) is the probability of pixel x (for equal probability; $p(x) = 1$), is the mean of the distribution (for an interval of ± 0.5, $\mu = 0$).

Consequently, the standard deviation of quanitization (measurement) σ^2_q is 0.289 pixels or approximately 0.3 pixels. This, in part, explains why sub-pixel accuracies are achievable despite the fact that the elementary measurement unit is a pixel. It also expresses the need to have sub-pixel pointing capability (Cosandier and Chapman, 1992).

The requirement for sub-pixel pointing in softcopy photogrammetric systems is also due to the fact that most objects to be measured (e.g., fiducials, reseau crosses, images of targets of ground control points, natural features used for relative orientation and pass and tie points) are represented by several pixels. This permits the operator to use all of the available local information to carry out the actual measurement. Table 2.9 illustrates specific cases of sub-pixel pointing when a symmetrical target (e.g., fiducial, target) is measured.

Since a typical fiducial mark has a cross section of 20 to 60 μm, then a scanning resolution of 10 μm will result in a pixel width of 2 to 6 pixels, respectively. Similarly, a scanning resolution of 15 μm will result in a pixel width of 1.5 to 4 pixels, respectively. Thus, a 15 μm scanning resolution would offer a 1.7 μm precision if a typical fiducial is 40 μm in cross section. This would be satisfactory for most measurements involving fiducials imaged on aerial photography.

Table 2.9 Feature Size Versus Subpixel Resolution

Feature diameter (pixels)	Estimated standard deviation (pixels)	Number of potential edge observations
60.0	0.010 (1/100)	164
30.0	0.014 (1/70)	84
15.0	0.019 (1/50)	44
7.5	0.030 (1/30)	17
4.0	0.065 (1/15)	8
3.0	0.110 (1/9)	6
2.0	0.140 (1/7)	4
1.0	0.289 (1/3)	2

2.6.3.4 Interior Orientation Comparison

Typical specifications for analytical interior orientation require a minimum of four corner or side fiducials (8 fiducials maximum) and a 2D affine transformation model to be used. Residual errors of ± 3 - 7 μm are normally accepted as meeting the specifications. Experience has shown that a 15 μm pixel resolution will permit the same results to be achieved using a softcopy system. While lower scanning resolutions (i.e., > 15 μm) are capable of achieving satisfactory interior orientation results, the reduced image quality would hamper subsequent steps. However, with the introduction of automated measuring techniques, such as template matching, more progress will be made in the area of precision related to automated systems.

2.6.3.5 Relative Orientation

It is in this, and subsequent operations, where the effect of the scanning resolution is most critical. This is due to the fact that points used for relative orientation are usually not pre-signalized and are often natural features covering small, symmetrical areas and possessing high-contrast characteristics. The reduced definition of these points on the imagery has the effect of reducing the measuring acuity of the operator. This problem is experienced in both analytical and softcopy systems and is generally only a question of the resolution of the imagery.

The difference between analytical and softcopy systems comes from the characteristics of the inherent transport systems. In the case of analytical systems, the measurement error is always influenced by variations in stage positioning errors and the encoders. For softcopy systems, the digital nature of the imagery eliminates this additional error and the only transport error is introduced as a constant at the time of scanning.

On the radiometric side, the illumination and magnification can be used on the analytical plotter to enhance the image locally. Such local enhancement can be achieved in a non-linear manner on softcopy systems and zooming is virtually unlimited (except for pixel size considerations). Tests have shown that a 15 μm scanning resolution (0.6 m on the ground for 1:5,000 photography) produces results comparable to a typical analytical plotter. In fact, when images are acquired over urban areas a scanning resolution of 20 μm produces acceptable results due to the availability of better-defined objects.

2.6.3.6 Absolute Orientation

Both analytical and softcopy systems produce similar results when a 15 μm scanning resolution is used for the softcopy systems and targeted ground control points are used. The same statement is true for marked ("pugged") points since the drill size is 40 to 60 μm and the measurement acuities of both systems are, therefore, comparable. As with relative orientation, a lower scanning resolution (e.g., 20 μm) could be used if targets and "pugged" points are well defined.

2.6.3.7 Exterior Orientation

The process of exterior orientation uses the same algorithms for both softcopy and analytical methods. In these cases, ray intersections for ground points are used from all available photographs/images. Similar statements as those made for absolute orientation can be applied to the exterior orientation process. Again, results have shown that a 15 μm scanning resolution produces results that are competitive with those obtained from analytical methods. Superior results can be obtained with softcopy systems when a 10 μm scanning resolution is used, however, little improvement is seen if the scanning resolution is further refined.

2.6.3.8 Restitution

Model restitution is the process that is most affected by the choice of scanning resolutions. This is due to the fact that many of the features to be captured are not clearly defined as in the case of ground control targets or other man-made features. However, this problem is also encountered with analytical photogrammetric systems. In both cases, there is about a 1.5 - 2x reduction in acuity for features that are not well defined and that do not have good background detail. Given that these measurements are made in stereo viewing mode, the problem is somewhat compounded. Using the Nyquist theorem that states that a feature must be represented by a sampling frequency of twice its frequency, then a 0.30 m feature must be represented by an equivalent 0.15 m sampling frequency. For 1:5,000 photography, this implies a sampling frequency of 15 μm if the 2x reduction in acuity is assumed. With a small safety margin introduced, the 10-15 μm sampling resolution is, again, suggested. This has been verified through block adjustments using both simulated and real data.

This implies that points and lines collected through softcopy restitution would have error budgets as:

$$\sigma_T = (\sigma_{AT}{}^2 + \sigma_{GC}{}^2 + \sigma_S{}^2 + \sigma_{DIG}{}^2)^{1/2} \quad \text{for a digitized point} \tag{2.6}$$

$$\sigma_T = (30^2 + 20^2 + 2^2 + 20^2)^{1/2} = \pm 41.3 \ \mu\text{m (20.7 cm at ground scale)}$$

where σ_{AT} is standard deviation of aerial triangulation (± 0.15 m), σ_{GC} is standard deviation of control point measurements, σ_S is standard deviation of scanning process, and σ_{DIG} is standard deviation of digitizing a point feature,

and

$$\sigma_T = (\sigma_{AT}{}^2 + \sigma_{GC}{}^2 + \sigma_S{}^2 + \sigma_{DIG}{}^2)^{1/2} \quad \text{for a digitized line} \tag{2.7}$$

$$\sigma_T = (30^2 + 20^2 + 2^2 + 25^2)^{1/2} = \pm 43.9 \ \mu\text{m (22.0 cm at ground scale)}$$

where σ_{AT} is standard deviation of aerial triangulation (± 0.15 m), σ_{GC} is standard deviation of control point measurements, σ_S is standard deviation of scanning process, and σ_{DIG} is standard deviation of digitizing a line feature.

2.6.3.9 Orthophoto Generation

The superiority of the softcopy system is realized at the phase of orthophoto generation. In fact, most users of analytical photogrammetric systems now use a softcopy solution for orthophoto production. The reason that softcopy orthophoto systems are better stems from the fact that the images can readily be enhanced in a digital form. In addition, tasks such as mosaicking and draping over DEMs are ideally suited for softcopy systems. Finally, the final product is produced in a digital format that permits ease of communication to other computers or for the production of hard copies using a printer/plotter.

2.6.4 Direct Georeferencing

With the advent of the GPS, it became possible to precisely determine the location of an exposure station at the instant of exposure. This was primarily due to the availability of the precise time offered by GPS (Mostafa *et al.*, 1998). In addition, the inclusion of data from an IMU introduced the possibility of observing the angular orientation of the camera at the instant of exposure. When such systems are used to the exclusion of control points on the ground, the process of direct georeferencing is invoked.

The physical relationship between a camera, an IMU, and a GPS antenna in a mobile mapping system (MMS) is shown in Figure 2.18, in which $\mathbf{r}_P^M$ indicates the coordinates of a point in the mapping coordinate system, and $\mathbf{r}_p^c$ indicates the coordinates of the same point in the camera coordinate system. Mathematically, these coordinates are related by

$$\mathbf{r}_P^M = \mathbf{r}(t)_{GPS}^M - \mathbf{R}(t)_b^M \mathbf{R}_c^b \left(\mathbf{r}_{GPS}^c - \mu_p^P \mathbf{r}_p^c\right) \tag{2.8}$$

where $\mathbf{r}_{GPS}^M$ are the coordinates of the GPS antenna, $\mathbf{R}_b^M$ is the rotation matrix between the IMU (body) and mapping coordinate frames, μ_p^P is the scale between the image and mapping frames for the point, $\mathbf{r}_{GPS}^c$ is the vector connecting the GPS antenna and the camera, and $\mathbf{R}_c^b$ is the rotation matrix between the camera and IMU frames. *t* indicates the time changing quantities, specifically the GPS position and IMU orientation.

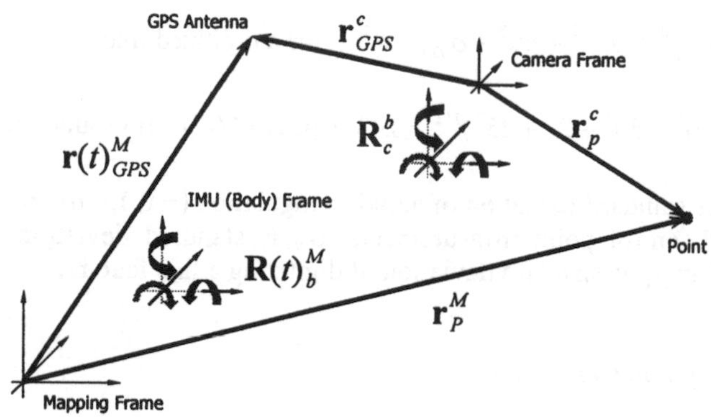

Figure 2.18 Relationships between Sensors in a Direct Georeferencing System (Ellum et al, 2002)

The significant terms in Equation 2.8 are $\mathbf{r}_p^c$, $\mathbf{r}_{GPS}^c$, and $\mathbf{R}_c^b$. The former term, i.e., the coordinates of the point in the camera coordinate system, has the effects of camera calibration embedded in it. The latter two terms describe the relative position and orientation between the navigation sensors and the camera. Determination of the $\mathbf{r}_{GPS}^c$ vector between the camera and the GPS antenna is known as lever-arm calibration and determination of the $\mathbf{R}_c^b$ rotation matrix between the camera and IMU is known as boresight calibration. It should be noted that some authors refer to both calibrations as the boresight calibration (or the calibration of the boresight parameters). However, to clearly demarcate between the two calibrations, the terminology presented above will be followed herein (Ellum *et al.*, 2002).

The effect of calibration errors on the total system error can be found by performing a first-order error analysis on Equation 2.8. The result, shown in Equation 2.9, illustrates that errors in calibration have the same effect as measurement errors. As a corollary, a more accurate calibration can mean that less accurate, and less expensive, navigation sensors are required.

Employing these equations with photogrammetric data captured using a mobile system can yield point coordinates of features with accuracies of 10-15 cm. Such accuracies are generally more than sufficient for most GIS applications.

$$
\begin{array}{lll}
\delta\mathbf{r}_P^m = & \delta\mathbf{r}(t)_{GPS}^M & \textit{GPS position error} \\
- & \delta\mathbf{R}(t)_b^M\,\mathbf{R}_c^b\left(\mathbf{r}_{GPS}^c - \mu_p^P\mathbf{r}_p^c\right) & \textit{IMU Attitude error} \\
- & \mathbf{R}(t)_b^M\,\delta\mathbf{R}_c^b\left(\mathbf{r}_{GPS}^c - \mu_p^P\mathbf{r}_p^c\right) & \textit{Calibration error in IMU/Camera misalignment angles} \\
- & \mathbf{R}(t)_b^M\,\mathbf{R}_c^b\delta\mathbf{r}_{GPS}^c & \textit{Calibration error in Camera/GPS offset} \\
+ & \mathbf{R}(t)_b^M\,\mathbf{R}_c^b\delta\mu_p^P\mathbf{r}_p^c & \textit{Scale error} \\
+ & \mathbf{R}(t)_b^M\,\mathbf{R}_c^b\mu_p^P\delta\mathbf{r}_p^c & \textit{Image point measurement and camera calibration error} \\
+ & \delta t\left[v(t) - \omega(t)\mathbf{R}_c^b\left(\mathbf{r}_{GPS}^c - \mu_p^P\mathbf{r}_p^c\right)\right] & \textit{Synchronisation error}
\end{array} \qquad (2.9)
$$

2.6.5 Photogrammetric Processing of Satellite Imagery

While the traditional photogrammetric method of relating object space to image space employs the collinearity equations, many new satellite sensors have complex imaging models that do not easily lend themselves to such equations (Chapman *et al*, 1991, Dowman and Dolloff, 2000). These modern sensors are very agile and can capture images under high dynamical situations. Satellites such as IKONOS, QuickBird and OrbView employ such sensors (see Table 2.3).

Several mathematical models have been introduced to attempt to rigorously model the imaging geometry of these sensors. One of the most successful models involves the use of rational polynomials or functions. These mathematical models first found success in conjunction with military satellites. The basic principle underlying these equations is that the complex imaging geometry can be represented by the ratio of two polynomial expressions:

$$r_n = a_1\,(X_i, Y_i, Z_i)\,/\,c_1\,(X_i, Y_i, Z_i)$$

$$c_n = b_1\,(X_i, Y_i, Z_i)\,/\,c_1\,(X_i, Y_i, Z_i) \qquad (2.10)$$

where, r_n, c_n, are the normalize row and column coordinates of a pixel, X_i, Y_i, Z_i are the coordinates of control points on the ground, and a_1, b_1, c_1 are the polynomial functions relating object space to image space.

A shortcoming of these was previously related to the inability to rigourously propagate the variances and covariances. In addition, the explicit satellite orbital parameters and related observations could not be directly entered into these expressions. Recently, these shortcoming have been overcome to a certain degree making them more attractive for use due to their flexible nature (Dowman and Dolloff, 2000).

2.7 REMOTE SENSING IN TELEGEOINFORMATICS

2.7.1 Imaging in Telegeoinformatics

It is frequently stated that a picture is worth a thousand words. The inclusion of image data in GISs is becoming increasing common as timely, up-to-date, complete and accurate data are expected to be in demand. As more users become aware of the potential of such information, there will be an inevitable increase in demand for current and new information and services.

The retail industry actively uses GISs to develop location models and sales strategies. Knowing the demographics of their potential clients enables retailers to select store locations. With the inclusion of image date especially from satellite, they will be able to keep up-to-date with the development changes in the vicinity of their existing and planned outlets. While the real estate has been using GIS data for some time, their full adoption has been somewhat delayed. Given the visual nature of residential dwelling and commercial building sales it would seem natural that the real estate sector would embrace this new opportunity. Casual browsing on the Internet reveals that the adoption of these new data sources is slowly being realized.

Many new automobiles are coming equipped with Automatic Vehicle Location (AVL) devices using digital maps. As satellite image date becomes more readily available and affordable, motorists will be more inclined to use these data sources to ensure their trips are pleasant and free of unforeseen detours and maintenance work along the roadways.

The rapid development of *Intelligent Transportation Systems* (ITSs) is another example of how GIS data, including satellite imagery, can participate in the modernization of the transportation sector. In fact, there are several departments of transportation in the U.S., Canada and Europe which are routinely using mobile mapping systems that employ GPS, IMU and imaging sensors. These systems determine asset location and highway parameters such as rutting, cracking, potholes and line-of-sight (Li *et al.*, 1994, Chaplin and Chapman, 1998)

Many groups in the transportation and shipping industry are routinely scheduling and monitoring their fleets using GIS data coupled with GPS. The use of satellite image data will further enhance their ability to increase the efficiency and safety of their drivers, fleets of vehicles and goods being shipped.

Applications in environmental assessment have been reported since the 1980s. The availability of short turn-around-time satellite image data will increase use of such information for environmental, search and rescue and disaster monitoring applications.

2.7.2 Mobile Mapping Technology and Telegeoinformatics

Technological advancements in positioning/navigation and imaging sensors that occurred in the 1990s, practically redefined the concept of airborne and land-based remote sensing and photogrammetric mapping. Mobile Mapping Systems (MMSs) can be defined as moving platforms upon which multi-sensor measurement systems have been integrated to provide 3D near-continuous positioning of both the platform's path in space and simultaneously collected geospatial data. A recent overview of the mobile mapping technology can be found in Grejner-Brzezinska *et al* (2002). The advent of first MMS in the early 1990s initiated the process of establishing modern, fully digital, virtually ground control-free remote sensing and photogrammetric mapping. By the end of the 1990s, Mobile Mapping Technology (MMT) has made remarkable progress, evolving from rather simple land-based systems to more sophisticated, real-time multi-tasking and multi-sensor systems, operational in land-based and airborne environments. Following the proliferation of GPS/INS integrated technology in the mid-1990s, the quality of direct platform orientation (DPO) reached the level of supporting even demanding airborne mapping. New, specialized systems, based on modern imaging sensors, such as CCD cameras, Lidar and hyper/multi-spectral scanners, are being developed, aimed at automatic data acquisition for GIS databases, thematic mapping, land classification, terrain modeling, etc.

MMT has been employed in a number of important real-life applications. Airborne MMSs have mainly been used for the same purposes as traditional aerial mapping technology. However, because they use digital imaging sensors and avoid or greatly reduce ground control requirements, MMSs provide much higher efficiency than traditional aerial mapping. Land-based MMSs, on the other hand, have mainly been used in highway and facility mapping applications in which traditional terrestrial mapping technology is either impossible or very inefficient to apply. Over the past decade, the uses of both airborne and land-based MMSs have been notably expanding in many new directions.

There are several advantages to using MMT in highway applications. Because it employs dynamic data acquisition, MMT can be directly used in highway-related applications such as traffic sign inventory, monitoring of speed limits and parking violations, and generation of road network databases. When laser technology is jointly applied, road surface condition inspection can also be achieved. As long as traffic velocity is less than approximately 70 km per hour, data acquisition can be performed without disturbing traffic flow. In addition, a single collection of data can be used to obtain diverse information for multiple purposes. Moreover, because data can be both collected and processed in a short time period, frequent repetition of road surveys and updating of databases are both possible and affordable.

Mobile GISs are one of the latest useful services that have come out of the inexpensive access capabilities provided by the new wireless Internet and mobile mapping capability. Its development has been stimulated by increasing demand for up-to-date geospatial information, along with technological improvements in hardware size, performance, power consumption and network bandwidth (Maguire, 2001). Mobile GIS applications can deliver mapping output in a number of different formats including text, image, voice, and video in field. Information can be requested from Internet servers using the wireless application protocol-based Wireless Markup Language (WML). Resulting maps can be displayed in the form of an embedded map (Maguire, 2001). Along with many others, GIS software market leaders AutoDesk, ESRI and Intergraph are currently developing software and system architecture for mobile GIS applications.

Compared with the traditional GIS applications, mobile GIS technology has two unique components: (1) portable mobile devices, such as wireless telephones and PDAs, serving as client terminals and (2) a wireless network. Wireless telephones can access data in three different ways: Wireless Application Protocols (WAP), General Packet Radio Service (GPRS), and I-Mode. Using special wireless services (such as GoAmerica) or a wireless modem, PDA users can access content through the Internet with Windows CE-based Internet Explorer or other WAP-capable browsers. Current applications of mobile GISs include field mapping, routing, tracking, data collection, and public safety. Two popular uses of the technology are routing and facility search (Srinivas *et al.*, 2001).

Shortcomings of the current mobile GIS technology include a limited number of service providers which offer wireless data transfer services and limited wireless bandwidth. The current transfer speed through CDMA (Code Division Multiple Access) WAP wireless telephones is about 9.6 kbps. The speed of GPRS WAP telephones is about 53.6 kbps. These transfer speeds would be unacceptable for use in transferring large spatial data sets, such as those found with remote sensing data. As the technology improves, mobile GIS will become much more accepted, especially when wireless data transfer speeds are greatly increased with the introduction of the third-generation communication systems, expected in the next few years.

Telegeoinformatics, as an emerging discipline with its roots in MMT, integrates the theory and applications of geospatial informatics (geoinformatics), telecommunication, and mobile computing technologies. Geoinformatics, as a foundation component of Telegeoinformatics, is based on GISs, which utilizes remote sensing and geolocation techniques such as GPS/INS, to gather both spatial and attribute information. Telegeoinformatics directly exploits geoinformatics, i.e., GIS, GPS, satellite remotely sensed imagery and aerial photography in a mobile computing environment, where the mobile computers are linked through a wireless communication network.

REFERENCES

Avery, T. E. and Berlin, G. L., 1992, *Fundamentals of Remote Sensing and Airphoto Interpretation*, 5 th ed., (Upper Saddle River: Prentice Hall).

Baker, J. C., O'Connell, K. M. and Williamson, R. A. (Eds.), 2001, *Commercial Observation Satellites: At the Leading Edge of Global Transparency,* (Bethesda: ASPRS and RAND).

Chaplin, B. and Chapman, M. A., 1998, A procedure for 3D motion estimation from stereo image sequences for a mobile mapping system, In *Proceedings of the ISPRS Commission III Symposium on Object Recognition and Scene Classification from Multispectral and Multisensor Pixels,* July 6-10, Columbus, Ohio, pp. 17 - 22.

Chapman, M.A., Tam, A. and Yasui, T., 1991, A rigorous approach for the estimation of terrestrial coordinates from digital stereo SPOT imagery, In *Proceedings of the 14th Canadian Symposium on Remote Sensing*, Calgary, Alberta, pp. 495-501.

Congalton, R. C., 1991, A review of assessing the accuracy of classifications of remotely sensed data, *Remoter Sensing of Environment*, **37**, pp. 35-46.

Cosandier, D. and Chapman, M. A., 1992, High precision target determination using digital imagery, In *Proceedings of the SPIE Symposium on Videometry*, Boston, pp. 111-122.

Dowman, I. and Dolloff, J. T., 2000, An evaluation of rational functions for photogrammetric restitution, *IAPRS*, 33 (B3), pp. 254-266 (CD-ROM).

Ellum, C., El-Sheimy, N. and Chapman, M. A., 2002, The calibration of image-based mobile mapping systems, unpublished manuscript, 22 pp.

ERDAS, 1999, *Field Guides*, 5 th ed., (Atlanta: ERDAS).

Green, K., Kempka, D. and Lackey, L., 1994, Using remote sensing to detect and monitor land-cover and land-use change, *Photogrammetric Engineering & Remote Sensing*, **60**(3), pp. 331-337.

Grejner-Brzezinska, D. A., Li, R., Haala, N. and Toth, C., 2002, Multi-sensor systems for land-based and airborne mapping: Technology of the future? *IAPRS*, 34 (2): pp. 31-42.

Jensen, J. R., 1996, *Introductory Digital Image Processing*, (Upper Saddle River: Prentice Hall).

Jensen, J. R., 2000, *Remote Sensing of the Environment—An Earth Resource Perspective*, 1 st ed., (Upper Saddle River: Prentice Hall).

Lillesand, T. M. and Kiefer, R. W., 2000, *Remote Sensing and Image Interpretation*, 4 th ed., (Toronto: John Wiley & Sons).

Li, R., Schwarz, K.-P., Chapman, M. A. and Gravel, M., 1994, Integrated GPS, INS and CCD cameras for rapid GIS data acquisition, *GIS World*, **7**(4), pp. 41-43.

Lo, C. P. and Yeung, A. K. W., 2002, *Concept and Techniques of Geographic Information Systems*, (Upper Saddle River: Prentice Hall).

Maguire, D., 2001, Mobile geographic services. *Map India 2001*, 11 pp.

Maune, D. (Ed.), 2001, *Digital Elevation Model Technologies and Applications: The DEM Users Manual*, (Bethesda: ASPRS).

Mostafa, M. M. R., Schwarz K. P. and Chapman, M. A., 1998, Development and testing of an airborne remote sensing multi-sensor system, In *Proceedings of the ISPRS WG II/1 Conference on Data Integration: Systems and Techniques*, Cambridge, UK, 13 -17 July, pp. 217-222.

Richard, J. A. and Jia, X., 2000, *Remote Sensing Digital Image Analysis*, 3 rd ed., (New York: Springer-Verlag).

Srinivas, U., Chagla, S. M. C. and Sharma, V. N., 2001, Mobile mapping: challenges and limitations, *Map India 2001*, 6 pp.

Wolf, P. R. and Dewitt, B. A., 2000, *Elements of Photogrammetry with Applications in GIS*, 3 rd ed., (Toronto: McGraw Hill).

CHAPTER THREE

Positioning and Tracking Approaches and Technologies

Dorota Grejner-Brzezinska

3.1 INTRODUCTION

Telegeoinformatics, as an emerging discipline, integrates the techniques and technologies of geoinformatics, telecommunications, and mobile computing. Geoinformatics, as a distinctive component of Telegeoinformatics, is based on Geographic Information Systems (GISs), remote sensing and geolocation techniques, such as the Global Positioning System (GPS), or in general, Global Navigation Satellite Systems (GNSS), to gather and process geospatial position and attribute information. Geospatial data can be also acquired by digitizing maps, or by traditional surveying methods. However, modern techniques fully rely on remote sensing supported by GNSS used for georeferencing of airborne and satellite imagery. Thus, GNSS, remote sensing, and GISs are considered fundamental geospatial technologies, supporting acquisition, analysis and distribution of geospatial data. Satellite radionavigation approaches are also widely used in mobile mapping, which rapidly transforms from emerging to well-established, fully digital and automated technology, serving as a primary data source of data on road networks, facilities and infrastructures. Georeferenced, remotely sensed data are subsequently processed using photogrammetric methods, and converted to intelligent information stored in GIS databases.

Mobile computing, as a next step towards real-time GISs, is a combination of Mobile Internet, wireless communication (such as cellular phone), and position location and information technologies. The key to successful mobile computing is real-time positioning of the mobile user, enabling a large number of applications including wireless mapping, emergency response, fleet tracking, traveler information services, agriculture and environmental protection, and location-based marketing, just to name a few. For example, if the mobile user is asking for directions to the closest post office, the cellular service provider should be able to locate the user, forward his/her coordinates to the Location-Based Service (LBS) provider, where the actual navigation instructions are generated, and send the driving directions back to the user. In order to respond to the mobile user's request, not only must his/her location be found, but also a complete knowledge of the mobile terminal's environment and its realistically predicted change of location must be known. Thus, an efficient combination of a tracking technique with GISs, which provides the location information/attributes, is the key to the success of any LBS/application. GISs, in essence, can serve not only as a resource depository, but also as an instruction and decision clearinghouse, enabling mobile network-based decision-making.

The major focus of this chapter is on positioning and tracking techniques supporting LBSs, which require position information (e.g., from GPS) and location attributes provided by GIS databases. Tracking is commonly defined as a combination of positioning and telemetry providing a full tracking solution for mobile vehicles, where position is computed and then transmitted via a communications network to a control center. However, a positioning system maintaining a continuous log of the object's trajectory is also referred to as a tracking system. Several positioning techniques known as *radiolocation* are presented, with a special emphasis on the satellite-based positioning technology, i.e., GPS. It will be shown that virtually all major LBSs are GPS-supported, if not directly through the mobile GPS user unit, then through at least time synchronization or positioning of the cellular network base stations. Since Telegeoinformatics is focused on the fusion of GISs, telemetry and GPS as a primary tracking technique, more in-depth information about GPS is provided, with a special emphasis on fundamental aspects of positioning with GPS, GPS signal structure, achievable positioning accuracies, modes of positioning with GPS, and GPS modernization. In addition, a Russian counterpart of GPS, the Glonass satellite system, and the forthcoming European system, Galileo, are briefly presented. Next, positioning methods based on cellular networks are discussed. Finally, an inertial navigation concept and other indoor and outdoor tracking techniques are presented.

3.2 GLOBAL POSITIONING SYSTEM

GPS is an example of a single technology that profoundly changed contemporary navigation, surveying and mapping techniques. As a result of progressive innovation and a significant drop in the price of the equipment in the last decade, GPS technology currently supports a variety of applications ranging from precise positioning, cadastral mapping and engineering, to environmental and GIS surveys. GPS can be found on spaceships, aircraft and land-based mapping vehicles. It provides navigation means to airplanes, ships, transportation fleets, emergency cars, golf carts, etc., and new applications, such as LBSs, spring up every day.

3.2.1 Definitions and System Components

The NAVSTAR (NAVigation System Timing And Ranging) GPS is a satellite-based, all-weather, continuous, global radionavigation and time-transfer system, designed, financed, deployed and operated by the U.S. Department of Defense (DOD). The concept of NAVSTAR was initiated in 1973 through the joint efforts of the U.S. Army, the Navy and the Air Force. The first GPS satellite was launched in 1978, and in 1993 the system was declared fully operational. GPS technology was designed with the following primary objectives:

- Suitability for all classes of platforms (aircraft, ship, land-based and space), and a wide variety of dynamics;

- Real-time positioning, velocity and time determination capability;
- Availability of the positioning results on a single global geodetic datum;
- Restricting the highest accuracy to a certain class of users (military);
- Redundancy provisions to ensure the survivability of the system;
- Providing the service to an unlimited number of users worldwide;
- Low cost and low power users' unit.

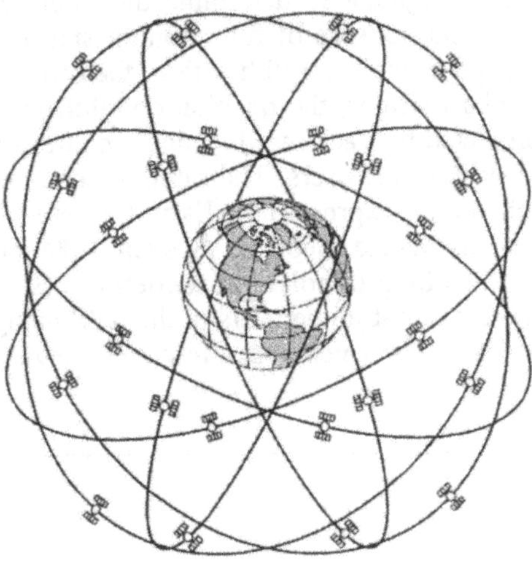

Figure 3.1 The GPS Constellation (Rizos, 2002b)

The nominal GPS Operational Constellation consists of 24 satellites (21 plus three spares) that orbit the Earth at the altitude of ~20,000 km in 12 hours (Figure 3.1). The satellites approximately repeat the same track and configuration once a day, advancing by roughly 4 minutes each day. They are placed in six nearly circular orbital planes, inclined at about 55 degrees with respect to the equatorial plane, with nominally four satellites in each plane. This configuration assures the simultaneous visibility of five to eight satellites at any point on Earth. The current constellation (as of September 2002) consists of 28 Block II/IIA/IIR satellites. Since February 22, 1978 – the launch of the first GPS Block I satellite – the system evolved through several spacecraft designs, focused primarily on the increased design life, extended operation time without a contact from the Control System (autonomous operation), better frequency standards (clocks), and the provision for the accuracy manipulation, controlled by DOD. Block I satellites are referred to as the original concept validation satellites; Block II satellites are the first full-scale operational satellites (first launch in February 1989); Block IIA satellites are the second series of operational satellites (first launch in November 1990), and Block IIR satellites, the operational replenishment satellites carried the GPS into the 21st century (first launch in January 1997). The Block IIF, the follow-on satellites, is currently in the design phase, with the first launch planned for 2006. The information about the current status of the constellation can be found, for example, at the U.S. Coast Guard Navigation Center (2002).

GPS consists of three fundamental segments: *Satellite Segment*, i.e., the satellite constellation itself, the *User Segment*, including all GPS receivers used in a variety of civilian and military applications, and the *Control Segment*, responsible for maintaining the proper operability of the system (Figure 3.2). Since the satellites have a tendency to drift from their assigned orbital positions, primarily, due to so-called orbit perturbations caused by the Earth, Moon and planets gravitational pull, solar radiation pressure, etc., they have to be constantly monitored by the Control Segment to determine the satellites' exact location in space. The Control Segment consists of five monitor stations, each checking the exact altitude, position, speed, and overall health of the orbiting satellites 24 hours a day. Based on these observations, the position coordinates and clock bias, drift and drift rate can be predicted for each satellite, and then transmitted to the satellite for the re-transmission back to the users. The satellite position is parameterized in terms of predicted ephemeris, expressed in Earth-Centered-Earth-Fixed (ECEF) reference frame, known as World Geodetic System 1984 (WGS84). The clock parameters are provided in a form of polynomial coefficients, and together with the predicted ephemeris are broadcast to the users in the GPS navigation message. The accuracy of the predicted orbit is typically at a few-meter level.

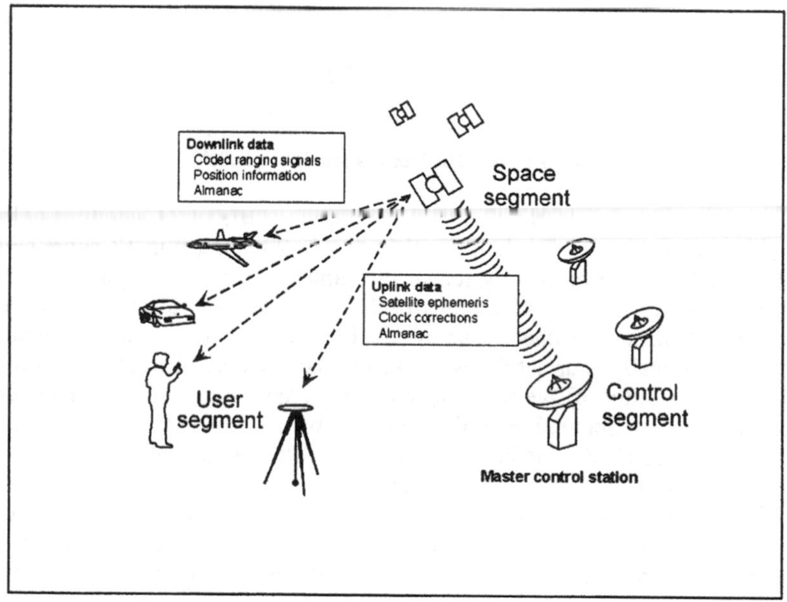

Figure 3.2 GPS Elements (Rizos, 2002b)

3.2.2 GPS Signal Structure

The signal generated by a GPS satellite oscillator contains three primary components: pure sinusoidal waves or carriers (L1 and L2 with frequencies of

54×10.23 MHz and 120×10.23 MHz, respectively),[1] Pseudo-Random Noise (PRN) codes and the navigation message (Figure 3.3). There are two PRN codes, precise P(Y)-code, superimposed on L1 and L2 carriers, and coarse-acquisition C/A-code, superimposed on L1 carrier. All signals transmitted by GPS satellites are coherently derived from a fundamental frequency of 10.23 MHz, as shown in Table 3.1. The frequency separation between L1 and L2 is 347.82 MHz or 28.3%, and it is sufficient to permit accurate dual-frequency estimation of the ionospheric group delay affecting the GPS observables (see Section 3.2.3.1.1).

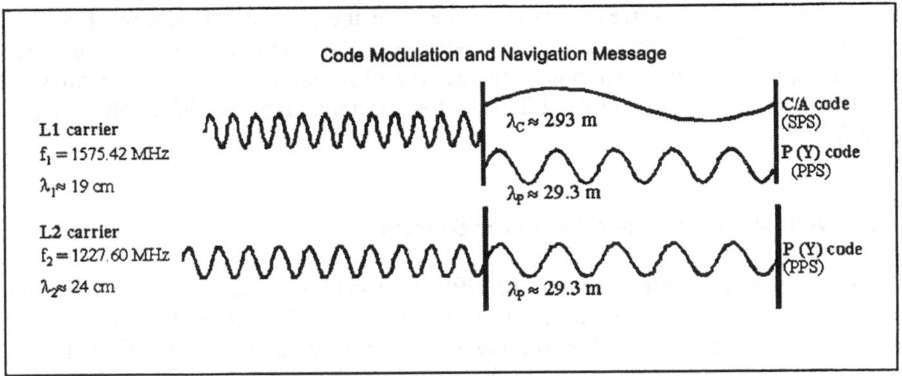

Figure 3.3 GPS Signal Characteristics

Table 3.1 Basic Components of the GPS Satellite Signal

Component	Frequency [MHz]	Ratio of fundamental frequency f_o	Wavelength [cm]
Fundamental frequency f_o	10.23	1	2932.6
L1 Carrier	1,575.42	154	19.04
L2 Carrier	1,227.60	120	24.45
P-code	10.23	1	2932.6
C/A code	1.023	1/10	29326
W-code	0.5115	1/20	58651
Navigation message	50×10^{-6}	1/204,600	N/A

Each satellite transmits a C/A-code and a unique one-week long segment of P-code, which is the satellite's designated ID, ranging from 1 to 32. PRN is a very complicated digital code, i.e., a sequence of "on" and "off" pulses that looks almost

[1] f_o =10.23 MHz is the fundamental frequency of GPS. The GPS satellites also transmit an L3 signal at 1381.05 MHz (135×10.23 MHz), associated with their dual role as a nuclear burst detection satellite as well as S-band telemetry signal.

like random electrical noise. This carrier modulation enables the measurement of the signal travel time between the satellite and the receiver (user), which is a fundamental GPS observable (see sections 3.2.3.2 and 3.2.4). An access to the C/A code is provided to all users and is designated as the Standard Positioning Service (SPS). Under the *Anti-Spoofing* (AS) policy imposed by DOD, the additional W-code is implemented to encrypt the P-code into the Y-code, available exclusively to the military users, and designated as Precise Positioning Service (PPS). PPS guarantees positioning accuracy of at least 22 m (95 percent of the time) horizontally, and 27.7 meters vertically, while guaranteed positioning accuracy of SPS is 100 m (95 percent of the time) horizontally, and 156 meters (95 percent of the time) vertically. However, most of the time the practical accuracy of SPS is much higher (see Section 3.2.4.2). Under AS, the civilian receivers must use special signal tracking techniques to recover observables on L2, since no C/A (civilian) code is available on L2 (see, for example, Hofman-Wellenhof *et al.*, 2001).

3.2.3 GPS Observables and the Error Sources

There are two fundamental types of GPS observables: *pseudorange* and *carrier phase,* both subject to measurement errors of systematic and random nature. For example, systematic errors due to ionosphere or troposphere delay the GPS signal, and cause the measured range to be different from the true range by some systematic amount. Other errors, such as the receiver noise, are considered random. The following sections provide an overview of the GPS observables and the primary error sources.

3.2.3.1 Systematic Errors

Satellite orbital errors (errors in predicted ephemeris), reference station position errors (in differential/relative positioning), satellite and receiver clock errors, and effects of propagation media represent bias errors, which have to be properly accommodated before positioning solutions can be obtained. The biases can be either modeled mathematically and accounted for in the measurement model (such as tropospheric correction), or a special observation and data reduction techniques must be applied to remove their effects. For example, dual frequency observables can be used to mitigate or remove the effects of the ionospheric signal delay, and differential GPS or relative positioning can be used to mitigate the effects of imperfectness of the satellite and receiver clock or broadcast ephemeris errors.

3.2.3.1.1 Errors Due to Propagation Media

The presence of free electrons in the *ionosphere* causes a nonlinear dispersion of electromagnetic waves traveling through the ionized medium, affecting their speed, direction and frequency. The largest effect is on the signal speed, and as a result, GPS pseudorange is delayed (and thus, is measured too long), while the phase advances (and thus, is measured too short). The total effect can reach up to 150 m, and is a function of the Total Electron Content (TEC) along the signal's path, the frequency of the signal itself, the geographic location and the time of observation, time of the year and the period within the 11-year sun spot cycle (the last peak in ionospheric activity was 2001). The effects of the ionosphere can be removed from the GPS observable by combining dual-frequency data in so-called iono-free linear combination, and by relative processing as will be explained later (see also Hofman-Wellenhof *et al.*, 2001).

The *troposphere* is a nondispersive medium for all frequencies below 15 GHz, thus phase and pseudorange observables on both L1 and L2 frequencies are delayed by the same amount. Since the amount of delay is not frequency dependent, elimination of the tropospheric effect by dual-frequency observable is not possible, but is rather accomplished using mathematical models of the tropospheric delay. The total effect in zenith direction reaches ~2.5 m, and increases with the cosecant of the elevation angle up to ~25 m at 5° elevation angle. The tropospheric delay consists of the *dry* component (about 90% of the total tropospheric refraction), proportional to the density of the gas molecules in the atmosphere and the *wet refractivity*, due to the polar nature of the water molecules. In general, empirical models, which are functions of temperature, pressure and relative humidity, can eliminate the major part (90-95%) of the tropospheric effect from the GPS observables.

In addition, GPS signals can experience *multipath*, which is a result of an interaction of the upcoming signal with the objects in the antenna surrounding. It causes multiple reflections and diffractions, and as a result, the signal arrives at the antenna via direct and indirect paths. These signals interfere with each other, resulting in an error in the measured pseudorange or carrier phase, degrading the positioning accuracy. The magnitude of multipath effect tends to be random and unpredictable, varying with satellite geometry, location and type of reflective surfaces in the antenna surrounding, and can reach 1-5 cm for the carrier phases and 10-20 m for the code pseudoranges (Hofman-Wellenhof *et al.*, 2001). Properly designed choke ring antennas can almost entirely eliminate this problem for the surface waves and the signals reflected from the ground.

3.2.3.1.2 Selective Availability (SA)

Another source of errors in the measured range to the satellites is Selective Availability (SA). SA is the DOD policy of denying to non-military GPS users the full accuracy of the system. It is achieved by dithering the satellite clock (called delta process) and degrading the navigation message ephemeris (called epsilon process). The effects of SA (that are highly unpredictable) can be removed with encryption keys or through differential techniques (see section 3.2.4). SA levels were set to zero on May 2, 2000 at 04:00 UT (Universal Time). For more information on GPS error sources the reader is referred to Hofman-Wellenhof *et al.,* (2001) and Lachapelle, (1990).

3.2.3.2 Mathematical Models of Pseudorange and Carrier Phase

Pseudorange is a geometric range between the transmitter and the receiver, distorted by the propagation media and the lack of synchronization between the satellite and the receiver clocks. It is recovered from the measured *time difference* between the epoch of the signal transmission and the epoch of its reception by the receiver. The actual time measurement is performed with the use of the PRN code. In principle, the receiver and the satellite generate the same PRN sequence. The arriving signal is delayed with respect to the replica generated by the receiver, as it travels ~20,000 km. In order to find how much the satellite's signal is delayed, the receiver-replicated signal is delayed until it falls into synchronization with the incoming signal. The amount by which the receiver's version of the signal is delayed is equal to the travel time of the satellite's version (Figure 3.4). The travel time, Δt (~0.06 s), is converted to a range measurement by multiplying it by the speed of light, c.

There are two types of pseudoranges: C/A-code pseudorange and P-code pseudorange. The precision of the pseudo-range measurement is partly determined by the wavelength of the chip in the PRN code. Thus, the shorter the wavelength, the more precise the range measurement would be. Consequently, the P-code range measurement precision (noise) of 10-30 cm is about 10 times higher than that of the C/A code. The pseudorange observation can be expressed as a function of the unknown receiver coordinates, satellite and receiver clock errors and the signal propagation errors (Equation (3.1)).

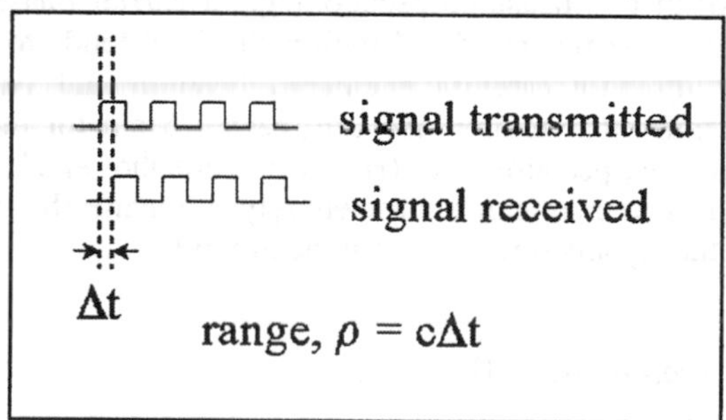

Figure 3.4 Principles of Pseudorange Measurement Based on Time Observation

$$P_{r,1}^{s} = \rho_r^s + \frac{I_r^s}{f_1^2} + T_r^s + c(dt_r - dt^s) + M_{r,1}^s + e_{r,1}^s$$

$$P_{r,2}^{s} = \rho_r^s + \frac{I_r^s}{f_2^2} + T_r^s + c(dt_r - dt^s) + M_{r,2}^s + e_{r,2}^s \quad (3.1)$$

Where

$$\rho_r^s = sqrt\left[(X^s - X_r)^2 + (Y^s - Y_r)^2 + (Z^s - Z_r)^2\right] \quad (3.2)$$

$P_{r,1}^s, P_{r,2}^s$: pseudoranges measured between receiver *r* and satellite *s* on L1 and L2

ρ_r^s : geometric distance between satellite *s* and receiver *r*

$\frac{I_r^s}{f_1^2}, \frac{I_r^s}{f_2^2}$: range error caused by ionospheric signal delay on L1 and L2

dt_r : the *r-th* receiver clock error (unknown)

dt^s: the *s-th* satellite clock error (known from the navigation message)

c : the vacuum speed of light

$M_{r,1}^s, M_{r,2}^s$: multipath on pseudorange observables on L1 and L2

T_r^s : range error caused by tropospheric delay between satellite *s* and receiver *r* (estimated from a model)

$e_{r,1}^s, e_{r,2}^s$: measurement noise for pseudorange on L1 and L2

X^s, Y^s, Z^s: coordinates of satellite *s* (known from the navigation message)

X_r, Y_r, Z_r: coordinates of receiver *r* (unknown)

f_1, f_2 : carrier frequencies of L1 and L2

Carrier phase is defined as a difference between the phase of the incoming carrier signal and the phase of the reference signal generated by the receiver. Since at the initial epoch of the signal acquisition, the receiver can measure only a fractional phase, the carrier phase observable contains the initial unknown *integer ambiguity, N*. Integer ambiguity is a number of full phase cycles between the receiver and the satellite at the starting epoch, which remains constant as long as the signal tracking is continuous. After the initial epoch, the receiver can count the number of integer cycles that are being tracked. Thus, the carrier phase observable (φ) can be expressed as a sum of the fractional part measured with mm-level precision, and the integer number of cycles counted since the starting epoch. The integer ambiguity can be determined using special techniques referred to as *ambiguity resolution algorithms*. Once the ambiguity is resolved, the carrier phase observable can be used to determine the user's location. The carrier phase can be converted to a phase-range observable, Φ (Equation (3.3)), by multiplying the measured phase, φ, with the corresponding wavelength λ. This observable is used in applications where the highest accuracy is required.

$$\Phi_{r,1}^s = \rho_r^s - \frac{I_r^s}{f_1^2} + T_r^s + \lambda_1 N_{r,1}^s + c(dt_r - dt^s) + m_{r,1}^s + \varepsilon_{r,1}^s$$

$$\Phi_{r,2}^s = \rho_r^s - \frac{I_r^s}{f_2^2} + T_r^s + \lambda_2 N_{r,2}^s + c(dt_r - dt^s) + m_{r,2}^s + \varepsilon_{r,2}^s \qquad (3.3)$$

Where:

$\Phi_{r,1}^s, \Phi_{r,2}^s$: phase-ranges (in meters) measured between station *r* and satellite *s* on L1 and L2

$N^s_{r,1}, N^s_{r,2}$: initial integer ambiguities on L1 and L2, corresponding to receiver r and satellite s

$\lambda_1 \approx 19$ cm and $\lambda_2 \approx 24$ cm are wavelengths of L1 and L2

$m^s_{r,1}, m^s_{r,2}$: multipath error on carrier phase observables on L1 and L2

$\varepsilon^s_{r,1}, \varepsilon^s_{r,1}$: measurement noise for carrier phase observables on L1 and L2.

Another observation that is sometimes provided by GPS receivers, and is primarily used in kinematic applications for velocity estimation, is *instantaneous Doppler* frequency. It is defined as a time change of the phase-range, and thus, if available, it is measured on the code phase (Lachapelle, 1990).

Equation (3.2), which is a non-linear part of Equations (3.1) and (3.3), requires Taylor series expansion to enable the estimation of the unknown user coordinates. In addition, Equations (3.1) and (3.3) can be solved for other parameters, such as user clock error. Secondary (nuisance) parameters in the above equations are satellite and clock errors, tropospheric and ionospheric errors, multipath, and integer ambiguities. These are usually removed by differential (relative) GPS processing (see section 3.2.4.1.2), by empirical modeling (troposphere), or by processing of dual frequency signals (ionosphere). As already mentioned, ambiguities must be resolved prior to users' position estimation.

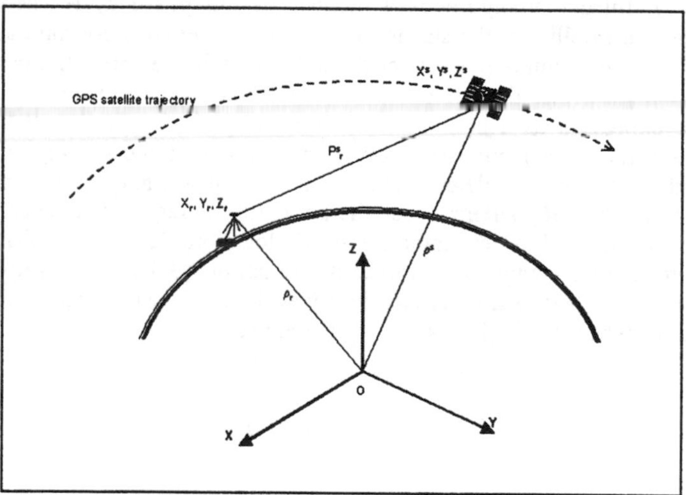

Figure 3.5 Range from Satellite s to Ground Receiver r

3.2.4 Positioning with GPS

The main principle behind positioning with GPS is triangulation in space, based on the measurement of a range (pseudorange or phase-range) between the receiver and the satellites (Figure 3.5). Essentially, the problem can be specified as follows: given the position vectors of GPS satellites (such as ρ^s of satellite *s* in Figure 3.5) tracked by a receiver *r*, and given a set of range measurements (such as P_r^s) to these satellites, determine a position vector of the user, ρ_r . A single range measurement to a satellite places the user somewhere on a sphere with a radius equal to the measured range. Three simultaneously measured ranges to three different satellites place the user on the intersection of three spheres, which corresponds to two points in space. One of them is usually an impossible solution that can be discarded by the receiver. Even though there are three fundamental unknowns (coordinates of the user's receiver), the minimum of four satellites must be simultaneously observed to provide a unique solution in space (Figure 3.6), as explained next.

As already mentioned, the fundamental GPS observable is the signal travel time between the satellite and the receiver. However, the receiver clock that measures the time is not perfect and may introduce an error to the measured pseudorange (even though we limit our discussion here to pseudoranges, the same applies to the carrier phase measurement that is indirectly related to the signal transit time, as the phase of the received signal can be related to the phase at the epoch of transmission in terms of the signal transit time). Thus, in order to determine the most accurate range, the receiver clock correction must be estimated and its effect removed from the observed range. Hence, the fourth pseudorange measurement is needed, since the total number of unknowns, including the receiver clock, is now four. If more than four satellites are observed, a least squares solution is employed to derive the optimal solution.

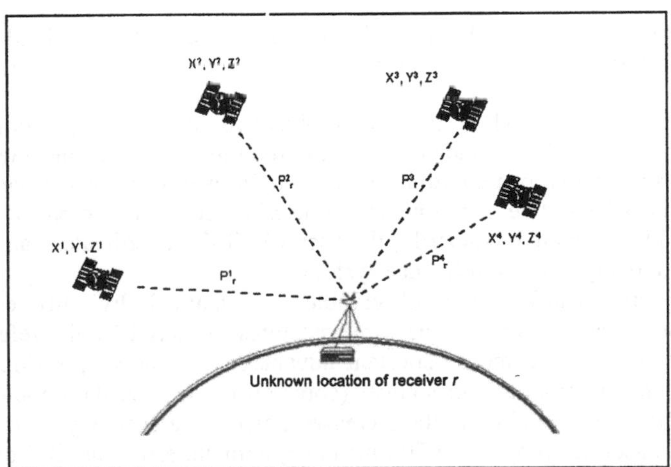

Figure 3.6 Determination of Position in Space by Ranging to Multiple Satellites

3.2.4.1 Point vs. Relative Positioning

There are two primary GPS positioning modes: *point positioning*, or *absolute positioning*, and *relative positioning*. However, there are several different strategies for GPS data collection and processing, relevant to both positioning modes. In general, GPS can be used in static and kinematic modes, using both pseudorange and carrier phase data. GPS data can be collected and then post-processed at a later time, or processed in real time, depending on the application and the accuracy requirements. In general, post-processing in relative mode provides the best accuracy.

3.2.4.1.1 Point (Absolute) Positioning

In point, or absolute positioning, a single receiver observes pseudoranges to multiple satellites to determine the user's location. For the positioning of the moving receiver, the number of unknowns per epoch equal to three receiver coordinates plus a receiver clock correction term. In the static mode with multiple epochs of observations there are three receiver coordinates and *n* receiver clock error terms, each corresponding to a separate epoch of observation *1* to *n*. The satellite geometry and any unmodeled errors will directly affect the accuracy of the absolute positioning.

3.2.4.1.2 Relative Positioning

The relative positioning technique (also referred to as differential GPS, DGPS) employs at least two receivers, a reference (base) receiver, whose coordinates must be known, and the user's receiver, whose coordinates can be determined relative to the reference receiver. Thus, the major objective of relative positioning is to estimate the 3D baseline vector between the reference receiver and the unknown location. Using the known coordinates of the reference receiver and the estimated ΔX, ΔY and ΔZ baseline components, the user's receiver coordinates in WGS84 can be readily computed. Naturally, the user's WGS84 coordinates can be further transformed to any selected reference system.

A relative (differenced) observable is obtained by differencing the simultaneous measurements to the same satellites observed by the reference and the user receivers. The most important advantage of relative positioning is the removal of the systematic error sources (common to the base station and the user) from the observable, leading to the increased positioning accuracy. Since for short to medium baselines (up to ~60-70 km) the systematic errors in GPS observables due to troposphere, satellite clock and broadcast ephemeris errors are of similar magnitude (i.e., they are spatially and temporally correlated), the relative positioning allows for a removal or at least a significant mitigation of these error sources, when the observables are differenced. In addition, for baselines longer than 10 km, the so-called ionosphere-free linear combination must be used (if dual frequency data are available) to mitigate the effects of the ionosphere (see Hofman-Wellenhof *et al.*, 2001).

The primary differential modes are (1) single differencing mode, (2) double differencing mode, and (3) triple differencing mode. The differencing can be performed between receivers, between satellites and between epochs of observations. The single-differenced (between-receiver) measurement, $\Phi_{i,j}^{k}$, is obtained by differencing two observables to the satellite *k*, tracked simultaneously by two receivers *i* (reference) and *j* (user): $\Phi_{i,j}^{k} = \Phi_{i}^{k} - \Phi_{j}^{k}$ (see Figure 3.7). By differencing observables from two receivers, *i* and *j*, observing two satellites, *k* and *l*, or simply by differencing two single differences to satellites *k* and *l*, one arrives at the double-differenced (between-receiver/between-satellite differencing) measurement: $\Phi_{i,j}^{k} = \Phi_{i}^{k} - \Phi_{j}^{k} - \Phi_{i}^{l} + \Phi_{j}^{l} = \Phi_{i,j}^{k} - \Phi_{i,j}^{l}$. Double difference is the most commonly used differential observable. Furthermore, differencing two double differences, separated by the time interval $dt = t_2 - t_1$, renders triple-differenced measurement, $\Phi_{i,j}^{k,l}(dt) = \Phi_{i,j}^{k,l}(dt_2) - \Phi_{i,j}^{k,l}(dt_1)$, which in case of carrier phase observables effectively cancels the initial ambiguity term. Differencing can be applied to both pseudorange and carrier phase. However, for the best positioning accuracy with carrier phase double differences, the initial ambiguity term should be first resolved and fixed to the integer value. Relative positioning may be performed in static and kinematic modes, in real time (see the next section) or, for the highest accuracy, in post-processing. Table 3.2 shows the error characteristics for between-receiver single and between-receiver/between-satellite double differenced data.

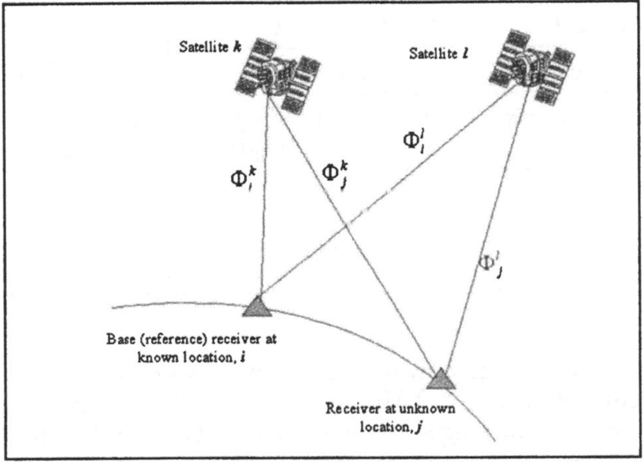

Figure 3.7 Between-Receiver Phase-Range Double Differencing

3.2.4.1.3 DGPS Services

As explained earlier, the GPS error sources are spatially and temporally correlated for short to medium base-user separation. Thus, if the base and satellites locations are known (satellite location is known from broadcast ephemeris), the errors in the measurements can be estimated at specified time intervals and made available to the nearby users through a wireless communication as *differential corrections*.

These corrections can be used to remove the errors from the observables collected at the user's (unknown) location. This mode of positioning uses DGPS services to mitigate the effects of the measurement errors, leading to the increased positioning accuracy in real time. DGPS services are commonly provided by the government, industry and professional organizations, and enable the users to use only one GPS receiver collecting pseudorange data, while still achieving superior accuracy as compared to the point-positioning mode. Naturally, in order to use a DGPS service, the user must be equipped with an additional hardware capable of receiving and processing the differential corrections. In an alternative implementation of DGPS, the reference station broadcasts the observables (carrier phase and pseudorange) instead of the differential corrections. In that case, the user unit has to perform the relative positioning, as described in Section 3.2.4.1.2.

Table 3.2 Differencing Modes and Their Error Characteristics

Error source	Single Difference	Double Difference
Ionosphere	Reduced, depending on the baseline length	Reduced, depending on the baseline length
Troposphere	Reduced, depending on the baseline length	Reduced, depending on the baseline length
Satellite clock	Eliminated	Eliminated
Receiver clock	Present	Eliminated
Broadcast ephemeris	Reduced, depending on the baseline length	Reduced, depending on the baseline length
Ambiguity term	Present	Present
Noise level w.r.t. one-way observable	Increased by $\sqrt{2}$	Increased by 2

DGPS services normally involve some type of wireless transmission systems. They may employ VHF or UHF systems for short ranges, low-frequency transmitters for medium ranges (beacons) and L-band or C-band geostationary satellites for coverage of entire continents, which is called Wide Area DGPS (WADGPS). WADGPS involves multiple GPS base stations that track all GPS satellites in view and, based on their precisely known locations and satellite broadcast ephemeris, estimates the errors in the GPS pseudo-ranges. This information is used to generate pseudorange corrections that are subsequently sent to the master control station, which uploads checked and weighted corrections to the communication geostationary satellite, which in turn transmits the corrections to the users. The positioning accuracy of WADGPS, such as OmniSTAR™ (OmniSTAR, 2002), is at the sub-meter level.

Example DGPS services include Federal Aviation Administration (FAA)-supported satellite-based Wide Area Augmentation System (WAAS), ground-based DGPS services, referred to as Local Area DGPS (LADGPS), such as U.S.

Coast Guard and Canadian Coast Guard services, or FAA-supported Local Area Augmentation System (LAAS). LADGPS supports real-time positioning typically over distances of up to a few hundred kilometers, using corrections generated by a single base station (Rizos, 2002a). WAAS and LAAS are currently under implementation, with a major objective of supporting aviation navigation and precision approach (Federal Aviation Administration, 2002; U.S. Coast Guard, 2002). The accuracy of WAAS was expected at ~7.6 m at 95 percent of the time, but it is already significantly better than these specifications (~2 m and less horizontal RMS) (Lachapelle *et al.*, 2002). WAAS is a public service and any user equipped with the appropriate receiver may have an access to it.

Another approach gaining popularity in a number of countries is establishing local networks of Continuously Operating Reference Stations (CORS) that normally support a range of applications, especially those requiring the highest accuracy in post-processing or in realtime (though the real-time support is currently rather limited). Government agencies, such as National Geodetic Survey (NGS), Department of Transportation (DOT) or international organizations, such as International GPS Service (IGS), deploy and operate these networks. Normally, all users have a free access to the archived data that can be used as a reference (base data) in carrier phase or range data processing in relative mode. Alternatively, network-based positioning using carrier-phase observations with a single user receiver in real time can be accomplished with local specialized networks, which can estimate and transmit carrier phase correction (see, e.g., Raquet and Lachapelle, 2001).

3.2.4.2 How Accurate is GPS?

The positioning accuracy of GPS depends on several factors, such as the number and the geometry of the observations collected, the mode of observation (point vs. relative positioning), type of observation used (pseudorange or carrier phase), the measurement model used, and the level of biases and errors affecting the observables. Depending on the design of the GPS receiver and the factors listed above, the positioning accuracy varies from 10 meters with SA turned off (about 100 m with SA turned on) for pseudorange point positioning, to better than 1 centimeter, when carrier phases are used in relative positioning mode. In order to obtain better than 10 m accuracy with pseudoranges, differential positioning or DGPS services must be employed.

The geometric factor, Geometric Dilution of Precision (GDOP), reflects the instantaneous geometry related to a single point. Normally, more satellites yield smaller DOP value, and GDOP of six and less indicates good geometry (a value around 2 indicates a very strong geometry). Other DOP factors, such as Position DOP (PDOP), Vertical DOP (VDOP) or Relative DOP (RDOP), the last one related to the satellite geometry with respect to a baseline, can be also used as the quality indicators. These DOPs are normally computed by GPS receivers and provided in real time as the quality assessment. Formulas for computing various geometric factors are provided in Hofman-Wellenhof *et al.*, (2001).

Other factors affecting the GPS positioning accuracy depend on: (1) whether the user is stationary or moving (static vs. kinematic mode), (2) whether the positioning is performed in real time or in post-processing, (3) the data reduction algorithm, (4) the degree of redundancy in the solution, and (5) the measurement noise level. The currently achievable GPS accuracies, provided as two-sigma, corresponding to 95 percent confidence level, are summarized in Table 3.3 (Rizos, 2002b). The lower bound of the relative positioning accuracy listed in Table 3.3 cannot be stated with precision, as it depends on several hardware and environmental factors, as well as the survey geometry, among others (the symbol → indicates the increase of the values listed). Thus, the accuracy levels listed in Table 3.3 should be understood as the best achievable accuracy.

Table 3.3 Currently Achievable GPS Accuracy (Rizos, 2002b)

Positioning mode			
Point positioning (pseudorange)		Relative positioning	
PPS	1-5 m	Static survey (carrier phase)	2 mm (→) plus 1 ppm[2] (up to < 0.1 ppm)
SPS, SA off	4-10 m	Kinematic survey (carrier phase)	5 mm (→)
SPS, SA on	0-100 m	DGPS services (pseudorange)	50 cm (→)

3.2.5 GPS Instrumentation

Over the past two decades, the civilian as well as military GPS instrumentation evolved through several stages of design and implementation, focused primarily on achieving an enhanced reliability of positioning and timing, modularization and miniaturization. In addition, one of the most important aspects, especially for the civilian market, has been the decreased cost of the receivers, as the explosion of GPS applications calls for a variety of low-cost, application-oriented and reliable equipment. By far, the majority of the receivers manufactured today are of the C/A-code single frequency type. However, for the high precision geodetic applications the dual frequency solution is a standard. Even though the civilian and military receivers, as well as application-oriented instruments have evolved in different directions, one might pose the following question: *Are all GPS receivers essentially the same, apart from functionality and user software?* The general answer is, yes, all GPS receivers support essentially the same functionality blocks, even if their implementation differs for different types of receivers.

The following are the primary components of a generic GPS receiver (Figure 3.8): *antenna* and *preamplifier*, *radio-frequency (RF) front-end section*, *a signal tracker block*, *microprocessor*, *control/interface unit*, *data storage device*, and *power supply* (Langley, 1991; Parkinson and Spilker, 1996; Grejner-Brzezinska, 2002). Any GPS receiver must carry out the following tasks:

---

[2] ppm – part per million

- Select the satellites to be tracked based on GDOP and Almanac;[3]
- Search and acquire each of the GPS satellite signals selected;
- Recover navigation data for every satellite;
- Track the satellites, measure pseudorange and/or carrier phase;
- Provide position/velocity information;
- Accept user commands and display results via control unit or a PC;
- Record the data for post-processing (optional);
- Transmit the data to another receiver via radio modem for real-time solutions (optional).

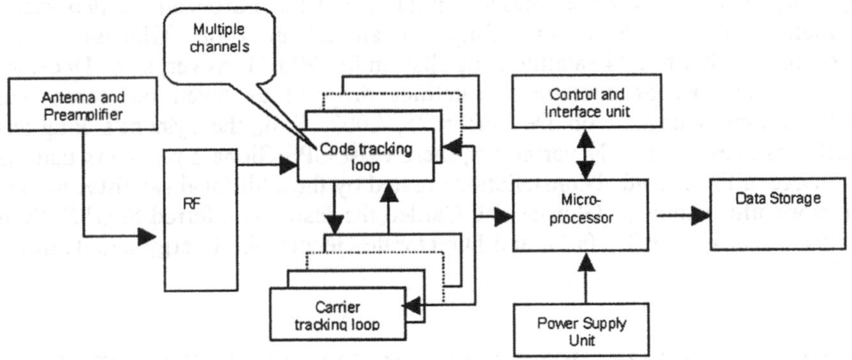

Figure 3.8 Basic Components of a GPS Receiver (Grejner-Brzezinska, 2002)

An important characteristic of the RF section is the number of channels, and hence the number of satellites that can be tracked simultaneously. Older receivers had a limited number of channels (even as little as one), which required sequencing through satellites in order to acquire enough information for 3D positioning. Modern GPS receivers are based on dedicated channel architecture, where every channel tracks one satellite on L1 or L2 frequency.

3.2.6 GPS Modernization and Other Satellite Systems

The current plan of GPS modernization is focused on improving the quality of civilian uses of GPS, primarily through the implementation of a new code on L2 frequency and a new civilian signal on L5 frequency of 1227.6 MHz (Spilker and Van Direndonck, 2001). In addition, a new M-code (encrypted) will be implemented exclusively for military use, ensuring that military and civilian users will have entirely separate signals and codes. Consequently, AS policy will be abandoned (SA has already been turned off). The new dual-frequency civilian tracking capability will be available on the Block IIR GPS satellites scheduled for

[3] A set of parameters included in the GPS satellite navigation message that is used by a receiver to predict the approximate location of a satellite. The almanac contains information on all of the satellites in the constellation.

launch starting 2003 (Rizos, 2002c), while the other improvements are planned for Block IIF satellites that are scheduled for launch starting in 2006.

In terms of GPS instrumentation, hardware miniaturization, improvements in reliability, faster sampling rates, lower noise and more multipath resistance, more real-time operations and ubiquitous dual-frequency measurement capability are expected (Rizos, 2002c). The current trend of integrating GPS with other sensors such as inertial, vision systems, laser scanners, and pseudolites will continue to serve more specialized, customized applications. It is expected that the increasing number of CORS-based local services will serve real-time users.

Other GNSS that are complementary to GPS are the Russian Glonass system, originally developed for the Russian military, and the European Galileo system (civilian), expected to become fully operational in 2008. Glonass became operational with a full 24-satellite constellation in 1996. However, as of December 2002, the number of operational satellites dropped to seven but three more satellites were launched on December 25, 2002. Still, the system's long-term stability is questionable. Nevertheless, there exist GPS/Glonass receivers that take advantage of the extended constellation created by the additional satellites in view. For more information on Glonass and Galileo the reader is referred to *GPS World*, *Galileo's World,* GIBS (2002), and Directorate-General for Energy and Transport (2002).

3.3 POSITIONING METHODS BASED ON CELLULAR NETWORKS

LBSs are made possible through a suitable relationship between the cellular service providers, cellular networks and mobile users' terminals, which work in synch, to locate the user (mobile terminal), and then to transfer the position data either upon the request or as a continuous stream. The major issue is to locate the user with a required accuracy and limited latency. A popular approach to finding the user's location is a *radiolocation* technique, which uses parameters of radio signal that travels between the mobile user and reference (base) stations to derive the user's location. The signal parameters most commonly used in radionavigation are: angle of arrival, time of arrival, signal strength and signal multipath signature matching. The time of arrival and the signal strength can be directly converted to the range measurements.

The most popular technique of finding the user's location is triangulation, based either on angular or distance observations (or some combination of both), between the mobile terminal and the base stations. The base stations in the radiolocation techniques are either cellular service towers (cellular network) or GPS satellites. Thus, in general, the technologies for finding the user's location in LBSs can be divided into *network-based* or *satellite-based* (currently primarily GPS-based) systems. Another classification is based on the actual device that performs the positioning solution, i.e., mobile user or the base station (control center), leading to *mobile terminal (user)-centric*, *network-centric*, or *hybrid solutions*. In the network-centric systems, the user's position is determined by the base station and sent back to the user's set, while in the terminal-centric solution, the position computation is performed by the user's set.

Table 3.4 Cellular Network-Based Techniques Supporting LBSs

LBS Technique	Primary observable	Upgrade of the User Terminal or Network	Location Calculation and Control
GPS A-GPS	• Time (range) to multiple satellites • 3D location • Minimum of 3 ranges required for 2D positioning	• User terminal (GPS receiver, memory, software) • Non-synchronized networks may require an enhancement	Mobile Terminal
E-OTD	• Signal travel time difference between the user and the base stations • 2D location	• User Terminal (memory, software) • Base station time synchronization	Mobile Terminal
CGI-TA	• Cell ID • The accuracy does not meet the E-911[4] requirements • 2D location	• None	Network
TOA	• Signal travel time between the user and the base stations • 2D location	• Supports legacy terminals • Monitoring equipment at every base station	Network
TDOA	• Signal travel time difference between the user and the base stations • 2D location	• Network interconnection	Network
AOA	• Time (range) to multiple cell towers (minimum of three measurements is required) • 2D location	• Network interconnection • Antenna arrays to measure angles	Network
RSS	• Received signal strength • 2D location	• None	Network
Location/ Multipath Pattern Matching	• Multipath signature at the users location • 2D location	• None	Network
Hybrid system such as A-GPS+CGI	• GPS range • Cell ID • 3D or 2D	• Same as for GPS method	Mobile terminal plus Network

[4] E-911 (Enhanced 911) services. According to the Federal Communication Commission (FCC) mandate, the wireless carriers were supposed to provide the location of the emergency calls to Public Safety Answer Points (PSAPs) with the accuracy of 125 m for 67 percent of calls, and 300 m for 95 percent of calls by October 2001.

In this section, we present an overview of the three main location techniques used in LBSs: (1) mobile terminal (user)-centric, (2) network-centric, and (3) hybrid solutions (Caffery and Stuber, 1998; Hellebrandt and Mathar, 1999; Hein *et al.*, 2001; Abnizova, *et al.*, 2002; Andersson, 2002; Djuknic and Richton, 2002; Francica, 2002; SnapTrack, 2002). The summary characteristics of these methods are presented in Tables 3.4 and 3.5.

Table 3.5 Cellular Network-Based Techniques Supporting LBSs: Cost, Latency and Accuracy

LBS Technique	Total Cost[5]	Latency (TTFF)	Accuracy*
A-GPS GPS	Moderate	< 10 s up to 60 s (cold start)	High (5-10 m)
E-OTD	High	< 10 s	Moderate-to-high
CGI-TA	Low	< 10 s	Low Depends on cell size
TOA	High	< 10 s	Moderate-to-high
TDOA	High	< 10 s	Moderate-to-high
AOA	Low	< 10 s	Low-to-moderate
RSS	Low	< 10 s	Moderate
Location/ Multipath Pattern Matching	Moderate	< 10 s	Moderate
Hybrid system (A-GPS + CGI)	Moderate	< 10 s	High

* The levels of accuracy are defined using the 95% CEP (Circular Error Probable) as follows:
- High level equals to 95% CEP within 50 m
- Moderate level equals to 95% CEP within 300 m
- Low level equals to 95% CEP greater than 300 m (Pietila and Williams, 2002).

3.3.1 Terminal-Centric Positioning Methods

The *terminal-centric methods* rely on the positioning software installed in the mobile terminal (see Table 3.4). They are further divided into:

- GPS method;
- Network Assisted GPS (A-GPS);

[5] Total cost includes handset, infrastructure and maintenance; for details, see SnapTrack (2002).

- Enhanced Observed Time Difference (E-OTD); this method can also be used in the *network-centric* mode, according to Andersson (2001).

The *GPS method* uses ranging signals directly from a number of GPS satellites and provides instantaneous point-positioning information, with the accuracy of 5-50 m, depending on the availability of GPS signals. *A-GPS* uses an assisting network of GPS receivers that can provide information enabling a significant reduction of the time-to-first-fix (TTFF) from 20-45 s, to 1-8 s. For example, the timing and navigation data for GPS satellites may be provided by the network, which means that the receiver does not need to wait until the broadcast navigation message is read. It only needs to acquire the signal to compute its position almost instantly. For the timing information to be available through the network, the network and GPS would have to be synchronized to the same time reference. In essence, the assistance data make it possible for the receiver to make the time measurements (equivalent to ranges) to GPS satellites without having to decode the actual GPS message, which significantly speeds up the positioning process. According to Andersson (2001) the assistance data is normally broadcast every hour, and thus it has a very little impact on the network's operability. More details on positioning with GPS can be found in section 3.2.4 of this chapter.

Another terminal-centric solution is the *E-OTD*, which measures the time of the signal arrival from multiple base stations (within the wireless network) at the mobile device. The time differences between the signal arrivals from different base stations are used to determine the user's location with respect to the base stations, provided that the base stations' coordinates are known and the base stations send time-synchronized signals. For the positioning and timing purposes, the base stations might be equipped with stationary GPS receivers. Thus, the base stations in E-OTD serve as reference points, similar to GPS satellites. However, this method is not subject to limitations in signal availability affecting GPS. The positioning accuracy of E-OTD is about 100-125 m. Since E-OTD requires monitoring equipment at virtually every base station, it adds to the cost of LBSs.

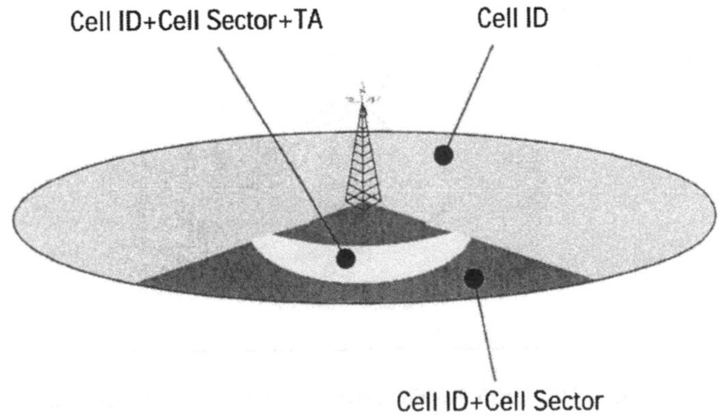

Figure 3.9 Cell-ID with Cell Sector and TA (SnapTrack, 2002)

3.3.2 Network-Centric and Hybrid Positioning Methods

The main *network-centric methods* are:

- Cell Global Identity with Timing Advance (CGI-TA);
- Time of Arrival (TOA);
- Uplink Time Difference of Arrival (TDOA);
- Angle of Arrival (AOA);
- Location (Multipath) Pattern Matching;
- Received Signal Strength (RSS).

CGI uses the *cell ID* to locate the user within the cell, where the cell is defined as a coverage area of a base station (the tower nearest to the user). It is an inexpensive method, compatible with the existing devices, with the accuracy limited to the size of the cell, which may range from 10 – 500 m (indoor micro cell), to an outdoor macro cell reaching several kilometers (Andersson, 2002). CGI is often supplemented by the Timing Advance (TA) information that provides the time between the start of a radio frame and the data burst (Figure 3.9). This enables the adjustment of a mobile set's transmit time to correctly align the time, at which its signal arrives at the base. These measurements can be used to determine the distance from the user to the base, further reducing the position error (SnapTrack, 2002).

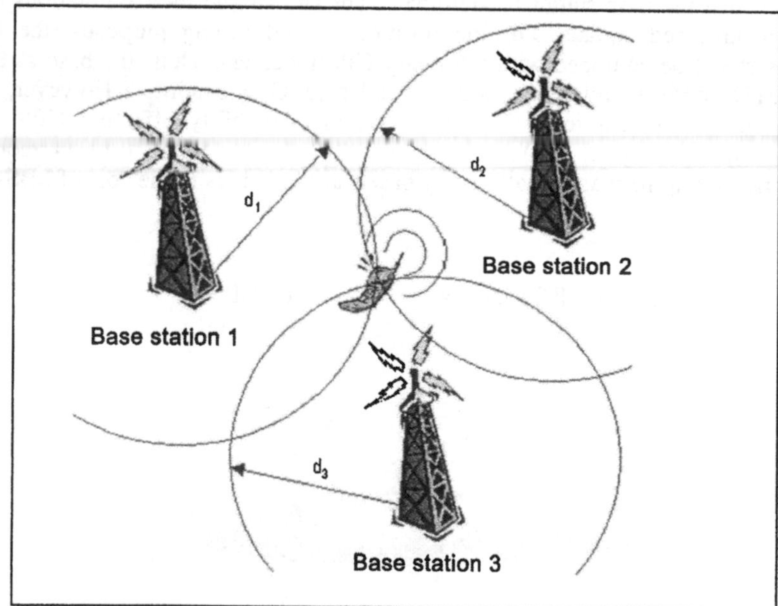

Figure 3.10 Triangulation of the User's Position Based on the Distance Measurements to Three Base Stations (Cell-Towers). The latitude and longitude of the user are obtained as the intersection of three circles centered at the towers, with radii of d_1, d_2 and d_3

TOA is based on the travel time information (equivalent to a range) between the base station and the mobile terminal. In essence, the user's location can be found by triangulation, at the intersection of three (or more) arcs centered at the tower locations, with radii equal to the measured distances (Figure 3.10). The actual observation is the signal travel time, $t_U^r - t_B^t$, which is converted to a distance by multiplying it by the speed of light, where t_U^r, and t_B^t denote the time of signal arrival at the user (*U*) and the time of signal transmission at the base (*B*), respectively. For a higher accuracy, the signal delay corrections, such as the tropospheric correction, might be applied. The basic observation equation (Equation (3.4)), which is used to determine the user's location, represents the measured distance from the user to a base station as a function of the coordinates of $U(x_U, y_U)$ and $B(x_B, y_B)$. Since the coordinates of the base station are known, linearized equations (similar to Equation (3.4)) written for multiple base stations allows for the estimation of the user's coordinates by the least squares adjustment. It should be mentioned here that the base station time synchronization required by this method might need an additional hardware and software support to achieve the required timing accuracy. It is achieved with the use of the so-called Location Measurement Units (LMUs) placed at known locations. LMUs, similarly to the mobile user, receive the signal from the surrounding towers. This information combined with LMUs known position coordinates enables the estimation of the clock offsets between pairs of base stations.

$$d_{UB} = (t_U^r - t_B^t)c = \sqrt{(x_U - x_B)^2 + (y_U - y_B)^2} \tag{3.4}$$

The concept of *TDOA* is similar to E-OTD, however, in TDOA the time of the user's signal arrival is measured by the network of base stations that observe the apparent arrival time differences (equivalent to distance differences) between pairs of sites. Since each base station is usually at a different distance from the caller, the signal arrives at the stations at slightly different times. The receivers, synchronized by an atomic clock (provided, for example, by GPS), send the user's voice call and timing data to the mobile switch, where the times are compared and computed to generate the coordinates (latitude and longitude) of the caller. To calculate the distance difference between the two base stations, a hyperbola is defined, with each base station located at one of its foci. The intersection of the hyperbolas defined by different pairs of base stations determines the 2D location of the mobile terminal (Balbach, 2000; Hein *et al.*, 2000; Hein *et al.*, 2001). A minimum of three stations must receive the signal to enable the user's location estimation as an intersection of two hyperbolas. The basic observation equation (Equation (3.5)), which is used to determine the user's location, measures the distance from the user to two base stations (A, B).

$$\begin{aligned} d_{UB} - d_{UA} &= (t_U^r - t_B^t)c - (t_U^r - t_A^t)c \\ &= \sqrt{(x_U - x_B)^2 + (y_U - y_B)^2} - \sqrt{(x_U - x_A)^2 + (y_U - y_A)^2} \end{aligned} \tag{3.5}$$

The *AOA* method is based on the observation of the angle of signal arrival by at least two cell towers. The towers that receive the signals measure the direction of the signal (azimuth) and send this information to the AOA equipment, which determines the user's location by triangulation using basic trigonometric formulas. The accuracy of AOA is rather high but may be limited by the signal interference and multipath, especially in urban areas. Much better and more reliable results are obtained by combining AOA with TOA (Deitel *et al.*, 2002).

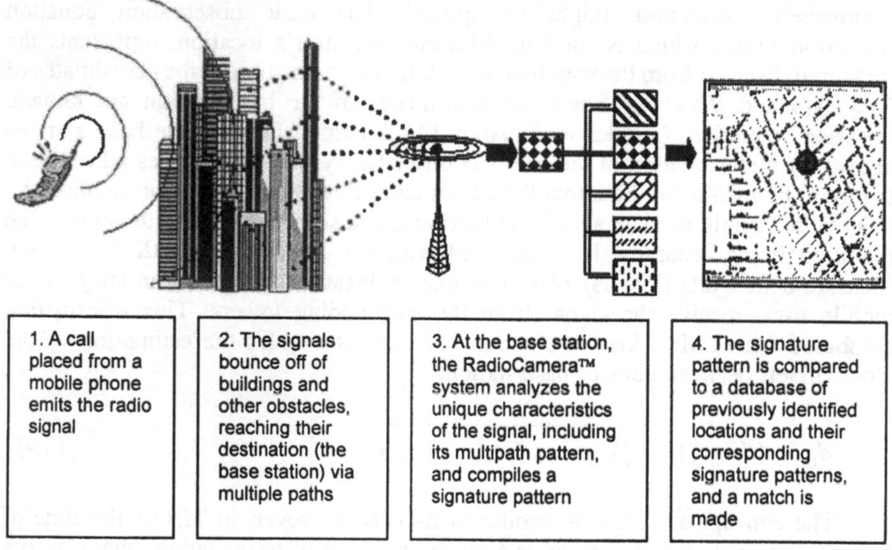

1. A call placed from a mobile phone emits the radio signal
2. The signals bounce off of buildings and other obstacles, reaching their destination (the base station) via multiple paths
3. At the base station, the RadioCamera™ system analyzes the unique characteristics of the signal, including its multipath pattern, and compiles a signature pattern
4. The signature pattern is compared to a database of previously identified locations and their corresponding signature patterns, and a match is made

Figure 3.11 The Location Pattern Matching process (U.S. Wireless, Corp., 2002)

The *Location (Multipath) Pattern Matching* method uses multipath signature in the vicinity of the mobile user to find its location. The user's terminal sends a signal that gets scattered by bouncing off the objects on its way to the cell tower. Thus, the cell tower receives a multipath signal and compares its signature with the multipath location database, which defines locations by their unique multipath characteristics (Figure 3.11). An example implementation, developed by the U.S. Wireless Corp. (2002), uses the location pattern matching technology (RadioCamera™) by measuring the radio signal's distinct radio frequency patterns and multipath characteristics to determine the user's location. With this method, the subscribers do not need any special updates to the mobile terminals to access the services, and wireless carriers do not need to make the infrastructure investments to offer LBSs.

Another network-centric method, applied in LBSs, is the software only approach, where no additional hardware on the cell towers (base stations) or the mobile phones (terminals) is required. The method is based on the signal strength model observed for the area. By merging the information about the actual *Received Signal Strength (RSS)* with the existing (mapped) signal data, the system can predict the user's location. An important feature of this technology is that it can determine the location of any digital cell phone or wireless device without any modification or add-ons or enhancements to the wireless carrier's existing network. By simply analyzing the existing RX (received) signal level (dBm) from multiple base stations to a standard wireless phone or device, the actual location of the phone can be calculated in seconds. The basic mathematical model is the relationship between the signal strength and the distance between the mobile station and the base. As in other ranging techniques, the user is located on a circle around the base with a radius equal to the distance measured using the signal strength. The prototype system based on RSS called GeoMode™, developed by Digital Earth Systems, has been tested in several metropolitan regions where it demonstrated location accuracy of 20-50 meters for 92% of the time (McGeough, 2002). In more challenging environments such as Manhattan, accuracy better than 100 meters for 80% of all stationary tests was obtained.

Several positioning methods presented in this section are based on the time measurement, such as the time of the user's signal arrival recorded at the base stations (TOA), or the time differences between the signal arrival from multiple base stations recorded by the mobile set (E-OTD, CGI-TA), or the signal travel time between GPS satellites and the mobile user. These time measurements can be converted to a range measurement (or the range difference), enabling the user's position determination by a common method of triangulation. Clearly, TOA or CGI-TA have an advantage over TDOA by working with the existing Global System for Mobile communications (GSM) mobiles (discussed in the following section), but may require significant investments in the supporting infrastructure (this is especially true with TOA). CGI-TA is rather inexpensive, as the cell information is already built into the networks. The E-OTD and TDOA methods require an extensive infrastructure support; moreover, E-OTD needs also customized handsets at the users' end (SnapTrack, 2002).

In general, one may argue that the user is less in control when the determination of his/her location is placed entirely within the network, as opposed to the mobile device. Perhaps the most autonomous is the GPS method and A-GPS, which rely on mobile devices that have an integrated GPS receiver. Clearly, GPS is the most accurate method of locating the mobile user; however, its accuracy may be limited by interference, jamming, strong multipath and losses of signal lock under foliage, overpasses or in urban canyons, as well as other factors. However, with the upcoming GPS modernization (see section 3.2.6) bringing the new, stronger civilian signal, and providing additional redundancy provisions, it is expected that GPS use in LBSs will only increase.

It should be mentioned that the existence of a variety of positioning technologies, without a standardized method, may pose a problem to both the users and the providers, as the user is only covered in the area serviced by his/her provider and may not be covered elsewhere, if another provider uses different location-identification method. It is rather difficult to define a single best technology, as each has its own advantages and disadvantages. Perhaps hybrid solutions offer the best choice, as they normally combine highly accurate with highly robust methods, resulting in multiple inputs improving both the robustness and the coverage. One example hybrid solution, listed in Tables 3.4 and 3.5, based on a combination of a handset-based GPS method with a network-based CGI-TA to cover GPS losses of lock, should offer a solution more reliable to the one offered by each technique alone. Other hybrid solutions are, for example, AOA plus

TDOA, called Enhanced Forward Link Triangulation (E-FLT), also called Enhanced Forward Link Time Difference, E-OTD plus A-GPS, and AOA plus RSS. Clearly, a selection of the positioning technique supporting LBSs should be guided by the following issues, depending on the needs of the potential users, applications covered, and the compliance with the existing standards:

- Required positioning accuracy;
- Frequency of the positioning update;
- Constant tracking versus positioning upon request;
- Maximum allowable latency;
- Compliance with U.S. FCC E-911;
- Compatibility with the users' terminals.

In summary, any wireless method of position determination is subject to errors. The major error sources include multipath propagation, Non-Line of Sight (NLOS), and multiple access interference. Multipath affects primarily AOA and RSS, but can also affect the time-based methods. Under NLOS the arriving signal is reflected or diffracted, and thus takes a longer path as compared to the direct LOS signal, affecting primarily the time measurement. Co-channel interference is common to all cellular systems, where users share the same frequency band. The multiple access interference can significantly affect the time measurement. In order to obtain satisfactory positioning performance, steps must be taken to mitigate the effects of the error sources (Caffery *et al.*, 1998).

3.3.3 GSM and UMTS Ranging Accuracy

GSM[6] is a widely used mobile communication standard also known as 2G (second generation), while Universal Mobile Telecommunication System (UMTS) is its emerging 3G (third generation) counterpart (Pietila and Williams, 2002; MobileGuru, 2002). GSM, ranging from ~450 MHz to ~2000 MHz, uses Time- and Frequency-Division Multiple Access (TDMA/FDMA), while UMTS is a Code Division Multiple Access (CDMA) system, operating on a carrier frequency of about 2 GHz. TDMA/FDMA means that the signal bandwidth is divided into frequency slots, each one further subdivided into time slots. The FDMA part involves the division by frequency of the (maximum) 25 MHz bandwidth into 124 carrier frequencies spaced 200 KHz apart. One or more carrier frequencies of the radio frequency part of the spectrum designated for mobile communication are

---

6 In 1982 the Conference of European Posts and Telegraphs (CEPT) formed a study group called the Groupe Spécial Mobile (GSM), to study and develop a public land mobile system that would meet the following criteria: good subjective speech quality, low terminal and service cost, support for international roaming, ability to support handheld terminals, support for range of new services and facilities, spectral efficiency, and ISDN compatibility. In 1989, GSM responsibility was transferred to the European Telecommunication Standards Institute (ETSI). Phase I of the GSM specifications was published in 1990, followed by commercial service initiated in 1991. Currently, hundreds of GSM networks are operational on every continent, and the acronym GSM now stands for Global System for Mobile communications (Wireless KnowHow, 2002).

assigned to each base station. Each of these carrier frequencies is then divided in time, using a TDMA scheme (Wireless KnowHow, 2002).

CDMA supports synchronized networking while GSM networks are generally unsynchronized (Pietila and Williams, 2002). Also, CDMA is able to support more calls in the same spectrum, as it dynamically allocates the bandwidth. Wideband CDMA (W-CDMA) has been selected for the third generation (3G) of mobile telephone systems in Europe, Japan and the United States. CDMA's signal structure is similar to the one used in satellite navigation techniques, i.e., channels are allocated on the same frequency and separated with codes. This approach may suffer from near-far effects, where a transmitter close to the receiver will effectively jam the signal from the distant transmitters. A solution to overcome this problem is to implement an Idle Period Down-Links (IPDL) of the base stations, meaning that a base station will have brief periods of no transmission, allowing all receivers in the vicinity to receive the signal. It should be mentioned that in the W-CDMA standardization, the positioning methods based on time difference are Idle Period Down Link - Observed Time Difference of Arrival (IPDL-OTDOA) and Advanced Forward Link Triangulation (A-FLT, also called Advanced Forward Link Time Difference). A-FLT and E-FLT mentioned earlier are essentially the same algorithms, according to Djuknic and Richton (2002), except, E-FLT covers the legacy handsets.

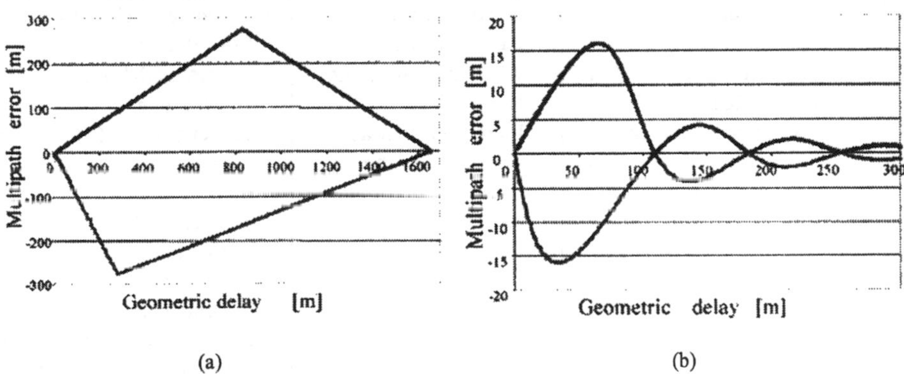

Figure 3.12 Multipath Error Envelops for (a) GSM and (b) UMTS (Hein, 2001)

The signal structure of a mobile communication system is rather complex, and different communication standards, such as GSM or UMTS, will differ with respect to the mode of access, frequency, time slot, power, etc. Several of these aspects are important also from the positioning standpoint, and would affect the final accuracy of the user's location, determined with the ranging signals. For example, GSM and UMTS suffer from different levels of multipath, as shown in Figure 3.12 (Hein, 2001). Multipath and other measurement errors (noise, tropospheric delay, and time synchronization error) accumulate in the user's location accuracy, and their effect is increased by the Horizontal Dilution of Precision (HDOP) factor, which reflects the geometry of the user-base configuration (for more information on DOP factors, see section 3.2.4.2). Table 3.6

shows examples of the expected positioning accuracy with GSM and UMTS while Table 3.7 lists advantages and disadvantages of stand-alone GSM and UMTS, and their combination with GNSS (Hein *et al.*, 2000; Hein *et al.*, 2001).

Table 3.6 GSM and UMTS Positioning Error Budget (Hein, 2001)

Error source	GSM [m]	UMTS [m]
Measurement noise	270	18
Multipath	0-250	0-17
Troposphere	0.3-3	0.3-3
Network/handset synchronization	3-6	3-6
Oscillator error	7.5	7.5
Total error (1 sigma)	270-380	19-26

Table 3.7 Advantages and Disadvantages of Stand-Alone GSM and UMTS, and Their Combination with GNSS (Hein *et al.*, 2001)

	GSM	UMTS/CDMA	GSM/UMTS/CDMA and GNSS
Pros	• No upgrade to the existing infrastructure if CGI or RSS is used • Limited increase of accuracy if ranging methods are used	• No addi tional hardware needed while system design is optimized • Increased accuracy when using ranging methods, as compared to CGI or RSS	• A dvanced technology, low integration effort to integrate GNSS into GSM phone • Reductio n in TTFF, GNSS receiver up-time reduced (wireless assisted GNSS) • Further re duction in TTFF and signal tracking enabled under bad conditions when common oscillators and microprocessors used (tight integration)
Cons	• Accuracy might be poor for CGI and RSS • Accuracy strongly depends on location if ranging technique is used • Position availability is always dependent on the access to the network	• Accuracy strongly depends on location • Position availability is always dependent on the access to the network	• Integrated GNSS and GSM units may be relatively large • Network service needed for wireless assisted GNSS • Much development needed to reach the tight integration level

3.4 OTHER POSITIONING AND TRACKING TECHNIQUES: AN OVERVIEW

3.4.1 Inertial and Dead Reckoning Systems

The name 'dead reckoning' is an abbreviation of 'deduced reckoning,' which means that the present location is deduced or extrapolated from a known prior position, modified by the known (observed) direction of motion and the velocity. Thus, in order to find the position coordinates using a Dead Reckoning System (DRS), the starting location and orientation must be known, since the system is sensitive only to a change in the direction and the apparent distance traveled derived from the continuously measured speed. An example of DRS is an *inertial navigation system* (INS). INS consists of multiple *inertial measurement units* (IMUs), such as *accelerometers* and *gyroscopes*, each one being a separate DRS.

Inertial navigation utilizes the inertial properties of sensors mounted on a moving platform, and provides self-contained determination of instantaneous position and other parameters of motion, by measuring a specific force, angular velocity, and time. Two primary types of IMUs are *accelerometers*, which sense linear acceleration in the inertial frame (i.e., fixed non-rotating frame), and *gyroscopes*, which sense the inertial rotational motion (angular rates, angular increments or total angular displacements from an initial known orientation). In principle, INS requires no external information except for initial calibration (initialization and alignment), including externally provided 3D position, velocity and attitude. A stationary self-alignment is routinely performed if no external velocity/attitude data are available. The IMU errors, however, grow with time. Therefore, INS should be recalibrated periodically (by performing Zero Velocity Update (ZUPT), or by using an external aid such as, for example, GPS) to maintain reliable navigation quality. In the stand-alone mode, an INS results are primarily affected by the initial sensor misorientation, accelerometer biases and gyroscope drifts, causing a time-dependent positioning error.

Accelerometers use a known mass (proof-mass) attached to one end of a damped spring, which is attached to the accelerometer housing. Under no external acceleration condition, the spring is at rest and exhibits zero displacement. An external force applied to the housing will cause its acceleration, however, due to inertia, the proof-mass will lag behind, resulting in a displacement. The displacement of the mass and extension/compression of the spring is proportional to the acceleration of the housing (Allen *et al.*, 2001; Jekeli, 2001). Since, according to Einstein's principle of equivalence, accelerometers do not sense the presence of gravitational field (but can sense the reaction due to the gravitational forces), external gravity information must be provided to obtain navigation information. In inertial navigation, the velocity and position are obtained through real-time integration of the governing differential equations (equations of motion), with accelerometer-measured specific force as an input. More details on inertial navigation can be found in Allen *et al.* (2001), Grejner-Brzezinska (1999), Grejner-Brzezinska (2001a and b) and Jekeli (2001).

Any spinning object tends to keep its axis pointed in the same direction (so-called gyroscopic inertia, or rigidity in space), and if a force is applied to deflect its orientation, it responds by moving at right angles to the applied force. *Gyroscopes* (or gyros) are mechanisms displaying strong angular momentum characteristics, capable of maintaining a known spatial direction through an appropriate torque control, since inertially referred rate of angular momentum is proportional to the applied torque. Three mutually orthogonal gyroscopes can facilitate a three-dimensional inertially non-rotating Cartesian frame, if they are mounted on a gimbaled platform maintaining the gyros orientation in space (space-stable system). Consequently, the gyroscopes maintain the knowledge of the orientation of the inertial platform, upon which orthogonal triad of accelerometers is mounted. Alternatively, gyros can be mounted directly on the vehicle (strapped down to the vehicle). In this case, since there is no gimbal platform performing the torque to maintain the gyroscopes' orientation, the torque is applied mathematically directly to the gyros. Since the physically or mathematically applied torque is proportional to the gyroscope's inertially referenced angular motion, it can be used to calculate the relative angular orientation between the gyro's initial and present spatial direction. The systems with no moving parts (no gimbaled platform), where the instrumentation of a reference frame is not facilitated physically but rather analytically, are referred to as strapdown INS. Since the lack of gimbaled structure allows for reduction in weight, size, power consumption, and ultimately cost, they are the primary modes of inertial navigation in a number of land and airborne applications.

A gyroscope-based DRS is called a *gyrocompass*. Thus, its directive action is based on the mechanical laws governing the dynamics of rotating bodies. The essential part of a gyrocompass consists of a spinning wheel (gyroscope), mounted in a way allowing freedom of movement about three mutually perpendicular axes. Essentially, a gyroscope becomes a gyrocompass if it can be controlled so that its axis of spin is aligned parallel with the true meridian (north-seeking gyroscope), under the influences of the Earth's rotation and gravity. As the Earth rotates, gravitational force attempts to change the gyroscope's axis of rotation. The resulting motion of the axis of the gyroscope at right angles to the applied force causes it to move to keep the alignment with the Earth's axis of rotation (Navis, 2002).

Another DRS, which can measure the distance traveled, is an *odometer* (wheel counter), based on the concept of counting wheel turns, starting at a known location. The measuring accuracy of odometers may vary, and typically ranges from 0.1 to 0.01 percent. The main factor that degrades the accuracy of the wheel counter measurement is the error in the scaling factor, which is affected by several conditions, one of them being unevenness of the surface on which the distance is measured (Da and Dedes, 1995). Modern odometers normally have digital displays (control panel), memory storage, built-in calculators for automatic measurements of areas and volumes, instant unit conversion, and can automatically include wheel radius for accurate wall-to-wall measurements. If in addition to the distance traveled the application requires heading and attitude data, an odometer can be coupled with vertical and directional gyroscopes in a self-contained gyroscope package. This kind of DRS is capable of providing full orientation and positioning information.

3.4.1.1 What Are the Errors in Inertial Navigation?

Errors in inertial navigation are functions of the following factors: (1) initial condition errors including the alignment process, (2) errors in gravitational attraction compensation, (3) errors in coordinate transformation, (4) time-dependent accelerometer and gyroscope errors, and possibly errors from external navigation aids used, and finally, (5) errors excited by the dynamics of the vehicle. As a consequence, the INS-determined vehicle trajectory will diverge from the actual path, depending primarily on the quality of the IMU sensors and the mission duration. For example, a high-reliability and medium accuracy strapdown INS, such as Northrop Grumman LN100 (based on Zero-lock™ Laser Gyro (ZLG™) and A-4 accelerometer triad, gyro bias of 0.003deg/h, accelerometer bias of 25 μg) demonstrates the positioning quality of 1.48 km/h CEP (Circular Error Probable rate, at 50% probability level) in the stand-alone navigation mode (Litton Systems, Inc., (1994). Table 3.8 shows representative error characteristics of a medium quality unaided inertial navigator.

Table 3.8 INS Performance Error Characteristics Assuming 4-8 min Alignment (May, 1993)

Position	1.48 km/h (CEP)
Velocity	0.76 m/s (RMS)
Heading	0.1 deg (RMS)
Pitch and roll	0.05 deg (RMS)
Angular rate	0.04 deg /s (RMS)

3.4.2 Digital Compass

A *digital compass* is another device used for orientation tracking in navigation, guidance and vehicle compassing. It is a solid-state device capable of detecting the Earth's weak magnetic field, whose circuit board includes the basic magnetic sensors and electronics to provide a digital indication of heading. The achievable accuracy in heading measurement is typically at the level of 1-5 degrees. The device is sensitive to tilt, and any tilt greater than 10-15 degrees will create directional errors. Most advanced systems include compensation for hard iron distortions, ferrous objects, and stray fields. Other applications besides navigation and compassing cover attitude reference, satellite antenna positioning, platform leveling, and integration with other devices, such as GPS or laser range finder.

3.4.3 Additional Location Tracking Systems

The location techniques presented in this section represent other positioning sensors based on various physical media employed, commonly used in location tracking. These systems are most commonly used for indoor positioning and tracking, but some of them can be also used outdoors.

3.4.3.1 Acoustic (Ultrasonic) Tracking

An *ultrasonic tracker* utilizes high frequency sound waves (approximately 20,000 Hz) to locate objects either by triangulation of several transmitters, time-of-flight (TOF) method, or by measuring the signal's phase difference between the transmitter and the receiver (phase-coherence method). The TOF method measures the time of travel between a transmitter and the receiver, which multiplied by a speed of sound provides an absolute distance measurement. The phase coherence method provides the phase difference between the sound wave at the receiver and the transmitter, which can be converted to a change in distance, if the signal's wavelength (frequency) is known. Since this method is sensitive to a change in distance only, the initial distance to the target must be known. More details on the ultrasonic trackers can be found in Allen *et al.*, (2001). A single transmitter/receiver pair provides a range measurement between the target and the fixed point. Three distance measurements provide two solutions, one of which is normally discarded as impossible. Thus, to estimate the 3D position coordinates, minimum of three range observations between the known locations of the transmitters and the target are needed. This concept is similar to the GPS-based triangulation. An inherent problem of an ultrasonic tracker is a signal travel delay, due to slow speed of sound (331 m/s at 0° C, and varies with temperature and pressure).

3.4.3.2 Magnetic Tracking

Magnetic trackers use magnetic fields, such as low frequency AC fields or pulsed DC fields, to determine 3D location coordinates, attitude and heading relative to the transmitter. Typically, three orthogonal triaxial coils generating the source magnetic fields are used at the transmitter and the receiver. A magnetic field is generated when current is applied to the transmitter coil. At the receiving end, a time varying magnetic field induces a voltage with a magnitude proportional to the area bounded by the coil and the rate of change of the field. This voltage varies with a cosine of the angle between the direction of the field lines and the axis of the coil. From the induced voltage level, the information about the distance from the transmitter to the receiver and the axis-alignment between them can be extracted. The distance estimation is based on the fact that the magnetic field strength decreases with a third power of distance and with the cosine of the angle between the axis of the receiving coil and the direction of the magnetic field. Subsequently, in order to find the distance between the receiver and the transmitter, the voltage induced at the receiver is compared to the known voltage of the transmitted signals. The orientation can be found through the comparison of the strength (voltage) of the induced signals (Allen *et al.*, 2001).

Magnetic tracking systems are subject to error, primarily due to distortions of their magnetic fields by conducting objects or due to other electromagnetic fields in the environment, and generally, the error increases with the transmitter-receiver distance. If there is a metal object in the vicinity of the magnetic tracker's transmitter or receiver, the transmitter signals are distorted and the resulting position/orientation measurements will contain errors. One possible solution is the system calibration based on a map of distortions (calibration table), from which a correction term can be derived. More details on magnetic tracking can be found in Allen *et al.* (2001), Raab *et al.* (1979), and Livingston (2002).

3.4.3.3 Optical Tracking

Optical tracking systems also referred to as image-based systems, make use of light to measure angles (ray direction) that are used to find the position location. The essential parts of an optical system are the target and the detector (sensor). These systems rely on a clear LOS between the detector and the target. Detectors can be in the form of Charged Coupled Device (CCD)-based cameras, video cameras, infrared cameras, or lateral-effect photodiodes. Targets can be active, such as light-emitting diode or infrared-emitting diode, or passive, such as mirrors or other reflective materials, or simply natural objects (Allen *et al.*, 2001). Detectors are used to observe targets and to derive position and orientation of a target from multiple angular observations (multiple detectors). For example, a single point on a 2D detector imaging plane provides a single ray defined by that point and the center of projection; two 2D points allow 3D target positioning and additional points are required for orientation. According to Allen *et al.*, (2001) "in order to determine orientation, multiple targets must be arranged in a rigid configuration. Then the relative positions of the targets can be used to derive orientation." One possible configuration of an optical system is a set of three 1D sensors, each one narrowing the location of a target to a plane. Intersection of three planes provides coordinates of a target in a system-defined reference frame. Alternatively, two 2D sensors, each one determining a line, allow for location of a target in 3D.

In general, the image-based tracking systems provide high positioning accuracy and resolution, but these are a function of the type of sensors used (primarily its angular resolution), distance between the target and the sensor, specific application and the environment (outdoor vs. indoor). Also, the system geometry has an impact on its position location accuracy. For example, fixed sensors can observe moving targets, and moving sensors can observe fixed targets. An important aspect is also the selection of the reference frame, in which the optical measurements are made (Allen, *et al.*, 2001). More details on optical target location can be found in, Allen *et al.* (2001), Blais *et al.*, (2002), and Beraldin *et al.* (2000).

Another type of optical tracking systems is based on laser ranging, which provides range measurements to active or passive targets. To measure a distance, the TOF of a laser beam from a transmitter to the target and back is measured (see, for example, Blais *et al.*, (2000) and ACULUX (2002). This method is well suited for measuring distances from several meters to a few hundreds of meters, and even considerably longer distances, and thus, it is suitable for both outdoor and indoor applications. The accuracy of the distance measured ranges from micrometers for short-range devices, to a decimeter-level for very long-range systems.

In the long-range laser scanning systems placed on moving platforms (airborne or spaceborne), the laser beam is oriented and its projection center located using an integrated GPS/INS system (see section 3.5). This allows for direct estimation of 3D coordinates of a target (Baltsavias, 1999; Grejner-

Brzezinska, 2001a and b). Similarly, in the image-based tracking methods, if the imaging sensor can be directly georeferenced by GPS/INS (or ground control points can be used), the absolute location coordinates of the points in the imagery in the selected global or local reference frame can be determined by photogrammetric methods. This approach is normally used in land-based or airborne systems suitable for mapping and GIS data acquisition.

3.4.3.4 Pseudolite Tracking

Pseudo-satellite or pseudolite (PL) can be regarded as a mini-satellite that can be used for autonomous navigation and positioning in the indoor or outdoor environments. The principle of pseudolite-based positioning systems is directly derived from the global positioning technology for outdoor environments. The system is able to triangulate the position of an object by accurately measuring the distances from the object to the array of pseudolites, whose location coordinates are known in a selected reference frame. Other applications of pseudolites include precision landing systems, such as LAAS, discussed in 3.2.4.1.3, outdoor navigation and other system augmentation (such as GPS augmentation). In outdoor applications, the most commonly used type of pseudolites is a GPS pseudolite. It is a ground-based transmitter, which sends a GPS-like signal to support positioning and navigation in situations where the satellite constellation may be insufficient. PLs are usually located on building rooftops, high poles, or any high location in the vicinity of the survey area, resulting in a relatively low elevation angle, as compared to GPS satellites. The majority of GPS pseudolites transmit signals on L1 carrier (1575.42MHz), and the more advanced systems can transmit also on L2 carrier (1227.6MHz). PLs can be designed to both receive and transmit ranging signals (transceivers), and thus can be used to self-determine their own location. With some firmware modification, standard GPS receivers can be used to track PL signals.

The ionospheric and tropospheric errors do not apply to most of the pseudolite applications (tropospheric errors apply only to the outdoor situation). However, the most important error sources are multipath and the near-far problem, where the PL transmitter can be very close to the receiving antenna, as compared to, for example, GPS satellites. The methods most commonly used to mitigate these problems are the proper transmitter and receiver design and the appropriate signal structure. For example, one possible method for eliminating the near-far problem is the technique of pulsing the pseudolite's signal, while a higher chipping rate (CR) can mitigate the undesired effect of multipath (Progri and Michalson, 2001). In addition, to mitigate multipath errors, helical antennas are usually employed for the transmission of the pseudolite signals. Other possible problems related to the use of pseudolites are: (1) any errors in PL location will have a significant impact on the receiving antenna coordinates due to the short distance between the receiver and the PL, (2) since PL is stationary, its location bias is constant, and its effect on position coordinates of the receiver depends on the geometry between the PL and the receiver, (3) a differential technique may eliminate fewer error sources, as opposed to the differential GPS, especially if significant range differences exist between the receiver and the PLs in the array.

Table 3.9 Primary Sensors of a Multisensor Tracking/Imaging System and Their Functionality (Grejner-Brzezinska, 2001a)

Primary sensor	Sensor Functionality
GPS	• Image geo-positioning in 3D • Time synchronization between GPS and INS • Image time-tagging • INS error control • Furnishes access to the 3D mapping frame through WGS84
INS	• Image orientation in 3D • Supports image georeferencing • Provides bridging of GPS gaps o Provides continuous, up to 256Hz, trajectory between the GPS measurement epochs • Supports ambiguity resolution after losses of lock, and cycle slip detection and fixing
Pseudolite Transmitter/trans-ceiver	• Primary positioning functions identical to GPS • Supports GPS constellation during weak geometry (urban canyons)
CCD Camera	• Collects imagery used to derive object position o Two (or more) cameras provide 3D coordinates in space
Laser Range Finder	• Supports feature extraction from the imagery by providing precise distance (typical measuring accuracy is about 2-5 mm)
LIDAR* **(airborne systems)**	• Source of DSM/DTM[7], also material signature for classification purposes Supports feature extraction from the imagery
Multi/hyper spectral sensors (airborne systems)	• Spectral responses of the surface materials at each pixel location • Wealth of information for classification and image interpretation
Voice recording, touch-screen, barometers, gravity gauges	• Attribute collecting sensors (land based systems)

*LIDAR – Light Detection and Ranging

[7] DSM – Digital Surface Model; DTM – Digital Terrain Model

For more information on pseudolites and their applications, the reader is referred to Baltrop *et al.* (1996), Elrod and Van Direndonck (1996), Wang *et al.* (2001), Progri and Michalson (2001), Progri *et al.* (2001), and Grejner-Brzezinska *et al.* (2002).

3.5 HYBRID SYSTEMS

Virtually all position location techniques presented here display some inherent limitations related either to the associated physical phenomenon, system design specifications, or application environmental constraints. Consequently, no single technique/sensor can provide complete tracking information with continuously high performance and reliability. However, the sensors/techniques can be integrated with each other to provide redundancy, complementarity and/or fault-resistance, rendering more robust positioning/tracking systems. One example of a commonly used hybrid or integrated system is an inertial-optical hybrid system, in which the optical system supports calibration of inertial errors during the slow motion of the system, where the optical sensor's performance is the best. During the rapid motion, the inertial system performs better than the optical, thus complementary behavior of both systems renders more reliable and accurate performance.

Another example is a GPS/INS system, which is often combined with additional imaging sensors, such as frame CCD or video. Integrated GPS/INS systems are commonly used in positioning and navigation, as well as in direct georeferencing of imaging sensors in mobile mapping and remote sensing. GPS contributes its high accuracy and long-term stability (under no losses of lock), providing means of error estimation of the inertial sensors. GPS-calibrated INS provides reliable bridging during GPS outages, and supports the ambiguity resolution after the GPS lock is reestablished. The effective positioning error level depends on systematic and random GPS errors as amplified by satellite geometry. Well-calibrated, GPS-supported INS provides precise position and attitude information between the GPS updates and during GPS losses of lock, facilitating immunity to GPS outages, and continuous attitude solution. In general, using a GPS-calibrated, high to medium accuracy inertial system, attitude accuracy in the range of 10-30 arcsec can be achieved (Schwarz and Wei, 1994; Abdullah, 1997; Grejner-Brzezinska, 1997). GPS/INS works well if GPS gaps are not too frequent and not excessively long. However, in case of urban canyon or indoor navigation, there is usually very limited or no GPS signal. Consequently, a PL array may be used to supplement the satellite signal (Wang *et al.*, 2001; Grejner-Brzezinska *et al.*, 2002).

In summary, any combination of GPS and INS functionality into a single integrated navigation system represents a fusion of dissimilar, complementary data, and should be able to provide a superior performance as opposed to either sensor in a stand-alone mode. In fact, integration of these two systems is often the only way to achieve the following goals (Greenspan, 1996):

- Maintaining a specified level of navigation during GPS outages;

- Providing a complete set of six navigational parameters (three positional and three attitude components) and high rate (higher than available from conventional GPS, i.e., > 20Hz);
- Reducing random errors in the GPS solution;
- Maintaining a GPS solution under high vehicle dynamics and interference.

Combination of positioning/orientation and imaging sensors renders a multisensor tracking/imaging system that can be designed for use either in outdoor or indoor environments (Bossler *et al.*, 1991; El-Hakim *et al.*, 1997; El-Sheimy and Schwarz, 1999; Behringer, 1999; You *et al.*, 1999a and 1999b; Grejner-Brzezinska, 2001a and b). Table 3.9 lists the example functionality of a multisensor system designed for mobile mapping and GISs. However, the sensor functionality can be generalized for other applications.

Table 3.10 Positioning Techniques: Summary of Characteristics

Location technique	Positioning method	Environment/ Availability	Accuracy (Highest achievable)*
Radiolocation:			
• GPS	Triangulation, 3D	Outdoor/global	High
• DGPS	Triangulation, 3D	Outdoor/global	High (mm to cm in relative positioning)*
• Pseudolite	Triangulation, 3D	Indoor /outdoor/local	High
• Time-based (TOA, TDOA, E-OTD)	Triangulation, 2D	Outdoor/indoor/local	Moderate-to-high
• AOA	Triangulation, 2D	Outdoor/indoor/local	Low-to-moderate
• RSS	Triangulation, 2D	Outdoor/indoor/local	Moderate
• Multipath pattern	Pattern matching	Outdoor/indoor/local	Moderate
• CGI-TA	Cell ID + distance to the base	Outdoor/indoor/local	Low
Inertial Navigation	Integration of accelerations, 3D	Outdoor/indoor/local	High-to-Low Errors grow with time
Acoustic (ultrasonic)	Triangulation, 3D Phase coherence, 3D	Indoor/outdoor/local	High (mm to cm for short distances)* Affected by NLOS
Optical	Triangulation, 3D	Indoor/outdoor/local	High (mm to cm for short distances)* Affected by NLOS
Magnetic	Triangulation, 3D	Indoor/outdoor/local	High (mm to cm)* Affected by magnetic field distortions

3.6 SUMMARY

Location and tracking techniques presented in this chapter are essential to geoinformatics, which is the foundation component of Telegeoinformatics. These techniques are used in remote sensing, providing geospatial information to GIS databases subsequently used in Telegeoinformatics applications. They are also a key component of real-time indoor and outdoor tracking for LBSs, such as emergency response, mapping, robot, fleet and personnel tracking, agriculture and environmental protection, or traveler information services. The main emphasis was put on GPS-based location technology; however other major techniques used in LBSs and in indoor/outdoor tracking were also presented. Moreover, the primary concepts of position determination using different media and observation principles were discussed. The summary of the position tracking techniques presented is provided in Table 3.10.

Position location and attributes provided by GIS databases, combined with wireless communication and Mobile Internet are crucial in Telegeoinformatics, where the user's location must be known in real time, enabling a number of applications mentioned above. These applications are expected to expand in the next few years, and will require an updated infrastructure, efficient tracking and communication techniques, as well as advanced optimization algorithms. These algorithms should be able to determine the shortest path to the object area, based on the GIS database information available via wireless communication, and the user's current location. GPS seems by far the most accurate and the fastest expanding positioning technology supporting active tracking for outdoor applications, which can be augmented for better performance and reliability by other techniques, such as INS. In addition, the forthcoming Galileo system brings a promise of the extended GNSS constellation, for better position location availability, reliability and accuracy.

REFERENCES

Abdullah, Q., 1997, Evaluation of GPS-Inertial Navigation System for Airborne Photogrammetry, presented at ACSM/ASPRS Annual Convention and Exposition, April 7-10, Seattle, WA.

Abnizova, I., Cullen, P., Taherian, S., 2002, Mobile terminal location in indoor cellular multi-path environment. Online. Available HTTP: <http:// www. wlan01. wpi.edu/proceedings/wlan69d.pdf> (accessed 11 November, 2002).

ACULUX, 2002, LaseRanger™ Noncontact Optical Ranging Sensor, User's Guide. Online. Available HTTP: <http://www.aculux.com/pdf/microsoft%20word%20-%20LR_232.PDF> (accessed 11November, 2002).

Allen, B. D., Bishop, G., and Welch, G., 2001, SIGGRAPH 2001 Course materials, *Tracking: Beyond 15 Minutes of Thought*. Online. Available HTTP: <http://www.cs.unc.edu/~tracker/media/pdf/SIGGRAPH2001_CoursePack_11.pdf> (accessed 12 December, 2002).

Andersson, C., 2001, Wireless Developer Network web page. Online. Available HTTP: <http://www.wirelessdevnet.com/channels/lbs/features/mobilepositioning.html> (accessed, 25 October, 2002).

Andersson, C., 2002, Mobile Positioning – Where You Want to Be! Online. Available HTTP: <http://www.wirelessdevnet.com/channels/lbs/features/mobilepositioning.html> (accessed 25 October, 2002).

Balbach, O., 2000, UMTS-Competing Navigation System and Supplemental Communication System to GNSS, Proceedings, ION GPS, Salt Lake City, pp. 519-527.

Baltsavias, E.P., 1999, Airborne laser scanning: basic relations and formulas, ISPRS Journal of Photogrammetry and Remote Sensing (54), pp. 199–214.

Barltrop K.J., J.F. Stafford and B.D. Elrod, 1996, Local DGPS with pseudolite augmentation and implementation considerations for LAAS. Proceedings, ION GPS, pp. 449-459, Kansas City, Missouri, September 17-20.

Behringer, R., 1999, Registration for Outdoor Augmented Reality Applications Using Computer Vision Techniques and Hybrid Sensors, *IEEE Virtual Reality*, pp. 244-251, Houston, TX, USA.

Beraldin, J.-A., Blais, F., Cournoyer, L., Godin, G., Rioux, M., 2000, Active 3D Sensing. Online. Available HTTP: <http://ai.iit.nrc.ca/cgi-bin/ftpsearch> (accessed 12 December, 2002).

Blais, F., Lecavalier, M., and Bisson, J., 2002, Real-time Processing and Validation of Optical Ranging in a Cluttered Environment. Online. Available HTTP: <http://www.icspat.com/papers/147mfi.pdf> (accessed 11 November, 2002).

Blais, F., Beraldin, J.-A., El-Hakim, S., 2000, Range Error Analysis of an Integrated Time-of-Flight, Triangulation, and Photogrammetric 3D Laser Scanning System, SPIE Proceedings, AeroSense, Orlando, Florida, Vol. 4035, April 24-28, pp. 236-247.

Bossler, J. D., Goad, C., Johnson, P., and Novak, K., 1991, GPS and GIS Map the Nation's Highway, *GeoInfo System Magazine*, March, pp. 26-37.

Caffery, J.J., Stuber, G.L., 1998, Overview of Radiolocation in CDMA Cellular Systems, *IEEE Communications Magazine*, Vol. 36 (4), April, pp. 38 –45.

Da, R. and Dedes, G., 1995, Nonlinear smoothing of dead reckoning data with GPS measurements, Proceedings of Mobile Mapping Symposium, The Ohio State University, May 24-26, published by ASPRS.

Deitel, H.M., Deitel, P.J., Nieto, T.R. and Steinbuhler, K., 2002, *Wireless Internet and Mobile Business: How to Program*, Prentice-Hall, Inc.

Directorate-General for Energy and Transport, GALILEO, European Satellite Navigation System, 2002. Online. Available HTTP: <http://europa.eu.int/comm/dgs/energy_transport/galileo/index_en.htm> (accessed, 12 December, 2002).

Djuknic, G.M. and Richton, R.E., Geolocation and Assisted-GPS, 2002. Online. Available HTTP: <http://www.lucent.com/livelink/090094038000e51f_White_paper.pdf> (accessed, 12 December, 2002).

El-Hakim, S.F. Boulanger, P. Blais, F. Beraldin, J.-A. and Roth, G., 1997, A mobile system for indoors 3-D mapping and positioning , Proceedings of the Optical 3-D Measurement Techniques IV, Zurich, September 29 - October 2, NRC 41548, pp. 275-282.

Elrod B.D. and Van Dierendonck A.J., 1996, Pseudolites, in: Parkinson, B.W. and Spilker, J.J. (Eds.), *Global Positioning System: Theory and Applications,* Vol. II, American Institute of Astronautics, Washington D.C., pp. 51-79.

El-Sheimy, N. and Schwarz, K.P., 1999, Navigating Urban Areas by VISAT – A Mobile Mapping System Integrating GPS/INS/Digital Cameras for GIS Application, *Navigation*, Vol. 45, No. 4, pp. 275-286.

Federal Aviation Administration, Satellite Navigation Product Teams, 2002. Online. Available HTTP: <http://gps.faa.gov/programs/waas/waas-text.htm> (accessed 12 December, 2002).

Francica, J., 2002, Location-based Services: Where Wireless Meets GIS. Online. Available HTTP: <http://www.geoplace.com/bg/2000/1000/1000spf.asp> (accessed, October 25, 2002).

GIBS, GPS Information and Observation System, 2002. Online. Available HTTP: <http://gibs.leipzig.ifag.de/cgi-bin/> (accessed 12 December, 2002).

Greenspan, R.L., 1996, GPS and Inertial Integration, Chapter 7 in *Global Positioning System: Theory and Applications*, Vol. II, Parkinson, B.W. and Spilker, J.J. (Eds.), American Institute of Aeronautics and Astronautics, Inc., pp. 187-220.

Grejner-Brzezinska D.A., 1997, High Accuracy Airborne Integrated Mapping System, *Advances in Positioning and Reference Frames*, Brunner, F.K. (Ed.), IAG Scientific Assembly, Rio De Janeiro, Brazil, September 3-9, Springer, pp. 337-342.

Grejner-Brzezinska D.A., 1999, Direct Exterior Orientation of Airborne Imagery with GPS/INS System: Performance Analysis, *Navigation*, Vol. 46, No. 4, pp. 261-270.

Grejner-Brzezinska D., 2001a, Mobile Mapping Technology: Ten Years Later, Part I, *Surveying and Land Information Systems*, Vol. 61, No.2, pp. 79-94.

Grejner-Brzezinska D., 2001b, Mobile Mapping Technology: Ten Years Later, Part II, *Surveying and Land Information Systems*, Vol. 61, No.3, pp. 83-100.

Grejner-Brzezinska D.A., 2002, GPS Instrumentation Issues, Chapter 10 in *Manual of Geospatial Science and Technology*, Bossler, J., Jensen, J., McMaster, R. and Rizos, C. (Eds), Taylor & Francis Books.

Grejner-Brzezinska D.A., Yi Y. and Wang J., 2002, Design and Navigation Performance Analysis of an Experimental GPS/INS/PL System, Proc. 2 nd Symposium on Geodesy for Geotechnical and Structural Engineering, Berlin, Germany, May 21-24, pp. 452-461; accepted for *Survey Review*.

Hein, G., Eissfeller, B., Öhler, V., and Winkel, J.O., 2001, Determining Location Using Wireless Networks, *GPS World,* Vol. 12, No. 3, pp. 26-37.

Hein, G., 2001, On the Integration of Satellite Navigation and UMTS, CASAN-1 International Congress, Munich, Germany. Online. Available HTTP: <http://ifen.bauv.unibw-muenchen.de/Aktuelles/IntSatNavUMTS-25-04-01.pdf> (accessed 25 October, 2002).

Hein, G., Eissfeller, B., Öhler, V., and Winkel, J.O., 2000, Synergies between Satellite Navigation and Location Services of Terrestrial Mobile Communication, Proceedings, ION GPS, Salt Lake City, pp. 535-544.

Hellebrandt, M. and Mathar, R., 1999, Location Tracking of Mobiles in Cellular Radio Networks, *IEEE Transactions on Vehicular Technology*, Vol. 48(5), September, pp.1558 –1562.

Hofman-Wellenhof, Lichtenegger, H. and Collins, J., 2001, *GPS Theory and Practice*, 5 th edition, Springer-Verlag Wien, New York.

Jekeli, C., 2001, Inertial Navigation Systems with Geodetic Applications, Walter de Gruyter, Berlin and New York, 352 pages.

Lachapelle, G., 1990, GPS observables and error sources for kinematic positioning, in *Kinematic Systems in Geodesy, Surveying, and Remote Sensing*, Schwarz, K. P. and Lachapelle, G. (Eds), IUGG, Springer-Verlag, New York.

Lachapelle, G., Ryan, S. and Rizos, C., 2002, Servicing the GPS user, in *Manual of Geospatial Science and Technology*, J. Bossler, J. Jensen, R. McMaster and C. Rizos (Eds), Taylor & Francis, London, New York, Berlin, Heidelberg, Paris, Tokyo, Hong Kong, Barcelona.

Langley R.B., 1991, The GPS Receiver: An introduction, *GPS World*, January, pp. 50-53.

Langley R.B., 1995, NMEA 0183: A GPS Receiver Interface Standard, *GPS World*, July, pp. 54-57.

Litton Systems, Inc., 1994, *LN-100G EGI Description*, Litton Systems, Inc., September.

Livingston, M., 2002, UNC Magnetic Tracker Calibration Research. Online. Available HTTP: <http://www.cs.unc.edu/~us/magtrack.html> (accessed 11 November, 2002)

May, M.B., 1993, Inertial Navigation and GPS, *GPS World*, September, pp. 56-66.

McGeough, J., 2002, Wireless Location Positioning Based on Signal Propagation Data. Online. Available HTTP: <http://www.wirelessdevnet.com/aroundtheweb.phtml?url=http://www.wirelessdevnet.com/library/geomode1.pdf> (accessed December 12, 2002).

MobileGuru, 2002, Second Generation (GSM). Online. Available HTTP: <http://www.mobileguru.co.uk/Mobile_Technology_Sec.html> (accessed 11 November, 2002)

Moffit, F. and Bossler, J., 1998, Surveying, 10th edition, Addison-Wesley.

Navis, 2002, Aids to Navigation, Gyrocompass. Online. Available HTTP: <http://www.navis.gr/navaids/gyro.htm> (accessed 30 December, 2002).

OmniSTAR, 2002, How it Works. Online. Available HTTP: <http://www.omnistar.com/howitworks.html> (accessed December 12, 2002).

Parkinson B. W. and Spilker J. J. (Eds.), 1996, *Global Positioning System: Theory and Applications*, American Institute of Aeronautics and Astronautics, Inc., Washington, D.C.

Pietila, S. and Williams, M., 2002, Mobile Location Applications and Enabling Technologies, Proceedings ION GPS, September 24-27, CD ROM.

Progri, I.F. and Michalson, W.R., 2001, An Alternative Approach to Multipath and Near-Far Problem for Indoor Geolocation Systems, Proceedings ION GPS, September 11-14, CD ROM.

Progri, I.F., Hill, J.M. and Michalson, W. R., 2001, An Investigation of the Pseudolite's Signal Structure for Indoor Applications, Proceedings ION 57th Annual Meeting, June 11-13, CD ROM.

Raab, F., Blood, E., Steioner, T. and Jones, H., 1979, Magnetic position and orientation tracking system, *IEEE Transactions on Aerospace and Electronic Systems*, Vol. 15, No. 5, pp. 709-718.

Raquet, J. and Lachapelle, G., 2001, RTK Positioning with Multiple Reference Stations, *GPS World*, Vol. 12, No. 4, pp. 48-53.

Rizos, C., 2002a, Making sense of the GPS technique, in *Manual of Geospatial Science and Technology*, Bossler, J., Jensen, J., McMaster, R., and Rizos, C. (Eds.), Taylor & Francis, London and New York.

Rizos, C., 2002b, Introducing the Global Positioning System, in *Manual of Geospatial Science and Technology*, Bossler, J., Jensen, J., McMaster, R., and Rizos, C. (Eds.), Taylor & Francis, London and New York.

Rizos, C., 2002c, Where do we go from here?, in *Manual of Geospatial Science and Technology*, Bossler, J., Jensen, J., McMaster, R., and Rizos, C. (Eds.), Taylor & Francis, London and New York.

Schwarz, K.P. and Wei, M., 1994, Aided Versus Embedded A Comparison of Two Approaches to GPS/INS Integration, Proceedings of IEEE Position Location and Navigation Symposium, Las Vegas, NE, pp. 314-321.

SnapTrack, 2002, Location Technologies for GSM, GPRS and WCDMA Networks. Online. Available HTTP: <http://www.snaptrack.com/Advantage/location_tech_9_01.pdf> (accessed 25 October, 2002).

Spilker, J. J. and Van Direndonck, A. J., 2001, Proposed new L5 civil GPS codes, *Navigation*, Vol. 48, No. 3, pp. 135-143.

U.S. Coast Guard Navigation Center, 2002, GPS General Information. Online. Available HTTP: <http://www.navcen.uscg.gov/gps/> (accessed 10 December, 2002).

U.S. Wireless Corp., 2002, Location Pattern Matching and The RadioCamera™ Network. Online. Available HTTP: <http://www.uswcorp.com/USWCMainPages/our.htm> (accessed 11 November, 2002).

Wang J., Tsujii, T., Rizos, C., Dai, L., and Moore, M., 2001, GPS and pseudo-satellites integration for precise positioning. *Geomatics Research Australasia*, Vol. 74, pp. 103-117.

Wireless KnowHow, 2002, GSM. Online. Available HTTP: <http://www.m-indya.com/mwap/gsm/whatis.htm> (accessed 20 December, 2002)

You, S., Neumann, U., and Azuma, R.T., 1999a, Hybrid Inertial and Vision Tracking for Augmented Reality Registration, *IEEE Virtual Reality*, pp. 260-267, Houston, TX USA.

You, S., Neumann, U., and Azuma, R.T., 1999b, Orientation Tracking for Outdoor Augmented Reality Registration. *IEEE Computer Graphics and Applications 19*, Nov/Dec, pp. 36-42.

CHAPTER FOUR

Wireless Communications

Prashant Krishnamurthy and Kaveh Pahlavan

4.1 INTRODUCTION

Wireless communications play an increasingly important role in the networking infrastructure of the world. It is predicted that in a few years, the number of computing and communications devices that will connect to the Internet without wires will far exceed traditional desktop computers and workstations. Consequently, the opportunity for pervasive computing is peaking at this time. In essence, wireless communications will provide the necessary groundwork for anywhere, anytime computing that becomes more relevant for Telegeoinformatics applications.

Recent trends in wireless communications are mostly in wireless data networks: over wide areas, Wireless Wide Area Networks (WWANs) and local areas, Wireless Local Area Networks (WLANs) and Wireless Personal Area Networks (WPANs). The former provides coverage and continuous connections over large areas, spanning cities, states and the entire country at vehicular speeds but at much smaller data rates ranging from 19.2 kbps for second generation services to 384 kbps for third generation services. WLANs are becoming increasingly popular for providing coverage in campuses, buildings and residences (home area networks). Recently there has been a lot of attention towards community or Neighbourhood Area Networks (NANs) or nanny networks using WLANs. WLANs can provide much larger data rates (between 2 and 54 Mbps) but are restricted to much smaller areas. WPANs are employed for ad hoc connectivity between different kinds of devices (not all traditional computers) such as PDAs, cellphones, printers, digital cameras, etc. In the future, it is expected that WWANs will co-exist with WLANs and WPANs in a seamless way so that wireless networked devices can utilize the best available network for the application at hand, creating a ubiquitous wireless network. This introduces several challenges related to the operational aspects of these networks. Although we can expect several types of devices such as laptops, palmtops, and cell phone-PDA combos to connect to this ubiquitous wireless network, in this chapter, we will refer to all of them as Mobile Stations (MSs).

Unlike wired networks, wireless networks have unique challenges that need to be addressed for correct and efficient operation. With the limited spectrum and bandwidth, innovative network design and physical layer approaches are required. The radio channel is a broadcast medium subject to interference, multipath and fading. As the medium is shared, the access to it has to be coordinated well. The devices connecting to wireless networks are usually portable or mobile, that is they operate with batteries, are lightweight and have limited resources like CPU power, memory, display size, etc. As devices are mobile, they need to be tracked to deliver voice calls or datagrams as need be. These result in complex network operations not seen in wired networks. Telegeoinformatics will be affected by these issues but it can also be potentially used to provide innovative solutions to some of these issues.

In this chapter we will provide an overview of existing and emerging wireless technologies. We will look at the latest trends in wireless communications, provide an overview of current and evolving technologies and applications, and discuss issues and challenges related to wireless communications in general. Wireless communications spans several different disciplines and to provide the reader with a good understanding, we discuss the technical and operational issues related to these systems in as much detail as possible. Some of the material is not directly applicable to Telegeoinformatics. However, at the end of most of the sections we discuss the impact of Telegeoinformatics on the contents of the section. The rest of the chapter is organized as follows. In Section 4.2, we classify wireless systems and discuss issues that are unique to wireless systems at the end. Section 4.3 discusses the radio propagation and physical layer aspects of wireless networks. Medium access issues are considered in Section 4.4. Section 4.5 considers network design and Section 4.6 the operational aspects like mobility, radio resources, power and security. We conclude this chapter with Section 4.7.

4.2 OVERVIEW OF WIRELESS SYSTEMS

In this section we classify wireless systems and architectures. Wireless systems can be classified in various ways as shown in Figure 4.1. The most common classifications are based on application or coverage (Pahlavan and Krishnamurthy, 2002), but some take into account the topology and mobility as well. The different classifications provided in Figure 4.1 are not mutually exclusive. For instance, there are voice-oriented, metropolitan-area, infrastructure, mobile wireless networks and data-oriented personal area, ad hoc, portable wireless data networks. Cell phones and palmtops are extensively used under all situations (even while driving at 60 mph on a highway) and need mobile connectivity. Laptops, cordless phones and digital cameras are simply portable. Desktop computers and workstations with wireless connectivity are stationary and cannot be used while moving. Point-to-point wireless links between corporations or residences and a network service provider fall under fixed wireless networks.

4.2.1 Classification of Wireless Networks

The majority of wireless networks today are WWANs based on *voice communications* (telephony) and common examples are the various kinds of cellular telephony systems. For Telegeoinformatics, data transfer is more important than voice communications. Circuit switched connections used for voice can also be used for data transfer, but the utilization of the channel is very inefficient. Over the latter part of the 1990s, the growth of the Internet has spurred *data-oriented* wireless systems that are overlaid on the cellular telephony systems. Most cellular telephony systems span entire nations and even across countries in Europe and fall under wireless *wide area* networks. Some of these networks are localized to urban areas and fall under *metropolitan area* networks. WWANs typically use *licensed spectrum* that cannot be employed by anyone else in a given geographic area. WWANs have a very structured backbone wired network with Base Stations (BSs) providing points of access to the backbone and serving geographic areas called *cells*. Consequently, they are called cellular systems as well. MSs make handoffs as they cross the boundaries of cells; that is, they switch from the current BS to a new BS for continuous connectivity. The existence of the structured infrastructure makes them fall under wireless systems with *infrastructure* topology.

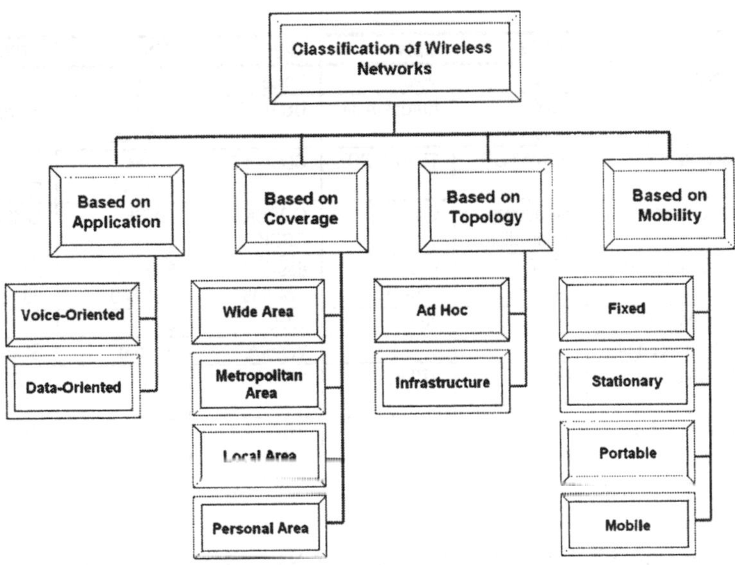

Figure 4.1 Classifications of Wireless Networks

The very first WWANs were the first generation (1G) analog cellular telephony systems such as the Advanced Mobile Phone System (AMPS) in the US. It is possible to overlay data traffic on such systems by borrowing unused voice channels. Cellular Digital Packet Data (CDPD) is one such example that was overlaid on AMPS (Taylor, 1997). CDPD uses the same spectrum and radio frequency (RF) infrastructure as AMPS but has its own protocols and architecture for supporting packet data applications. The 1G systems were gradually replaced (although not completely) in the 1990s by the second generation (2G) *digital* systems such as the Global System for Mobile (GSM) communications (Garg and Wilkes, 1999) that employs Time Division Multiple Access (TDMA) and the IS-95 system in the US that employs Code Division Multiple Access (CDMA) (Garg, 2000). Packet data oriented services are overlaid on the digital systems by grouping time slots or spread spectrum codes together and making them available on demand to MSs. The General Packet Radio Service (GPRS) (Kalden *et al.*, 2000) is such a data service overlay on GSM and has recently been deployed in Europe and the US. Sometimes these data overlays are referred to as 2.5G systems. Recently, efforts are on to migrate the 2G systems to third generation (3G) that employs only CDMA and can support both voice and data applications (3GPP, 3GPP2). 3G systems are expected to give a big boost to multimedia location services. The European version of 3G is called wideband CDMA (W-CDMA) or Universal Mobile Telecommunications System (UMTS). The North American version is called cdma2000 (Garg, 2002). Table 4.1 provides a summary of the characteristics and features of different wide area cellular wireless systems. Some of the details in the table are explained in subsequent sections.

Table 4.1 Some Wireless Wide Area Networks

System	AMPS	GSM	IS-95	CDPD	GPRS
Primary Application	Voice	Voice	Voice	Data	Data
Region	North America	Europe/Asia	USA/Asia	North America	Europe
Access Method	FDMA/FDD	TDMA/FDD	CDMA/FDD	TDMA DSMA	TDMA Reservation
Modulation Scheme	Frequency Modulation	GMSK	OQPSK/ QPSK	GMSK	GMSK
Frequency Bands (MHz)	869-894 824-849	935-960 890-915	869-894 824-849	869-894 824-849	935-960 890-915
Carrier Spacing (kHz)	30	200	1250	30	200
Bearer channels/carrier	1	8	Variable	Shared	Shared
Channel bit/chip rate (kb/cps)	Not applicable	270.833	1228.8	19.2	270.833
Data rates per user per trans. (kbps)	4.8 (circuit switched)	9.6 – 14.4 (circuit switched)	9.6 – 14.4 (circuit switched)	19.2	Up to 160

Contrary to WWANs, data-oriented WLANs have been built as extensions to existing Local Area Networks (LANs). MSs connecting through WLANs can use a simple infrastructure topology with Access Points (APs) instead of BSs or can connect to one another directly in ad hoc mode. Technologies that provide short range connectivity of diverse devices such as digital cameras and headsets to desktop and palm computers fall under WPANs. As these devices connect and disconnect as needed, they are said to have an *ad hoc* topology. WLANs and WPANs use *unlicensed spectrum* unlike the WWANs. The most common WLANs deployed today belong to the IEEE 802.11 standard for wireless LANs (IEEE 802.11) that has several sub-classes. WPANs are predominantly based on Bluetooth (see Bluetooth) (or the IEEE 802.15.1 standard) and to a far lesser extent on the HomeRF standard. Table 4.2 provides a summary of the characteristics and features of different local and personal area wireless systems. The details in the table are discussed in subsequent sections. We do not address WPANs based on infrared (IR) communications as they are typically restricted to line-of sight (LOS) situations.

The wide area systems typically support mobility while the local and personal area systems typically support only portability. However, as it is clear from Tables

4.1 and 4.2, wide area systems support much lower data rates than local area systems. The types and value of Location Based Services (LBSs) and Location based Computing (LBC) will be very different in these areas. WLANs and WPANs are better suited for a rich set of information services that are more granular (e.g., detailed to the level of a room) and consumer portal services (e.g., navigation in a mall or a museum). WWANs are likely to support fleet management, safety, telematics and a smaller set of information services and consumer portals. LBSs are discussed further in chapter 6.

In what follows, we examine some of the WWAN, WLAN and WPAN systems in more detail. In particular, we will consider wide area voice and data-oriented wireless systems that have infrastructure topology and local and personal area data-oriented systems that can have either a simplified infrastructure topology or ad hoc topology.

Table 4.2 Some Wireless Local and Personal Area Networks

Parameters/Standard	IEEE 802.11	IEEE 802.11b	IEEE 802.11a	HIPERLAN/2	HIPERLAN/1	Bluetooth
Status	Approved with products	Approved with products	Approved with products	In preparation	Approved but no products	Approved with products
Frequency Band	2.4 GHz	2.4 GHz	5 GHz	5 GHz	5 GHz	2.4 GHz
Modulation Scheme	DSSS with PSK, FHSS with FSK	DSSS with CCK	OFDM with PSK and QAM[1]	OFDM with PSK and QAM	GMSK	FHSS with FSK
Data Rate (Mbps)	1 - 2	1, 2, 5.5 and 11	6, 9, 12, 18, 24, 36, 54	6, 9, 12, 18, 24, 36, 54	23.5	Up to 0.732
Access method	CSMA/CA	CSMA/CA	CSMA/CA	TDMA/TDD	CSMA/CA	Polling

4.2.2 Wireless Network Architectures

A wireless network is far more complex than its wired equivalent. Figure 4.2 shows a schematic of the positioning of wireless networks today. Most wired networks have a fixed infrastructure that supports communications. For instance, when you lift a telephone handset, there are signals communicated to the local office. Switches in the local office communicate with other switches of the Public Switched Telephone Network (PSTN), across point-to-point links (fiber, copper, or even wireless) till the local office of the destination where the connection is completed to the destination phone (Carne, 1999). In the case of a data network, an Internet Protocol (IP) datagram from a host computer reaches a local router which examines the header to decide the next router for the packet. Eventually the packet gets routed through a Public Data Network (PDN) to a router attached to the

[1] Quadrature amplitude modulation that carries information in both phase and amplitude

destination network where it is picked up by the destination computer (Carne, 1999). In the case of wireless systems, certain extra entities are required to handle the wireless specific issues mentioned earlier in this chapter.

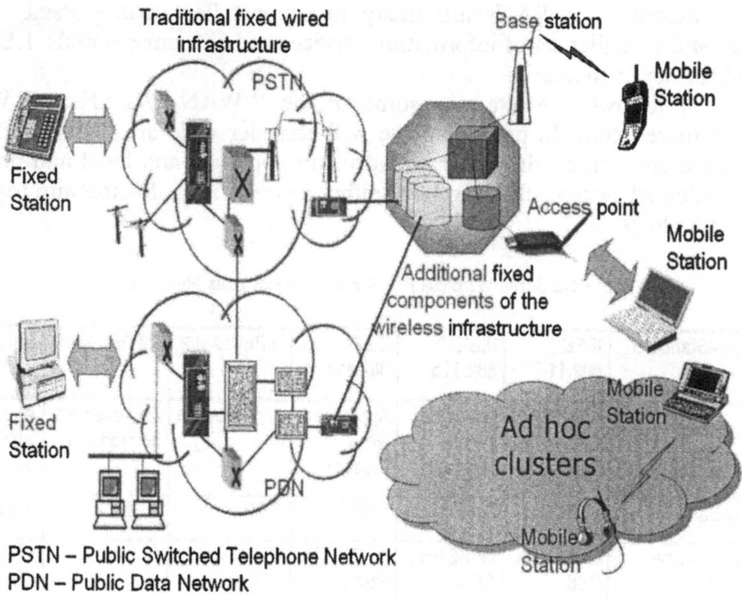

Figure 4.2 Positioning of Wireless Networks

First, as the spectrum is limited and regulated, transceivers can only transmit at limited power levels restricting their range. In order to provide coverage over large areas, in the infrastructure topology, several points of access to the fixed network (BSs or APs) are deployed. These are the physical radio transceivers (deployed by a service provider or an organization) that create the air interface, transmit and receive signals to and from MSs, and are involved in multiplexing on the link (medium access). Second, each system also has an allocated spectrum (licensed or unlicensed) and channelization (division of the total spectrum into frequency sub-bands). In some cases, as in WWANs shown in Table 4.1, there are separate frequency bands for the uplink and downlink. This is called Frequency Division Duplexing (FDD). In other cases, as in WLANs and WPANs, both uplink and downlink transmissions share the same frequency bands by transmitting at different points in time. This method is called Time Division Duplexing (TDD). Third, there are management entities and mobility-aware routers and switches for handling mobility, radio resources, power, billing and accounting, and security (Pahlavan and Krishnamurthy, 2002).

Typically, the network entity for managing radio resources is some kind of a radio controller. It manages the air interface and is involved in decisions related to the best RF carrier an MS should tune to, the transmit power levels it should use, whether the carrier/channel the MS is using is capable of providing acceptable quality, and when a handoff from one cell to another should be made. A mobile control center routes packets or calls to and from MSs and manages mobility by keeping track of the locations of MSs using some databases located in the *home area* and *visiting area* of MSs. Each MS is registered with its home area database (this is like the area code of a cellphone). Any incoming call or packet first needs to query this database to determine the current whereabouts of the MS (the visiting area – like the area code of the place where the MS is currently located). The home area database points to the visiting area database. The visiting area database keeps track of the MS with a finer granularity. The control center also ensures security via authentication centers and equipment registers to authenticate MSs and to prevent fraudulent/stolen devices from using the network and also performs accounting, billing, operations and maintenance. If we use location services as an example of Telegeoinformatics applications, WWANs are the focus today as they form the *carriers* of LBSs (see Chapter 6). In many cases, the WWAN carrier also hosts the wireless portals accessed by the end-user. The carriers already have an elaborate infrastructure to charge customers for enhanced services. WLANs and WPANs are becoming more popular as they support higher data rates. Note that in WLANs and WPANs there are few management entities. In ad hoc topologies, even the point of access is missing as MSs simply connect to one another as and when needed. We next consider three example architectures – GSM to present the most complexity, IEEE 802.11 that has a simple infrastructure and Bluetooth that has an ad hoc topology.

4.2.2.1 Example of a Complex Architecture: GSM

GSM (Goodman, 1997), the 2G standard for cellular telephony in use in many parts of the world is based on TDMA/FDMA. Each frequency carrier in GSM is 200 kHz wide and carries eight voice channels on eight time slots. In Europe, the uplink (MS to BS) uses the 890-915 MHz band and the downlink (BS to MS) the 935-960 MHz band. The modulation scheme is called Gaussian Minimum Shift Keying (GMSK) with optional frequency hopping. Circuit switched data at a maximum data rate of 14.4 kbps and Short Messaging Service (SMS) are supported. SMS consists of short alphanumeric messages that can be exchanged between MSs through the GSM network (Peersman and Cvetovic, 2000). Today, it is also possible to use multiple circuit switched connections to transfer data from 14.4 kbps to 57.6 kbps (using 4 time slots). A variety of the LBSs are carried via SMS today.

Figure 4.3 shows the reference architecture of GSM (Garg, 1999). Each link in the reference architecture has a reference name (e.g., the "O" interface between the management entities discussed below and some of the radio related entities). The Radio Subsystem consists of the MS and the Base Station Subsystem (BSS). It deals with the radio part of the GSM system. The BSS has two parts – a radio controller called the Base Station Controller (BSC) and a Base Transceiver System (BTS) (like a BS) that interfaces to the MS via the the air interface (called the U_m interface in GSM). The BSS contains parameters for the air interface such as GMSK modulation, status of carrier frequencies, and the channel grid as well as parameters of the A-interface (e.g., pulse code modulation signals at 64 kbps for a 4 kHz voice carried over Frame Relay). The BSC performs all Radio Resource Management (RRM) functions necessary to maintain radio connections to an MS, manages several BTSs, multiplexes traffic onto radio channels, handles intra-BSS handoff, reserves radio channels and frequencies for calls, and performs paging and transmits signaling data to the Mobile Switching Center (MSC). The BTS includes all hardware, transmitting and receiving facilities, antennas, speech coder and decoder, rate adapter and forms a radio cell (radius of 100m – 35km) or a cell sector (if directional antennas are employed).

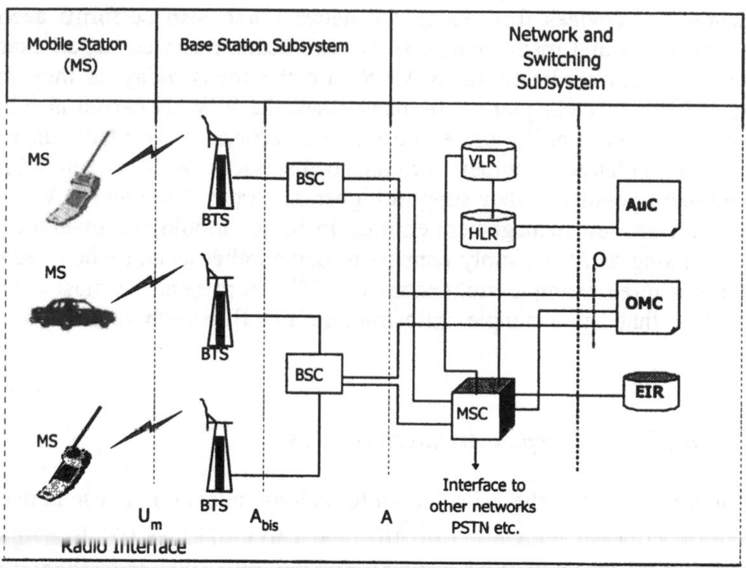

Figure 4.3 GSM Reference Architecture

The Network and Switching Subsystem is the heart of the GSM backbone. It provides connections to the standard public network, handoff related functions, functions for worldwide localization of users, support for charging, and accounting and roaming of users. It consists of the MSC which is the control center and the two types of databases: the Home Location Register (HLR) and a Visiting Location Register (VLR) that correspond to the home area and visiting area databases. The MSC is a high performance digital switch that is aware of mobility. It manages several BSCs. A Gateway MSC (GMSC) connects different service providers and networks like the PSTN, making use of signaling system 7 (SS-7) for signaling needed for connection set up, connection release, and handoff of connections. The MSC also handles call forwarding, multiparty calls, reverse charging and so on. The HLR is located in the *home network* of the MS. It stores all user relevant information: static information like the MS number, authentication key, subscribed services, and dynamic information like current Location Area (LA). For each MS there is exactly one HLR where the information is maintained. A VLR is associated with each MSC and it is a dynamic database that stores all information about active

MSs that are in its location. If a new MS comes into the location area, its information is copied from the HLR into the VLR. The Operation and Maintenance Center (OMC) monitors and controls all network entities using primarily SS-7 and handles traffic monitoring, status reports, accounting, billing, etc. The Authentication Center (AuC) contains algorithms for authentication and keys for encryption and is usually a special part of the HLR. The Equipment Identity Register (EIR) stores all device identifications and contains a blocked and stolen list and a list of valid and malfunctioning MSs.

The LA of today has a granularity of the cell size which could be anywhere between a few hundred meters and a few kilometers and the MS location is predictable to this granularity. As positioning techniques are implemented in these systems, additional entities like a "location control center" will have to be incorporated into the architecture (Koshima and Hoshen, 2000). As discussed in section 5, the position location information can be exploited for the operation of the network as well.

4.2.2.2 Example of a Simple Architecture: IEEE 802.11

In the case of WLANs, there are no databases and control centers and most of the functionality required for operation resides in the AP. The architecture of an IEEE 802.11 WLAN (IEEE 802.11) is shown in Figure 4.4.

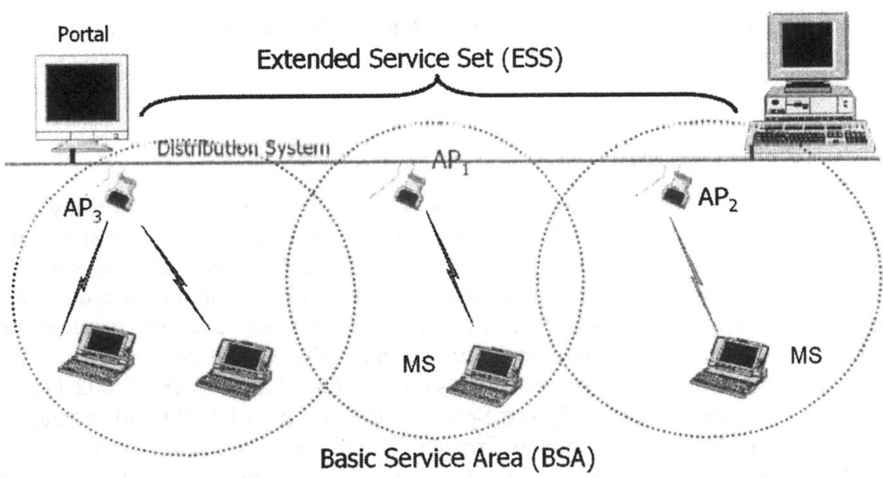

Figure 4.4 Architecture of IEEE 802.11

The AP is the point of access to the wired LAN (which is typically Ethernet, but could also be token ring). The wired LAN is called the distribution system. The AP thus provides access to distribution services via the wireless medium. A basic service area corresponds to the coverage area of one access point (it is the same as a cell). The Basic Service Set consists of the set of MSs controlled by one AP.

Several basic service sets are connected via the distribution system to form an Extended Service Set (ESS) (Crow *et al.*, 1997). The ESS is like a microscopic version of a cellular system and can cover a building or a campus. The AP in a WLAN broadcasts information related to the basic and extended service sets, registers MSs and enables communication between MSs or between an MS and the fixed network. Since packets are broadcast over LANs, mobility is not very complex. Every AP in the ESS can receive packets. Mechanisms to associate the MS to an AP, reassociate it to a different AP or dissociate it from an AP are used to determine which AP picks up packets intended for an MS. Typically there is no power control or complex RRM schemes in WLANs.

An important aspect of WLANs is that the organization that deploys the WLAN owns it unlike the cellular service providers. As the deployment of WLANs increases, it is likely that they will be the first candidates for limited-scale Telegeoinformatics applications in classrooms, campuses, malls, museums and other indoor areas owing to the low cost of deployment and ownership. A major problem for Telegeoinformatics with WLANs is the difficulty of position location within buildings (Pahlavan *et al.*, 1998 and Pahlavan *et al.*, 2002). Recently, there are efforts to use the WLAN infrastructure itself for positioning in indoor areas (Prasithsangaree *et al.*, 2002) as the Global Positioning System (GPS) does not work very well within buildings. While it is possible to deploy additional infrastructure for this purpose, using existing communications infrastructure is preferred for cost reasons. Location fingerprinting has been used in these approaches where information in a database of prior measurements of the received signal strength at different locations is matched with the values measured by an MS. See chapter 3 for more details.

4.2.2.3 Example of an Ad Hoc Topology: Bluetooth

Bluetooth (Bluetooth) is an example of a WPAN that considers networking in a Personal Operating Space (POS) – a 10m space around a person who is stationary or in motion. The idea here is to network "personal" devices such as sensors, cameras, handheld computers, audio devices, etc., with a short range around the person. The applications of Bluetooth range from being a cable replacement alternative to providing ad hoc connectivity and perhaps even connectivity to the fixed network. Bluetooth operates in the same 2.4 GHz bands as IEEE 802.11b. It employs Frequency Hopping Spread Spectrum (FHSS) with 1 MHz wide channels. The modulation scheme is Gaussian Frequency Shift Keying (GFSK) for a raw data rate[2] of around 1 Mbps on the air. In Bluetooth, a basic time slot is defined as 625 microseconds and a packet can occupy one, three or five slots. Sometimes a transmission is half a slot. The frequency is changed every packet and the frequency hopping sequence defines a piconet. The topology shown in Figure 4.5 is called a *scattered ad hoc* topology. The "cell" or "piconet" is defined by a *Master* device that controls the frequency hopping sequence as well as transmissions within

---

[2] The raw data rate does not account for overhead used in identifying time slots and packet headers.

its piconet. Multiple piconets with shared devices make a scatternet. There is no contention within a piconet (the devices take turns in transmitting) but there is interference between piconets. Piconets are created as and when necessary. When a Bluetooth device desires a connection, it scans the surroundings and tries to discover other Bluetooth devices in its vicinity. Other devices will or will not respond depending on how they are configured. The device initiating the connection is called the *Master*. Responding devices are *slaves* in the piconet. When the desired communication is completed, the devices disconnect thereby disbanding the piconet. It is possible for a device to be a slave in more than one piconet and perhaps a master in a piconet and a slave in another piconet (see Figure 4.5). It cannot be the master of two piconets as a piconet is defined by the master device. In Bluetooth, although the master device coordinates communications within the piconet, it does not perform mobility management, RRM or any other complex operations.

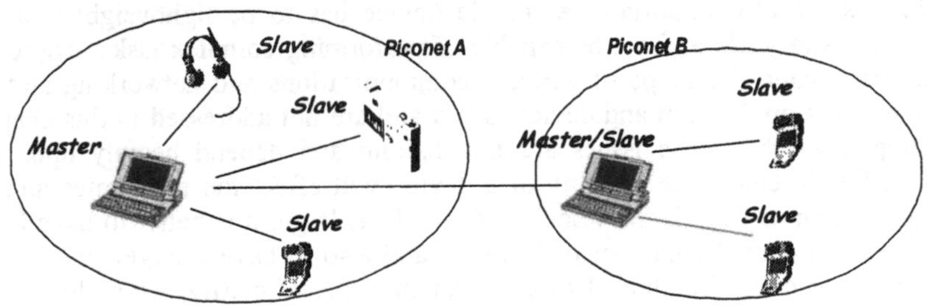

Figure 4.5 A Bluetooth Scatternet with Two Piconets

Bluetooth devices could potentially be deployed with positioning sensors to increase the granularity of position location in indoor areas. Infrared sensors have been used in the *active office project* (Ward *et al.*, 1997) where computers would autonomously change their functionality based on observations of people around them. Input from infrared sensors that were distributed throughout the environment, was used for this purpose. However, as of today it is not clear how the deployment of Bluetooth will evolve.

4.2.3 Issues and Challenges in Wireless Networks

Unlike wired networks, wireless networks face several issues and challenges. At the *physical layer*, the harshness of the radio channel leads to high bit error rates that need to be mitigated. This also has effects on higher-layer protocols like TCP. At the *Medium Access Control* (MAC) *layer*, since air is a shared broadcast medium, there is need for a simple decentralized medium access mechanism that does not sacrifice performance – throughput, delay and quality of service (QoS). There are *network design and deployment* issues because no single type of wireless access is available everywhere. Spectrum is scarce and needs to be reused. Coexistence of MSs that are transmitting in the same frequency bands at the same time leads to interference and degraded performance if the network design is not correct. *Radio resource and power management* is necessary to handle channel assignment/admission control, power control and handoff decisions. As the MSs operate on battery power, energy efficient design of software and protocols become necessary. *Mobility management* is important because of the need to use the MSs anywhere and anytime. Mobility management needs signaling and has its own QoS requirements. The broadcast nature of the wireless channel leads to *security* issues. There is need for correct entity authentication, confidentiality of signals and information on the air, management of keys required to perform encryption and so on. Lastly, there are *application level* issues in service discovery and data management. For instance, there must be mechanisms to ensure a coffee machine recognizes a PDA or vice versa. Important questions remain as to how data is maintained, where data should reside and how it can be efficiently accessed. Telegeoinformatics could play a very important role here as the position of the MS would often determine the answers to these questions. The form factor and capabilities are also important. A mobile device has to be lightweight, durable, have long battery life and yet be capable of performing complex tasks. These last issues are beyond the scope of wireless communications and networking and fall under wireless application and device issues and are not addressed in this chapter. We emphasize that these issues are not disjoint and depend heavily upon one another. For instance, the mobility of a device will affect the error rates and the network design as well. If the velocity of the MS is large, it is better to have larger cell sizes to reduce the number of handoffs and also to have a larger interleaving depth in error control coding. However, for the sake of clarity, we address these issues separately.

4.3 RADIO PROPAGATION AND PHYSICAL LAYER ISSUES

The physical layer of any digital communication system determines how exactly information (that includes packet headers and trailers) bits are converted into a form suitable for transmission and methods to detect this information at the receiver. In wireless networks, the lack of spectrum and the harsh radio channel make the selection of an appropriate transmission scheme (modulation and coding) an important issue. An understanding of radio propagation is essential for developing appropriate design, deployment and management strategies for any wireless network. In effect, it is the nature of the *radio channel* that makes wireless networks far more complicated than their wired counterparts. Radio propagation is heavily site specific and can vary significantly depending on the terrain, frequency of operation, velocity of the mobile terminal, interference sources, and other dynamic factors (see chapter 5 for more details). Telegeoinformatics could prove to be a valuable tool for modeling and understanding the radio propagation characteristics. Accurate characterization of the radio channel through key parameters and a mathematical model are important for predicting signal coverage, achievable data rates, specific performance attributes of alternative signaling and reception schemes, analysis of interference from different systems, and determining the optimum location for installing base station antennas. The reader is referred to Pahlavan and Levesque (1995) for an extensive discussion of radio propagation issues. In this section we briefly consider the important issues related to the physical layer.

4.3.1 Characteristics of the Wireless Medium

A signal transmitted over the air arrives at a receiver via many paths and this phenomenon is called multipath propagation. Especially at frequencies larger than 500 MHz, when the sizes of objects in the environment are larger than the wavelength of the carrier, it is possible to think of radio propagation as consisting of several discrete paths to the receiver. Some of these signals add constructively at the receiver and others destructively leading to random fading of the signal. Also, as the distance between the transmitter and receiver increases, the Received Signal Strength (RSS) gets smaller. The overall effects of the radio channel can be categorized broadly into large and small scale fading and each has several manifestations impacting different aspects of the communication system.

4.3.1.1 Large-Scale Fading

Large-scale fading refers to the macroscopic characteristics of the RSS. In free-space, the RSS decreases as the square of the distance, but in general, the presence of objects in the environment (including the earth) results in the mean value of the RSS decreasing as some exponent α of the distance d between the transmitter and the receiver. The value of α changes with the environment and is extremely site-specific. It also depends on the frequency of the carrier used in the system. The loss in signal strength is referred to as the *path-loss*. Over the last two decades, several measurement and modeling efforts have resulted in tabulation of path-loss for various frequencies and environments. There are several empirical models derived from these measurements, some of the popular ones being the Okumura-Hata model (Hata, 1980) for urban areas (used for frequencies around 900 MHz), the European Cooperative for Science and Technology (COST) model for micro-cellular areas (1900 MHz), and the partition-dependent and Joint Technical Committee (JTC) models for indoor areas (1900 MHz) (Pahlavan and Levesque, 1995). Path-loss models are used to predict the mean RSS as a function of distance from the transmitter, but the actual RSS has variability around the mean value (i.e., locations at the same distance from the transmitter have different RSS values). This variability is called "large scale fading" and occurs due to the nature of intervening obstacles between the transmitter and receiver. Thus, it is sometimes referred to as *shadow fading*. Measurements have shown that this variability has a log-normal distribution and it is not uncommon to refer to this effect as *log-normal shadow fading*. A common technique to overcome the effects of large scale fading is to introduce a fading margin in the link budget that will ensure that at least 95% of locations in a cell receive adequate signal strength.

4.3.1.2 Small-Scale Fading

Small-scale fading refers to how the signal varies microscopically, i.e., over small changes of time or space. Characterization of small-scale fading depends on the *scale* and typically the data symbol duration[3] is the scale of interest. It is also common to consider two separate manifestations of small-scale fading, namely variation in time and dispersion in time. The received signal amplitude varies with time because of multipath fading. This variation can be fast or slow compared to the symbol rate resulting in *fast fading* or *slow fading*, respectively. If the symbol rate is smaller than the rate of change of the amplitude, it is called fast fading. This can lead to unrecoverable errors in the received information. Slow fading is more common and results in burst errors that can be recovered using appropriate channel coding and interleaving schemes. The rate of change of the signal amplitude depends on the maximum *Doppler* frequency which in turn depends on the velocity of the MS and the carrier frequency. The larger the velocity (or frequency), the greater is the rate of change of the amplitude. The distribution of the received signal amplitude has also been characterized via measurements. In LOS situations, the amplitude has a *Ricean* distribution and in obstructed-LOS (OLOS) situations it has a good fit to the *Rayleigh* distribution (Pahlavan and Levesque, 1995). The effects of time variation are primarily increased bit error rates and packet error rates and some common mitigation techniques include using diversity (multiple copies of the signal), channel coding, and frequency hopping.

In the case of time dispersion, signals arriving via delayed paths interfere with the signals that have already arrived and are being detected at the receiver resulting in *Inter-Symbol Interference* (ISI) and irreducible errors. This problem is more acute in *wideband* systems where the symbol duration is much smaller than the multipath delays. In the frequency domain, we can think of the channel as being *frequency selective* (i.e., different frequency components of the spectrum face different attenuation). This situation is referred to as *frequency selective fading* as against *flat fading* where there is no inter-symbol interference. Time dispersion is characterized via the Root Mean Square (RMS) multipath delay spread of the channel. If the symbol duration is larger than the RMS delay spread, the time dispersion problem is severe. Alternatively, in the frequency domain, we can look at the coherence bandwidth (the range of frequencies in the channel that face identical attenuation). If the signal bandwidth is smaller than the coherence bandwidth, the time dispersion problem is less severe. In any case, time dispersion places a limit on the maximum data rates that can be supported on a wireless channel. In order to increase this limit, signal processing techniques like equalization, advanced modulation schemes like spread spectrum and Orthogonal Frequency Division Multiplexing (OFDM), or architectural techniques like directional antennas will have to be employed. In indoor areas, the RMS delay spread is between 30 and 300 ns (corresponding allowable symbol rates of 33 MSps and 3 MSps[4] respectively) and in outdoor areas, it can be as high as 5 μs restricting the symbol rates to 200 Kbps. Multipath delay spread also causes

---

[3] For example, at 1 Mbs and using a binary modulation scheme, the duration of one symbol is 10^{-6} s

[4] MSps = Mega Symbols per Second

physical limitations on the measurement of the time of arrival and angle of arrival of signals that impacts the accuracy of positioning systems (Pahlavan *et al.*, 1998).

4.3.1.3 Telegeoinformatics and Radio Propagation

Radio propagation characteristics are site specific as discussed before. Telegeoinformatics can be used for radio propagation modeling. Site-specific radio propagation prediction and modeling for wireless communication systems have employed geographic information systems (GISs) alone in a very limited manner. In (Lazarakis *et al.*, 1994), a GIS is employed to co-ordinate field data of radio signal strength with environmental and topographic information. The information is then used to compare statistical radio propagation path loss models with actual data. A procedure has been described there for incorporating radio channel measurements for a mobile radio environment into a GIS database. Ray tracing simulation schemes that predict the radio channel characteristics also employ GIS-based information (Catedra and Perez-Arriaga, 1999). Including a positioning system with mobile computing could make the process more automated and simpler.

4.3.2 Modulation and Coding for Wireless Systems

In selecting a modulation scheme for any system, issues that need to be considered are how it performs in an additive white Gaussian noise[5] (AWGN) channel (that is the best case scenario), how it performs in fading channels, what is its spectrum efficiency, how much power it consumes and how complex the transceiver will be. The performance measure is the bit error rate (BER or P_b) as a function of the energy per-bit to noise power spectral density ratio (E_b/N_0). The E_b/N_0 value is a measure of the "power requirements." There are tradeoffs between these issues that become relevant in selecting a modulation scheme (Sklar, 2001).

The most common modulation schemes are based on carrying information in the amplitude, phase, frequency, pulse position, pulse width, or code symbol. Depending on the number M of phases, frequencies, pulse positions etc., a symbol carries $\log_2 M$ bits (Proakis, 2000). In the case of Phase Shift Keying (PSK), as M increases, we increase the spectrum efficiency, i.e., we are able to stuff more bits/s in a Hz of bandwidth. However, the penalty is that the BER increases for the same E_b/N_0. Similarly, in the case of orthogonal Frequency Shift Keying (FSK), as M increases, the BER decreases for the same E_b/N_0 but the penalty is the reduction in spectrum efficiency. Similar tradeoffs exist for other modulation schemes. In general, it is possible to tradeoff between spectrum efficiency, power and BER to select the best modulation scheme. Certain modulation schemes are more bandwidth efficient than others. For example, in GSM and CDPD, a variation of FSK called *Minimum Shift Keying* (MSK) with a Gaussian filter (thus it is called GMSK) is used to reduce the bandwidth of a carrier. Pulse shaping techniques can be used to shape the bandwidth of a given modulation scheme. PSK is commonly

---

[5] Additive white Gaussian noise has a "flat" noise spectrum with average power spectral density of N_0.

employed in IEEE 802.11, IS-95, and HIPERLAN (HIgh Performance Radio LAN) (Wilkinson *et al.*, 1995) whereas FSK is employed in Bluetooth and the FHSS version of IEEE 802.11. Pulse position modulation is common in infrared systems.

Coding schemes are commonly employed to overcome the effects of time variation of the signal. Block codes and convolutional codes are commonly employed with interleaving in almost all cellular systems. For example, in CDPD, very powerful non-binary Reed-Solomon coding (Lin and Costello 1983) is used to correct up to 42 bit errors in a block of 378 bits. In CDMA systems, coding plays a very important role in reducing the effects of interference. In IEEE 802.11b systems, *complementary codes* in Complementary Code Keying (CCK) are used to increase the data rate by a factor of eleven without sacrificing the benefits of good error rate performance obtained with simple spread-spectrum.

In the case of time dispersion, equalization schemes are employed to overcome the ISI effects introduced by the channel. Equalization circuitry can be power hungry and today, advanced modulation schemes are employed to mitigate the effects of time dispersion. In IS-95 and other CDMA systems, Direct-Sequence Spread Spectrum (DSSS) is employed for multiple access. In DSSS, a symbol is split up into "chips" thereby spreading the spectrum and providing processing gain. A by-product of DSSS is that it can resolve or suppress multipath components with delays larger than a chip duration thereby mitigating ISI. Incidentally, DSSS is also employed in GPS for accurate positioning.

OFDM (van Nee and Prasad, 2000) is another scheme used to mitigate time dispersion. Here, data is sent on multiple narrow carriers that have bandwidths smaller than the coherence bandwidth of the channel. As the bandwidth is small, the symbol duration is large compared to the RMS delay spread. The carriers are spaced as close to one another as possible so as not to waste spectrum. Although OFDM is not a new technology, it has found new importance in applications like Digital Subscriber Line (DSL) modems where the channel is not uniform, for digital audio and video broadcast and finally WLAN applications (in IEEE 802.11a and HIPERLAN-2). Fast implementation of OFDM using Fast Fourier Transforms (FFT) is now possible. OFDM can be made adaptive to channel conditions as well as to improve the capacity of a carrier. Using OFDM for estimating the time-of-arrival of signals in WLANs is studied in Li *et al.* (2000). The mean position error using this technique was found to be between 3m and 7.5m depending on the nature of the radio channel.

4.4 MEDIUM ACCESS IN WIRELESS NETWORKS

Medium access (Gummala and Limb, 2000) basically decides who gets to transmit, how this decision is made and when they get to transmit. It involves multiplexing (how many MSs can share a single link) with the common frequency, time and code division multiple access (FDMA, TDMA, CDMA) schemes in circuit switched voice networks. In data networks (especially LANs), random access schemes like Carrier Sense Multiple Access with Collision Detection (CSMA/CD) in Ethernet (Carne, 1999) are used for simplicity and to accommodate bursty data traffic. Medium access also includes duplexing (how communication from MS_A to MS_B is separated from the communication from MS_B to MS_A) that was discussed previously in section 4.2.2. Medium access is impacted by architectures, whether it is an infrastructure with a centralized, fixed BS or AP, or ad hoc that is distributed and peer-to-peer. Channel characteristics, especially link quality and burst errors will impact the packet size to ensure a small packet error rate. As the medium is harsh and the loss of packets highly probable, most wireless MAC protocols need acknowledgments. MAC protocols (for wired or wireless networks) can be classified broadly into random access (such as ALOHA and CSMA), taking turns (such as polling and other round-robin access) and reservation schemes. We discuss some of these aspects below.

4.4.1 Medium Access Protocols for Wireless Voice Networks

In wireless voice networks, connections are circuit switched, i.e., an MS gets a dedicated channel for the duration of the connection and no other MS can access this channel during this period, whether or not the current MS is making full use of the channel. This is efficient for voice communications where two parties are likely to have conversations lasting for a significant amount of time. The conversations generate data continuously at a constant rate and they also have strict delay constraints making a dedicated channel efficient and beneficial. The channels can be a pair of frequency carriers (as in the case of AMPS), a pair of time slots (e.g., GSM) or a pair of spread spectrum codes (as in the case of IS-95) (Pahlavan and Levesque, 1995 and Pahlavan and Krishnamurthy, 2002). These are respectively called frequency, time and code division multiple access. Analog systems were primarily FDMA while 2G digital systems were based on TDMA and CDMA, both offering greater control over the types of services that could be provided to the end user. The limiting factor for all of these systems is interference that reduces system capacity. Using powerful channel coding and combining this with a robust modulation scheme has resulted in CDMA being the superior choice for 3G systems (Garg, 2002). Circuit switching is however inefficient for bursty data systems and the medium access mechanisms are significantly different.

4.4.2 Medium Access Protocols for Wireless Data Networks

4.4.2.1 Random Access Protocols

Random access protocols are usually preferred for bursty data services because of their simplicity. ALOHA, the earliest and simplest such protocol uses the "transmit whenever you want" approach for each MS. If the transmission is acknowledged, the MS simply proceeds with the next packet. Otherwise, it will have to retransmit lost packets. Packets are lost due to collisions between transmissions and thus, the ALOHA protocol has a very low maximum throughput (18%). Slotted versions are slightly better as transmission attempts can take place only at discrete points of time. Most practical random access protocols are based on some form of carrier sensing, which is an improvement over ALOHA. Here, MSs listen to the medium before transmitting to make sure that no competing MS is already transmitting, thereby reducing the number of collisions. Depending on the protocol, a variety of CSMA types exist. In non-persistent protocols, MSs back off for a random period when the medium is detected to be busy. In p-persistent protocols, they monitor the medium till it becomes idle and then transmit with a probability p. In the case of collisions, the random backoff time is doubled in binary exponential backoff schemes. In many cases, collisions are *detected* by changes in voltage in the medium and the utilization of the channel is improved. Collision detection is easier at baseband than at radio frequencies and is in fact quite impossible in wireless networks because of fades in signals and thus *collision avoidance* mechanisms are preferred.

While carrier sensing is suitable for wired data networks, there are problems with carrier sensing in the case of wireless data networks. The signal strength is a function of distance and location and is changed randomly due to path loss and shadow fading as discussed previously. Not all terminals at the same distance from a transmitter can "hear" the transmitter leading to the *hidden terminal problem*. Here, an MS_B that is within the range of the destination but out of range of a transmitter is referred to as a hidden node. Suppose MS_A transmits to the AP and MS_B cannot sense the signal. It is possible that MS_B may also transmit resulting in collisions and MS_B is called a "hidden terminal" with respect to MS_A. There are mechanisms for overcoming collisions due to hidden terminals. One mechanism called busy-tone multiple-access (BTMA) is an out of band signaling scheme. Any MS that hears a transmission will transmit a busy tone in an out of band channel that is then successively repeated so that all terminals in the network are informed of the busy nature of the channel. Another method uses control handshaking. MS_A sends a short request-to-send (RTS) packet to the AP which responds with a short clear-to-send (CTS) packet that is received by all MSs in its range. Upon receiving the CTS packet, MS_B defers the medium access till MS_A completes its transmission. The *exposed terminal problem* is the reverse of the hidden terminal problem. In this case, the exposed terminal is in the range of the transmitter but outside the range of the destination and unnecessarily backs off when the transmitter begins transmission. This results in low utilization of bandwidth. Solutions to this problem include proper frequency planning and selecting intelligent thresholds for the carrier sensing mechanism. Transmissions from ground level can be detected at a tower but not at the ground level in the case of WWANs. To overcome this problem, carrier sensing is done at the BS and the busy nature of the uplink is communicated to the MSs by the BS/AP on the downlink.

In summary, if simplicity demands a decentralized medium access protocol, CSMA or any of its variants is preferred. CSMA in wireless networks leads to the hidden terminal and exposed terminal problems. Collision detection in wireless networks is virtually impossible. Wireless systems that use CSMA are CDPD, IEEE 802.11 and HIPERLAN/1. CDPD employs carrier sensing as well as collision detection at the BS. IEEE 802.11 employs collision avoidance and the control handshaking mechanism to reduce the hidden terminal problem.

4.4.2.2 Taking Turns Protocols

Taking turns protocols are similar to token ring or bus in wired networks where a *token* is passed from one station to another so that they all get a fair chance to access the medium. There is no contention to access the medium and throughput and delay can be bounded. They are in general infeasible for wireless networks because errors in the recovery of tokens can render the system unstable and they have not been widely studied except for infrared systems.

Polling is another form of taking turns protocols where a centralized authority polls each MS for data and the MS can respond to the poll if it has anything to transmit. If the MS has nothing to transmit or it is inactive, the polling scheme consumes bandwidth unnecessarily. The advantage of polling is that it can guarantee bounded delays and throughput unlike random access schemes. Example systems that employ polling are the *Point Coordination Function* (PCF) in IEEE 802.11 and Bluetooth. Currently, PCF is optional in IEEE 802.11 and it is not widely deployed. We briefly discuss polling in Bluetooth below.

Bluetooth has a scattered ad-hoc topology as discussed in Section 4.2.2.3. A "cell" or "piconet" is defined by a Master device that controls the frequency hopping sequence and also the transmissions within its piconet. There is no contention within a piconet. The Master device is the device that initiates an exchange of data. A slave device is a device that responds to the Master. Slaves use the frequency hopping pattern specified by the Master and can transmit only in response to the Master. If the Master's transmission is corrupted, a slave cannot transmit on its own and has to wait for a repeated poll. The Master device also decides how many slots the slave can use for its packet (one, three or five). A Master device can simultaneously control seven slave devices. Although there is no contention within a piconet, there is interference between piconets because occasionally, the transmissions of devices may occur at the same frequency although they use frequency hopping. This is like CDMA using FHSS.

4.4.2.3 Reservation Protocols

This last category of medium access protocols bridges the space between random access schemes and taking turns schemes. The more common random reservation schemes employ statistical multiplexing of data on TDMA or CDMA systems. These are suitable for data services over existing voice networks (like GPRS over GSM). In these systems, a time slot (or code) can carry one voice packet and the idea is to use multiple such channels for data as and when necessary. The general method is to use a random access channel to "reserve" uplink channels. On the downlink, the BS can multiplex traffic intended for different MSs. The one disadvantage of such schemes is the signaling overhead and additional complexity. Several WWANs like Mobitex and GPRS use reservation. Some WLAN technologies like HIPERLAN/2 have also adopted reservation protocols.

The medium access in HIPERLAN/2 (Jonsson, 1999) for example is based on TDMA/TDD (Pahlavan and Krishnamurthy, 2002). As shown in Figure 4.6, each MAC frame starts with a Broadcast (BC) Phase where the downlink carries a *Broadcast Control Channel* (BCCH) and *Frame Control Channel* (FCCH). The status of the network and other announcements are contained in the BCCH and resource grants to the MSs are contained in the FCCH. This is followed by a Downlink (DL) Phase which carries more control information and data intended for MSs. MSs can either communicate through the AP (this is a mandatory mode) or directly after reserving a channel. In the Uplink (UL) Phase, MSs have to request capacity to transmit control or user data. The Direct Link (DiL) Phase is used for direct MS-MS communication. MSs request capacity from the AP and once this is allocated, they can then communicate directly. There is also a Random Access (RA) Phase that is used by MSs with zero capacity to obtain capacity for the UL phase or by new MSs entering the system or by MSs performing handoff.

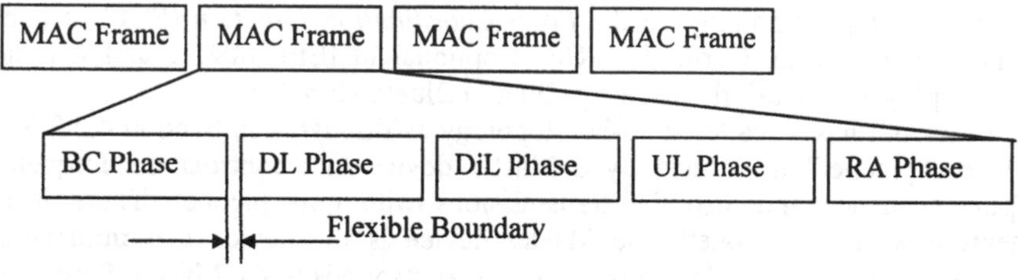

Figure 4.6 Basic MAC Frame Format in HIPERLAN/2

4.4.2.4 Impact on Telegeoinformatics

Most location-aware services are based on packet data. For information services, wireless portals and the like, random access schemes are acceptable. However, for real-time and multimedia services, time bounded medium access mechanisms such as polling or reservation might be more suitable. For the present, it is unlikely that medium access mechanisms adopted in various standards will change. Currently, standardization activities in the Task Group E of IEEE 802.11 are addressing changes to the medium access mechanism to support real-time services.

4.5 NETWORK PLANNING, DESIGN AND DEPLOYMENT

Design and deployment issues in wireless networks are extremely important and differ significantly from wired networks. In ad hoc networks, there is no architectural design as the topology continuously changes and the design issues are primarily at the protocol level. On the other hand, in infrastructure networks, the common approach is to deploy a cellular topology (frequency reuse is employed) based on some requirements. Initial deployments start with large cells and as demand increases, capacity enhancement techniques such as reuse partitioning, sectored cells and migration to next generation systems have been used. In this section we provide a brief overview of some of these aspects.

The *cellular concept* (MacDonald, 1979) basically deploys a large number of low-power BSs, each having a limited coverage area, and reuses the spectrum several times in the area to be covered to increase capacity. Capacity has traditionally meant voice traffic load in a cell. One measure of capacity has been the number of communication channels that are available (which could be a frequency carrier, a time slot or a spread spectrum code). In the case of wireless data networks, there has been little characterization of the traffic needs in a cell. The system design should also be such that performance measures like call blocking probability, handoff dropping probability and throughput satisfy the system requirements.

A primary factor in the design is the fact that reusing the spectrum leads to interference. The types of interference are different for different systems. For TDMA/FDMA based systems, co-channel interference (interference from signals transmitted by another cell using the same radio spectrum) and adjacent channel interference (interference from signals transmitted in the same cell with overlapping spectral sidelobes) are important and must be reduced. The transmit powers of BSs and MSs will influence the interference. The idea of clustering is used in TDMA/FDMA systems to reduce interference. Adjacent cells cannot use the same frequency carriers as the co-channel interference will be too severe. So the available spectrum is divided into chunks (sub-bands) that are distributed among the cells. Cells are grouped into clusters and each cluster of cells employs the entire available radio spectrum as shown in Figure 4.7. Here, the spectrum is divided into seven chunks labeled A through G. Each cell is allocated one of these chunks and the allocation is made so as to reduce the co-channel interference to the maximum allowed values[6]. Interference calculations are based on the path-loss and shadow fading discussed earlier. The cluster size N determines the co-channel interference and the number of channels allocated to a cell (capacity). The larger N is, the smaller is the co-channel interference and the smaller also is the number of channels available for a given cell (capacity reduces). Many preliminary calculations of the interference and design are based on a *hexagonal geometry* of cells (although cells are rarely of any regular shape) to keep the problem mathematically tractable. There has been a lot of research in this area to develop methods to improve capacity of the system. The use of directional antennas to *sector* the cells is a popular approach in practice to increase capacity by reducing interference.

In CDMA systems (Garg, 2000), all users transmit at the same time at the same frequency, but they use different codes. Thus there is interference from *within the cell* and interference from *outside the cell.* The measure of maximum allowed interference here is the value that provides a given "quality of signal." Usually this is the value that provides a frame error rate of 1% or less in CDMA systems. Also, power control, soft handoff and RSS threshold play a very important role in the design of CDMA systems. These issues are beyond the scope of this chapter.

---

[6] This maximum interference value depends on the system requirements. In voice networks, the maximum allowed value is expressed in terms of the minimum acceptable signal-to-interference ratio (SIR) based on the mean opinion score of voice quality. For AMPS, it is 18 dB and for GSM it is between 8 and 12 dB. In CDMA systems, the minimum SIR allowed is as small as 6 dB. For data networks, such a value is not useful and a more appropriate value is the error rate, throughput or delay.

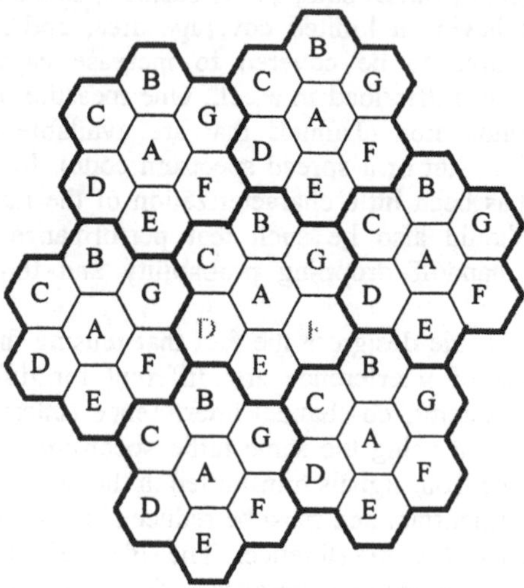

Figure 4.7 Hexagonal Cells with a Cluster Size of Seven

Wireless data networks such as CDPD and GPRS are overlaid on AMPS or GSM and as such they have to live with the design criteria used for those networks. The cell sizes, BS transmit power, and frequency allocations are all dictated by the underlying voice networks. Very little work has been done in designing WLANs although several optimization algorithms have been studied for the placement of APs within buildings. In the case of IEEE 802.11, initial deployments were based on the 915 MHz bands and there was only one channel, which meant that the co-channel interference could not be reduced by proper design. In the 2.4 GHz bands where most IEEE 802.11 networks now operate, there are three non-overlapping channels so that frequency reuse is possible. However, the operation of WLANs is different from WWANs especially those that are voice oriented because of the carrier sensing mechanism. More than co-channel interference, the exposed terminal problem becomes severe as an MS may simply back off upon sensing the transmission of another MS on the same frequency in a different cell. The carrier-sense thresholds have to be set intelligently to avoid this problem. In the 5 GHz bands, there are eleven non-overlapping channels further facilitating frequency reuse. Also, WLANs are deployed in buildings and three-dimensional planning is required. There is a need to take into account antenna patterns and building architecture. In WWANs, transmitter power control is a given and can be performed with different granularity. There are only three levels of transmit power at the AP in IEEE 802.11 WLANs and it is not clear what can be done at the MS to change its transmit power. This makes network planning and deployment more difficult in the case of WLANs. Another issue in the case of WLANs and WPANs is the nature of spectrum used by these systems. As the spectrum is unlicensed, many different technologies have to *coexist*. For instance, Bluetooth and IEEE 802.11b networks both operate in the 2.4 GHz bands and can potentially degrade the performance of one another.

Telegeoinformatics will prove to be very useful for network planning and design purposes. As discussed in Section 4.3.1.3, GISs have been used for characterizing radio propagation providing a better feel for the actual cell boundaries. Terrain data is also used in cellular network planning to statistically characterize path loss (Blaunstein, 2000) and obtain a first cut estimate of the size of a cell based on the location of the BS serving it. Combining this with traffic loads, frequencies and usage profiles in a comprehensive GIS would make the network planning process easier.

4.6 WIRELESS NETWORK OPERATIONS

So far, in this chapter, we have considered the architectural, physical layer, medium access and network design issues in wireless networks. Wireless networks are unique, as discussed in section 4.2.3 in terms of the resource limitations (bandwidth, battery, CPU, memory, etc.) and mobility. Consequently, network entities need special protocols and management functionality to address the problems arising from these unique characteristics. These management functions are as follows: *Radio Resources Management (RRM)* deals with power control, channel assignment and handoff decisions, *Power Management (PM)* deals with protocols and management for allowing MSs to save power, *Mobility Management (MM)* deals with location management and handoff management and *Authentication, Authorization, and Accounting (AAA)* deals with *security*, billing and access control of wireless networks. We look at these issues below.

4.6.1 Radio Resources Management

The scarcity of radio spectrum has resulted in frequency reuse as discussed in Section 4.5 to increase capacity in a serving area. This results in certain phenomena that need to be correctly addressed for proper operation of the wireless network. Firstly, signals from MSs operating in the coverage area of one BS cause interference to the signals of MSs operating in the coverage area of another BS on the uplink. Similarly, signals transmitted by one BS will interfere with the signals transmitted by another BS on the downlink. There is a need to reduce such an interference by properly controlling the transmit powers of MSs and BSs. Secondly, correctly controlling the transmit powers of mobile terminals can enhance their battery life and make mobile terminals lighter and handier to use. Since wireless terminals are mobile, they run on battery power which needs to be conserved as long as possible to avoid the inconvenience of requiring a fixed power outlet for re-charging. Most of this power is consumed during transmission of signals. Consequently, the transmit power of the mobile terminal must be made as small as possible. In turn, this requires reducing the coverage area of a cell so that the received signals are of adequate quality. Also, as MSs move, the ability to communicate with the current BS degrades and they will need to switch their connection to a neighboring BS, i.e., at some point during the movement, a decision has to be made to handoff from one BS to another. This decision will have to be made based on the expected future signal characteristics from several BSs that may be potential candidates for handoff, the capacities and available radio resources of such base stations, and interference considerations. For example, if an MS continues to communicate with a BS when it is deep into the coverage area of another BS, it will cause significant interference in some other cell that employs the same channel. Thirdly then, there is a need for the wireless network to keep track of the radio resources, signal strengths and other associated information related to communication between a mobile and the current and neighboring base stations (Pahlavan and Krishnamurthy, 2002).

RRM refers to the control signaling and associated protocols employed to keep track of relationships between signal strength, available radio channels, etc. in a system so as to enable an MS or the network to optimally select the best radio resources for communications. RRM handles three issues in general (a) BS and rate/channel assignment where the MS selects a BS upon powering up and is allocated a waveform or channel, (b) transmit power assignment and power control, and (c) handoff decision (including handoffs within a cell (intra-cell handoff) and handoff between cells (inter-cell handoff). In some technologies, other issues are part of RRM. For example, in HIPERLAN/2, there is an option for dynamic frequency selection whereby APs use RSS measurements to dynamically reconfigure their transmit powers and frequencies to reduce interference. In CDPD, the CDPD channels should be changed depending on incoming voice calls on the same frequencies (as voice has a higher priority) (CDPD Specs, 1995). This is called channel hopping. Both dynamic frequency selection and channel hopping are part of RRM. There may be dedicated channels for signaling RRM messages or they may be piggy backed on user data traffic. RRM control signals may be unidirectional (as in CDPD) or bi-directional (as in GSM). In the unidirectional case, the network broadcasts the RRM information and it is up to the MSs to make use of this information in handoff decisions and power control. In the bi-directional case, the network gets feedback from the MSs and can also instruct MSs to take appropriate actions like making a handoff or increasing the transmit power. The MS is always involved in RRM procedures. The equivalent of the BSS and sometimes the MSC is involved on the network side.

Telegeoinformatics has great potential to address RRM issues. The use of GPS technology in predictive radio channel resource allocation algorithms has been considered in (Chiu and Bassiouni, 1999). If the MS's location information is employed to reserve resources for it during handoff, the handoff blocking probability is shown to reduce via simulation. Teerapabkajorndet and Krishnamurthy (2001 & 2002), use the position information from GPS in making a handoff decision in WWANs. Considerable benefits are observed in terms of reducing the signaling load (even including the GPS error) on the network and increasing on the air throughput in these simulations.

4.6.2 Power Management

We distinguish between power control, power saving mechanisms, energy efficiency and radio resource management. By power control, we mean the algorithms, protocols, and techniques that are employed in a wireless network to dynamically adjust the transmit power of either the MS or the BS for reducing interference, with the by-product of elongating battery life.

Power-saving mechanisms are employed to save the battery life of a mobile terminal by explicitly making the mobile terminal enter a suspended or semi-suspended mode of operation with limited capabilities. This is however done in co-operation with the network, so as to not disrupt normal communications or provide the user with a perception of such a disruption even if there was one. The most common form of power saving mechanisms is to let the MS enter a sleep mode. Here, we briefly explain how this is done in CDPD (CDPD specs, 1995). The mechanisms are similar in most wireless networks. In CDPD, when an MS enters the sleep mode, the link layer is maintained in suspended mode and all timers are saved. If an MS does not transmit for a time denoted as T203, it is automatically assumed to have entered a sleep mode. The value of T203 is used to determine the "wake-up" time. Every T204 seconds the network broadcasts a list of identifiers of MSs (temporary addresses allocated to MSs when they power up) that have outstanding data packets to receive. Before entering the sleep mode an MS tracks the last T204 broadcast and wakes up at the next T204 to listen to the broadcast. If it has any pending packets, it will wake up to receive them and then go back to sleep. However, sometimes the MS may be powered down or shut without the network realizing it. So the network performs up to N203 broadcasts of the list of identifiers and if a MS does not wake up to receive packets, the network discards those packets. In some cases, packet forwarding is possible to a storage location.

Energy efficient design (Pahlavan and Krishnamurthy, 2002) is a new area of research that is investigating approaches to save battery life of a mobile terminal in fundamental ways such as in protocol design, via coding and modulation schemes, and in software. At the protocol level, the idea is to minimize transmissions and receptions that consume the most power. Transmissions that are not successful due to collisions or the harsh radio channel result in wasted power. Monitoring the channel for carrier sensing and receiving packets that are intended for other MSs also results in wasted power. Collision avoidance mechanisms are designed to reduce the number of collisions. To overcome the effect of bad channel conditions, some protocols have been suggested that transmit a short probe packet first and if a positive acknowledgment is received, the complete packet is transmitted. In terms of unnecessary receptions, in IEEE 802.11 a *virtual carrier sensing* is performed. Here the header of a transmission is decoded to determine the length of the transmission and for this period, the channel is not sensed. In HIPERLAN/1, a 34 bit hash value of the MS destination address is included in the header. If this hash value does not match, the MS discards the packet thereby eliminating unnecessary reception of the rest of the packet. Software approaches try to minimize power hungry operations like accessing the hard disk, using backlights of displays, etc.

It is possible to use the position information in reducing the power consumption in networks. A distributed algorithm for assigning paths for multi-hop routing of packets in ad hoc networks with a random deployment of mobile users traveling with random velocities has been developed based on GPS in (Rodoplu and Meng, 1999). Position information obtained from GPS is used to minimize the total energy consumed in the ad hoc network in this algorithm.

4.6.3 Mobility Management

The primary advantage of wireless communications is the ability to support tetherless access to a variety of services. Tetherless access implies that the MS has the ability to move around while connected to the network and continuously possess the ability to access the services provided by the system to which it is attached. This leads to a variety of issues because of the way in which most communications networks operate. Firstly, in order for any message to reach a particular destination, there must be some knowledge of where the destination is (location) and how to reach the destination (route). In static networks, where the end terminals are fixed, the physical connection (wire or cable) is sufficient to indicate the destination. In wireless networks, where the MS may be anywhere, there must be a mechanism to locate it in order to deliver the communication to it. *Location management* refers to the activities a wireless network should perform in order to keep track of where an MS is. The location of the MS must be determined such that there is a knowledge of which point of access (BS/AP) is serving the cell in which the MS is located. Secondly, once the destination is determined, it is not enough to assume that the destination will remain at the same location with time. When an MS moves away from a BS, the signal level from the current BS degrades and a handoff is made. *Handoff management* handles the messages required to make the changes in the fixed network to handle this change in the location during an ongoing communication. Location and handoff management together are commonly referred to as mobility management (Akyildiz *et al.*, 1998).

4.6.3.1 Location Management

Location management in general has three parts to it: location updates, paging, and location information dissemination. Location updates are messages sent by the mobile terminal regarding its changing points of access to the fixed network. These updates may have varying granularity and frequency. Each time the mobile terminal makes an update to its location, a database in the fixed part of the network has to be updated to reflect the new location of the mobile terminal. Whether or not there is a change in the location, the update message will be transmitted over the air and over the part of the fixed network. Since the updates are periodic, there will be some uncertainty in the location of the mobile terminal to something around a group of cells. In order to deliver an incoming message to the mobile terminal, the network will have to page the mobile terminal in such a group of cells. The paged terminal will respond through the point of access that is providing coverage in its cell. The response will enable the network to locate the terminal to within the accuracy of the cell in which it is located. Procedures can be then initiated to either deliver the packet or set up a dedicated communications channel for voice conversation. In order to initiate paging however, the calling party or the incoming message should trigger a location request from some fixed network entity. The fixed network entity will then access the database that will contain the most current location information related to the particular mobile terminal and use this information to generate the paging request, as well as deliver the message or set up a channel for the voice call. Location information dissemination refers to the procedures that are required to store and distribute the location information related to the mobile terminals serviced by the network.

The basic issue in location management is the tradeoff between the cost of the nature, number and frequency of location updates and the cost of paging (Wong and Leung, 2000). If the location updates are too frequent and the incoming messages few, the load on the network becomes an unnecessary cost, both in terms of the usage of the scarce spectrum as well as network resources for updating and processing of the location updates. If the location updates are few and infrequent, a larger area and thus a larger number of cells will have to be paged in order to locate the MS. Paging in all the cells where the MS is not located is a waste of resources. Also depending on the way paging is performed, there may be a delay in the response of the MS because the paging in the cell in which it is located might be performed much later than the cell in which it last performed its location update. For applications such as voice calls, this will result in unnecessary call dropping since the MS does not respond in a reasonable time. In the case of data networks, depending on the type of mobility management scheme implemented, packets might simply be dropped if the MS is not located correctly.

A definite possibility for improving location management is to use positioning information from GPS or other positioning systems. More information on location management can be found in chapter 5.

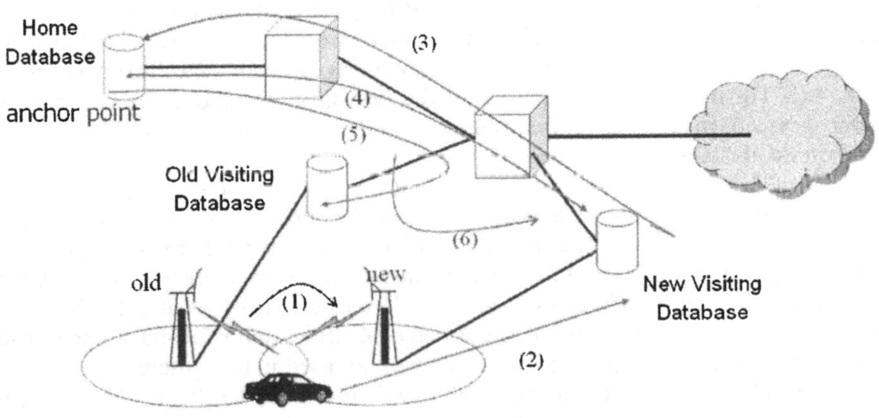

Figure 4.8 Steps in Mobility Management

4.6.3.2 Handoff Management

Handoff management has several steps to it and occurs when the MS switches its point of access during a connection. The decision mechanism or handoff control could be located in a network entity (as in AMPS) or in the MS (as in CDPD and IEEE 802.11) itself. These cases are called network controlled handoff (NCHO) and mobile controlled handoff (MCHO), respectively. In GSM/GPRS, information sent by the MS can be employed by the network entity in making the handoff decision. This is called mobile assisted handoff (MAHO). There are two types of databases in the network, the home database (compare with HLR in Figure 4.3) that also acts as an anchor and the visiting database (compare with VLR). Every MS is associated (registered) with a home database that keeps track of the profile of the MSs. The visiting database keeps track of all the MSs in its service area. The home and visiting databases communicate with each other during the handoff management process as described below. The steps in handoff management shown in Figure 4.8 are as follows. This is a generic description and the specifics in each technology will be different based on the implementation.

In the first step, a decision is made to handoff and handoff is initiated. This decision, as discussed above, may be made in the network by some entity with or without the help of the MS, or at the MS itself. The MS registers with the "new" visiting database via a handoff announcement message. This is the first information to a network entity in the case of a mobile controlled handoff. In the case of a network controlled or mobile assisted handoff, the new visiting database may already be aware or expecting this message. The new visiting database communicates with the home database to obtain subscriber profile and for authentication. This is the first information exchange between network entities about the changed location of the MS in MCHO. In the case of MAHO or NCHO, these entities may already be in communication. The home database responds to the new visiting database with the authentication of MS. If the MS is authenticated, in the case of circuit-switched connections, a pair of traffic channels that might be kept ready is allocated to the MS for continuing the conversation. In the case of packet data traffic, no such dedicated channels are required since the traffic is bursty. The two databases are updated for delivering new messages that may arrive to the MS. The new visiting database includes the MS in its list of terminals that are being serviced by it. The home database sends a message to the old visiting database to flush packets intended for and registration information related to the MS. This is because packets that may have been routed to the old visited network while the MS was making a handoff need to be dropped or redirected and the old visiting database needs to clear resources it had maintained for the MS since they are no longer required. The old visiting database flushes or redirects packets to the new visiting database and removes the mobile terminal from its list. Each of the above steps is important in order to correctly, securely, and efficiently implement handoff and release resources that are otherwise not used in the system.

Handoff management can impact the performance of LBSs and LBC. When an MS moves across cell/network boundaries and there is critical data in transit for LBC (required either from the MS or to be delivered to the MS), the handoff management procedures must be able to handle the requirements imposed by the data and application.

4.6.4 Security

Wireless access to the Internet is becoming pervasive with diverse mobile devices being able to access the Internet in recent times. While current deployment is small and the security risks appear to be low in these emerging technologies, the widely varying features and capabilities of wireless communication devices introduce several security concerns. The broadcast nature of wireless communications makes it susceptible to malicious interception and wanton or unintentional interference. At least, minimal security features are essential to prevent casual hacking into wireless networks. Since the advent of analog telephony, wireless service providers have suffered several billion dollars of losses due to fraud. In this section, we address security issues in general.

Security requirements for wireless communications are very similar to the wired counterparts but are treated differently because of the applications involved and potential for fraud. Different parts of the wireless network need security. Over the air security is usually associated with privacy of voice conversations. With the increasing use of wireless data services, message authentication, identification, and authorization are becoming important. The broadcast nature of the channel makes it easier to be tapped. Analog telephones were extremely easy to tap and conversations could be eavesdropped using an RF scanner. Digital systems of today are much harder to tap and RF scanners cannot be used directly. However, the circuitry and chips are freely available and it is not hard for someone to tap the signal itself.

A variety of control information is transmitted over the air in addition to the actual voice or data. These include call set-up information, user location, user ID (or telephone number) of both parties, etc. These should all be kept secure since there is potential for misusing such information. Calling patterns (traffic analysis) can yield valuable information under certain circumstances. In (Wilkes, 1995) various levels of privacy are defined for voice communications. At the bare minimum, it is desirable to have *wireline equivalent* privacy for all voice conversations. We commonly assume that all telephone conversations are secure. While wired communications are not entirely secure, it is possible to detect a tap on a wireline telephone. It is impossible to detect taps over a wireless link. To provide privacy that is equivalent to that of a wired telephone, for routine conversations it is sufficient to employ some sort of an encryption that will take more than simple scanning and decoding to decrypt. In order to alert wireline callers about the insecure nature of a wireless call that is not at all encrypted, a *"lack-of-privacy"* indicator may be employed.

Wilkes (1995) calls these two levels of security as levels one and zero respectively. Level-0 privacy is when there is no encryption employed over the air so that anyone can tap into the signal. Level-1 privacy provides privacy equivalent to that of a wireline telephone call, one possibility being encrypting the over-the-air signal. For commercial applications, a much stronger encryption scheme would be required that would keep the information safe for more than several years. Secret key algorithms with key sizes larger than 80 bits are appropriate for this purpose. This is referred to as Level-2 privacy. Encryption schemes that will keep the information secret for several hundreds of years are required for military communications and fall under Level-3 privacy. For wireless data networks, a bare minimum level would be to keep the information secure for several years. The primary reason for this is that wireless electronic transactions are becoming common. Credit card information, dates of birth, social security numbers, e-mail addresses, etc. can be misused (fraud) or abused (junk messages for example). Consequently, such information should never be revealed easily. A Level-2 privacy will be absolutely essential for wireless data networks. In certain cases, a Level-3 privacy is required. Examples are wireless banking, stock trading, mass purchasing, etc.

Most wireless networks today have only two security features, confidentiality and authentication. The former prevents eavesdropping or interception of information while the latter performs access control. An authentication center (see Figure 4.3) is used in WWANs for performing the necessary security algorithms. It is often collocated with the MSC. In WLANs, the AP can be manually configured to ensure confidentiality and authentication. For confidentiality, most wireless networks use a stream cipher derived from block ciphers with certain modes of operation. Authentication is based upon challenge-response protocols in WWANs and simple ability to decrypt in WLANs. Some public encryption algorithms used in today's wireless networks are RC-4 in CDPD and IEEE 802.11 and MILENAGE and KASUMI in 3G systems. Algorithms used in GSM are proprietary.

Security can impact LBSs and LBC. A major concern for LBSs is the privacy of location information. Similarly, LBC can be dealing with mission critical data that needs to be kept secure (confidential, authentic and so on). Security services in wireless networks need to be modified to account for these aspects as well.

4.7 CONCLUSIONS AND THE FUTURE

Telegeoinformatics involves location dependent computing and information delivery. Certainly, wireless communications is an important part of the infrastructure for Telegeoinformatics. Telegeoinformatics could also potentially be used to provide solutions to problems in wireless network operation. As discussed in this chapter, different technologies provide different coverage, data rates and services. They occupy different frequency bands and their availability is highly variable. In the future, it is expected that there will be *hybrid* networks – where there will be seamless roaming between WWANs and WLANs depending on availability, cost, service requirements, and so on. Recently, a variety of community WLANs for covering hot-spots are appearing in many countries providing high-speed wireless internet access (Schrick, 2002). Also, migration to a 3G wireless system is being implemented (Lucent Press Release, 2001) by service providers and manufacturers. In both cases, multi-rate data traffic (Nanda, 2000) is envisaged and in fact implemented in the case of WLANs. Service providers and manufacturers are also looking at integrating the two systems together into hybrid mobile networks. In such a case, knowledge of location and geographic information will become important in providing the infrastructure for Telegeoinformatics as well as to resolve the many issues (especially RRM, MM and security) associated with wireless communications itself.

REFERENCES

3GPP, Third generation partnership project website. Online. Available HTTP: <http://www.3gpp.org >.

3GPP2, Third generation partnership project-2 website. Online. Available HTTP: <http://www.3gpp2.org >.

IEEE 802.11, Website, tutorial and resources. Online. Available HTTP: <http://grouper.ieee.org/groups/802/11>.

Akyildiz, I.F., *et al.*, 1998, Mobility management in current and future communications networks, *IEEE Network Magazine*, July/August 1998, pp. 39-50.

Blaunstein, N., 2000, *Propagation in Cellular Networks*, (Boston: Artech House Publishers).

Bluetooth website. Online. Available HTTP: <http://www.bluetooth.com>.

Carne, E.B. 1999, *Telecommunications Primer*, Prentice Hall PTR, Upper Saddle River, N.J.

Catedra, M.F. and Perez-Arriaga, J., 1999, *Cell Planning for Wireless Communications*, (Boston: Artech House Publishers).

CDPD Specifications, Release in 1995.

Chiu, M-H. and Bassiouni, M., 1999, Predictive Channel Reservation for Mobile Cellular Networks Based on GPS Measurements, *Proceedings of ICPWC*.

Crow, B.P., Widjaja, I., Kim, L.G. and Sakai, P.T., 1997, IEEE 802.11 Wireless Local Area Networks, *IEEE Communications Magazine*, 35(9), pp. 116-126.

Garg, V.K. 2000, *IS-95 and CDMA2000*, (Upper Saddle River, N.J.: Prentice Hall).

Garg, V.K. 2002, *Wireless Network Evolution: 2G to 3G*, (Upper Saddle River, N.J.: Prentice Hall).

Garg, V.K. and Wilkes, J.E., 1999, *Principles and Applications of GSM*, (Upper Saddle River, N.J.: Prentice Hall).

Goodman, D.J., 1997, *Wireless Personal Communications Systems*, (Addison-Wesley).

Gummala, A.C. and Limb, J. 2000, Wireless Medium Access Control Protocols, *IEEE Communications Surveys and Tutorials*, 3(2), Second Quarter.

Hata, M., 1980, Empirical formula for propagation loss in land mobile radio services, *IEEE Transactions on Vehicular Technology*, VT 29(3), pp. 317 – 324.

Jonsson, M., 1995, HiperLAN2 – The Broadband Radio Transmission Technology Operating in the 5 GHz Frequency Band, *White paper*, HIPERLAN-2 Global Forum.

Kalden, R., Meirick, I. and Meyer, M., 2000, Wireless Internet access based on GPRS, *IEEE Personal Communications*, 7(2), pp. 8-18.

Koshima, H. and Hoshen, J., 2000, Personal locator services emerge, *IEEE Spectrum*, February 2000, pp. 41-47.

Lazarakis, F., Dangakis, K., Alexandridis, A.A. and Tombras, G.S., 1994, Field Measurements and Coverage Prediction Model Evaluation Based on a Geographical Information System, *Proceedings of PIMRC'94,* The Netherlands.

Lin, S. and Costello, D.J., 1983, *Error Control Coding*, (Prentice Hall).

Li, X., Pahlavan, K., Latva-Aho, M. and Ylianttila, M., 2000, Indoor geolocation using OFDM signals in HIPERLAN/2 wireless LANs, *Proc. PIMRC' 00*, September 2000.

Lucent Press Release, 2001, Verizon Wireless and Lucent Technologies demonstrate 3G CDMA2000 1xEV-DO high-speed wireless data technology. Online. Available HTTP: <http://www.lucent.com/press/1001/011031.nsb.html> (accessed October 2001).

MacDonald, V.H., 1979, The Cellular Concept, *The Bell System Technical Journal*, 58(1), pp. 15-41.

Nanda, S., Balachandran, K. and Kumar, S., 2000, Adaptation Techniques in Wireless Packet Data Services, *IEEE Communications Magazine*, January 2000.

Pahlavan, K. & Krishnamurthy, P., 2002, *Principles of Wireless Networks: A Unified Approach*, (New Jersey: Prentice-Hall PTR).

Pahlavan, K., Li, X. and Makela, J.P., 2002, Indoor geolocation science and technology, *IEEE Communications Magazine*, 40(2).

Pahlavan, K. & Levesque, A., 1995, *Wireless Information Networks*, (New York: John Wiley and Sons).

Pahlavan, K., Krishnamurthy, P. and Beneat, J., 1998, Wideband radio propagation modeling for indoor geolocation applications, *IEEE Communications Magazine*, April 1998, pp. 60-65.

Peersman, G. and Cvetkovic, S., 2000, The Global System for Mobile Communications Short Messaging Service, *IEEE Personal Communications*, June 2000, pp. 15-23.

Prasithsangaree, P., Krishnamurthy, P. and Chrysanthis, P., 2002, On Indoor Position Location with Wireless LANs, *Proceedings of PIMRC'02,* Lisbon, Portugal, September 2002.

Proakis, J.G., 2001, *Digital Communications*, 4th ed., (McGraw-Hill).

Rodoplu, V. and Meng, T.H., 1999, Minimum Energy Mobile Wireless Networks, *IEEE JSAC*, 17(8).

Schrick, B., 2002, Grass-roots wireless networks, *IEEE Spectrum*, June 2002, pp. 42-43.

Sklar, B. 2001, *Digital Communications: Fundamentals and Applications*, 2nd ed., (Prentice Hall).

Taylor, M.S., Waung, W. and Banan, M., 1997, *Internetwork Mobility: The CDPD Approach*, (Prentice Hall).

Teerapabkajorndet, W. and Krishnamurthy, P., 2001, Comparison of Performance of Location-Aware and Traditional Handoff-Decision Algorithms in Cellular Voice and Data Networks, *IEEE Vehicular Technology Conference '01*, Atlantic City, N.J., October 2001.

Teerapabkajorndet, W. and Krishnamurthy, P., 2002, Throughput Consideration for Location Aware Handoff in Mobile Data Networks, *Proceedings of PIMRC '02*, Lisbon, Portugal, September 2002.

Van Nee, R. and Prasad, R., 2000, *OFDM for Wireless Multimedia Communications*, (Boston: Artech House Publishers).

Ward, A., Jones, A. and Hopper, A., 1997, A New Location Technique for the Active Office, *IEEE Personal Communications*, October 1997.

Wilkes, J.E. 1995, Privacy and Authentication Needs of PCS, *IEEE Personal Communications*, August 1995.

Wilkinson, T.A., Phipps, T. and Barton, S.K., 1995, A Report on HIPERLAN Standardization, *International Journal on Wireless Information Networks*, 2, March 1995, pp. 99-120.

Wong, V.W.S. and Leung, V.C.M. 2000, Location management for next generation personal communication networks, *IEEE Network Magazine*, September/October 2000, pp.18-24.

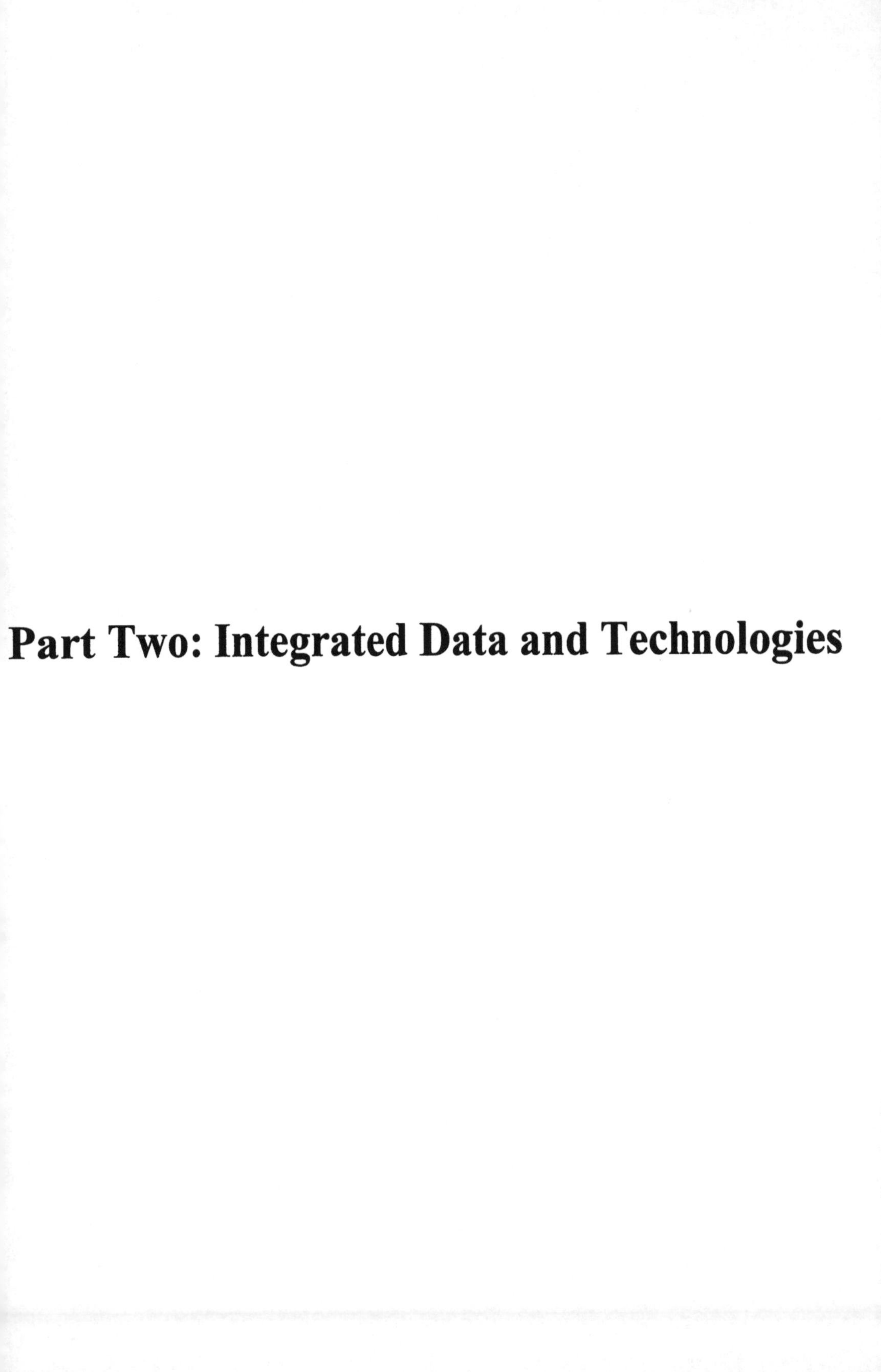

Part Two: Integrated Data and Technologies

CHAPTER FIVE

Location-Based Computing

Vladimir I. Zadorozhny and Panos K. Chrysanthis

5.1 INTRODUCTION

Modern Internet technologies abstracted away the concept of the real world. Web users have the illusion of one information storage, which is available for them anywhere in the world. Meanwhile, with rapid proliferation of wireless Internet, information about physical location of network devices cannot be disregarded anymore. In mobile commerce and Web applications over wireless Internet, location is crucial in presenting pertinent information and services to the customer. In this chapter, we introduce the concept of Location-Based Computing (LBC). LBC infrastructure comprises a distributed mobile computing environment where each mobile device is location-aware and wirelessly linked.

LBC enables Location-Based Services (LBSs) that combine the user's location with other information to provide appropriate and timely content for the user. LBC allows LBSs to utilize information about the current location of the mobile device to create a personal profile of each wireless mobile user. Such a profile may include traffic and weather information in the current location of the user, driving directions, entertainment, stores, wireless advertising, etc. As a result, the user can be provided with advanced LBSs such as destination guides, real-time turn-by-turn navigation, proximity searches, location-based billing, etc.

LBC infrastructure involves interoperating location-aware mobile applications exchanging large amounts of spatially oriented data in a volatile distributed environment. It utilizes and significantly extends such technologies as interoperable distributed processing and distributed data management. Location-based interoperability is addressed in public releases of major location interoperability standards, such as the Geography Markup Language (GML), recently published by the OpenGIS Consortium (OpenGIS, 2002). The location-based interoperability is closely related to the research and development on interoperable distributed systems, although for a considerable long period of time problems of distributed interoperability and LBC have been explored separately. One of the goals of recent work on location-based interoperability protocols is to integrate techniques from different areas in distributed location-based interoperable infrastructures. Adaptive LBC explores efficient implementation of interoperability between wireless and mobile applications that adjusts in real time to match varying capabilities and constraints of wireless mobile environments. A simple example of adaptive LBC is location-based routing in wireless networks. Location-based routing supports more scalable and efficient algorithms for discovering routes between mobile hosts.

The chapter is organized as follows. Section 2 overviews basic LBC infrastructure and outlines its major prerequisites. Section 3 considers location-based interoperability and its relation to open distributed processing. Section 4 addresses the issues of location-based data management in LBC infrastructure. Section 5 addresses advanced methods of adaptive LBC. Section 6 considers location-based routing as an example of adaptive LBC in wireless networks.

5.2 LBC INFRASTRUCTURE

LBC infrastructure is based on a general mobile computing environment (see Figure 5.1) in which the network consists of stationary and mobile stations, or mobile hosts (MHs). Specialized stationary hosts called *Base Stations* or *Mobility Support Stations (BS/MSSs)* are equipped with wireless communications capabilities enabling the MHs to connect to them, and through them, to the high-speed fixed network. The logical or physical area served by a single BS is called *a cell.* A *Mobile Positioning Center (MPC)* hides the positioning method from a developer of LBSs.

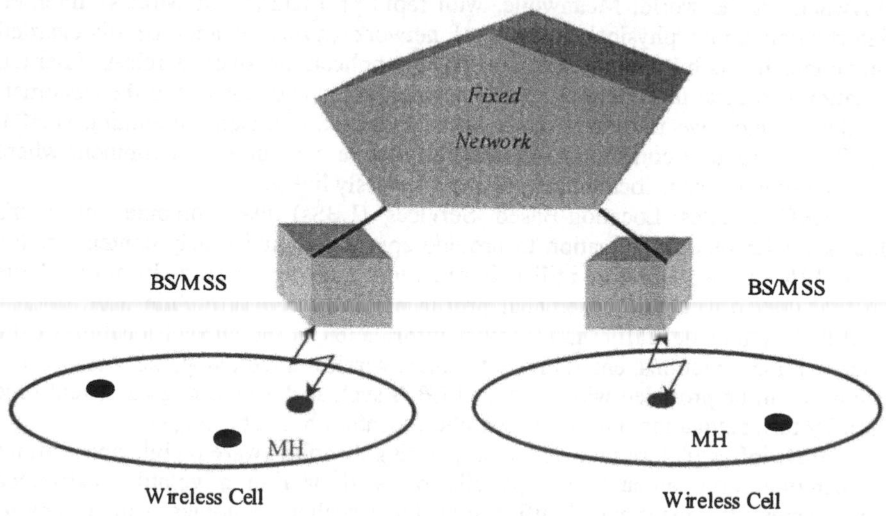

Figure 5.1 LBC Infrastructure: Wireless Mobile Environment

When a mobile device (e.g., phone) is activated or entering the cell, it connects to the network through the BS. A central database receives a signal acknowledging that the mobile device has entered the cell. Thus, the network operator always knows which cell the user is in so that data is sent to the right destination. For this reason, methods of precise position measurement of the mobile devices become a major prerequisite for LBC. One way to find a position of a mobile device in a wireless access network infrastructure is using base station antennas as reference points. All cells have a unique Cell Global Identity (CGI). The accuracy of the positioning depends on the size of the cell, which may be between 100 meters and 35 kilometers. To increase the accuracy, this method can be combined with timing advance procedure that measures the time it takes for the signal to go from the mobile host to the base station. Using timing advance values, the mobile system can calculate in which sector of the cell a mobile station is located. We refer readers to Chapter 4 for more detailed consideration of the wireless mobile infrastructure.

The position of a mobile device can also be computed using a satellite navigation system, such as the Global Positioning System (GPS) controlled by the US Department of Defence, or the Russian Global Navigation Satellite System (GLONASS). GPS provides specially coded satellite signals that can be processed in a GPS receiver, enabling the receiver to compute position, velocity and time. We refer readers to Chapter 3 for further reading on satellite navigation systems and network-based positioning.

Methods of spatial analysis developed in Geographic Information Systems (GISs) meet another important prerequisite of LBC infrastructure. GISs are hardware and software systems that are used to create, manage, analyze, and display spatial information. Evolving from Computer Aided Design packages, map display systems, and database and spreadsheet software, GIS processes allow the user to perform complex spatial analysis. Geographic data is an important aspect of any LBC framework. Integration of GISs with GPS also allows for the addition of real-time analysis capabilities to existing GISs. Chapter 2 gives a detailed overview of the GIS components and functionalities.

Any effort to build LBC infrastructure needs to utilize and significantly extend existing technologies.

- LBC infrastructure should utilize methods and techniques of *interoperable distributed processing* for interoperable network environments taking into account spatial information.
- Efficient implementation of LBC infrastructure is challenging. It is crucial to implement LBC in the most efficient way so it scales with millions of complex location-based data intensive applications. For this purposes, LBC infrastructure should utilize advanced methods of *distributed data management*.
- LBC infrastructure should also explore various optimization opportunities while considering the various trade-offs. For example, when a vector data set is sent from a server to a mobile device, should the entire topological information also be transmitted, or should the topology be constructed in the mobile device? Thus, LBC infrastructure should be highly *adaptive* to the unexpected behavior of a distributed environment.

5.3 LOCATION-BASED INTEROPERABILITY

One characteristic of the LBC infrastructure is a high level of interoperability between mobile applications and services. Interoperability is the ability of distributed systems to interact with each other and to dynamically integrate their functionalities, and hence to perform collaborative operations. Principles of interoperability allow users to share data and applications across distributed networks, varying processing platforms and vendor brands. In this section, we address the problem of supporting location-based applications with high interoperability requirements. This means that new mobile applications should also be able to interoperate with other mobile applications. For example, the next generation wireless environments are expected to support a variety of Mobile Web Services built from data services that are potentially located anywhere in the world (Sophica, 2001). The mobile services should be integrated into future mobile portals and made interoperable. Interoperable Mobile Web Services will allow users to combine potentially unrelated applications into composite services, which they can access via wireless devices. For instance, a user might link the address of his/her next appointment with driving directions and current traffic conditions to determine estimated driving time.

Issues of location-based interoperability are tightly related to research and development on *interoperable distributed systems*. One way to support system interoperability in distributed environments is through standard APIs and protocols that allow systems to interact with each other (*control, or invocation interoperability*). In distributed processing, as well as in any other software technology, standards also allow users to port applications across different distributed products. Currently, there is no single interoperable distributed systems standard accepted by all vendors of distributed systems. A visible effort in this area is Common Object Request Broker Architecture (CORBA). Control interoperability protocols enable higher levels of interoperability, such as *data interoperability* and *semantic interoperability*. Distributed applications typically involve processing heterogeneous data extracted from diverse data sources. *Data interoperability* standards, like Microsoft's Universal Data Access (UDA) (MSDN), offer a transparent and seamless ability to access and modify data across diverse data sources. Both CORBA and Microsoft's UDA are predominantly syntactic by nature and ignore application semantics. Methods of *semantic interoperability* take into account more refined application-specific semantic content of information that users create enabling more meaningful data exchange and reuse.

One of the goals of recent work on location-based interoperability protocols is to integrate distributed interoperability techniques in a distributed location-based interoperable infrastructure. Below we focus on CORBA as a major open distributed processing architecture, which, however, lacks the location-based computing capabilities. We also introduce two main protocols that allow distributed systems to receive location information from the network and implement invocation interoperability for LBC. They include Location Interoperability Forum (LIF, 2001), and Wireless Application Protocol Forum (WAP). After that we introduce two XML-based general-purpose specification languages for describing and exchanging location information: Geography Markup Language (GML, 2001) and Point of Interest Exchange Language (POIX). GML and POIX form a basis for data interoperability in LBC.

5.3.1 Open Distributed Processing and LBC

CORBA is a most well-known standard for open distributed processing, proposed by the Object Management Group (OMG), a consortium of leading companies in the area of information technology. OMG is moving forward in establishing a standard architecture for distributed object systems. CORBA defines a platform-independent communication infrastructure of object-oriented software framework and provides network-transparent mechanisms by which objects make requests and receive responses. Object Request Broker (ORB) supports interoperability between applications on different machines in heterogeneous distributed environments and interconnects multiple object systems. ORB allows clients to request operations supported by remote servers on different machine types, under different operating systems, and in different programming languages.

Figure 5.2 represents a generalized Object Management Architecture. ORB supports interaction between different computational agents (processes with their own address spaces) in a distributed environment. It assumes a common object model that allows us to interpret those agents in terms of objects and object operations. The operations are defined by signatures in *Interface Definition Language (IDL)* with C++-like syntax. The operations are requested by a client object. A client can also act as a server exporting its own interface to other clients. Thus, any client can also be a server.

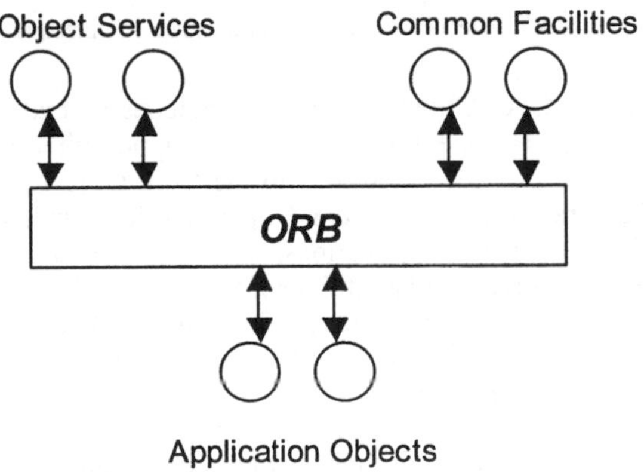

Figure 5.2 Object Management Architecture

ORB hides the details of remote calls, such as communication protocols, network topologies, platforms, underlying software, etc. Technically it is implemented by providing a client with an object reference that is properly bound to a target object (server). ORB is responsible for the server location, activation, passing parameters to the server and delivering results to the client. This functionality is supported by stub operations on the client side and skeleton operation on the server side. The compiler generates stubs and skeletons from IDL specifications. The stub invocation is made in a client programming language, which is ensured by proper mapping of IDL into the client programming language.

The general procedure of request processing is as follows (see Figure 5.3). After being invoked, a stub encodes (marshal) the request data (name of operation and actual parameters, including object references) in a Common Data Representation (CDR), such as General Inter-ORB Protocol that represents IDL data as octet streams. The client part of ORB sends the CDR buffer to the proper server. After receiving the CDR buffer, the server part of ORB determines a target object and passes the CDR buffer to the corresponding skeleton. The skeleton decodes (unmarshals) the request data and invokes the operation. After the operation has been executed, the skeleton marshals the result in the CDR buffer and sends it back to stub. The stub unmarshals the result and delivers it to the client.

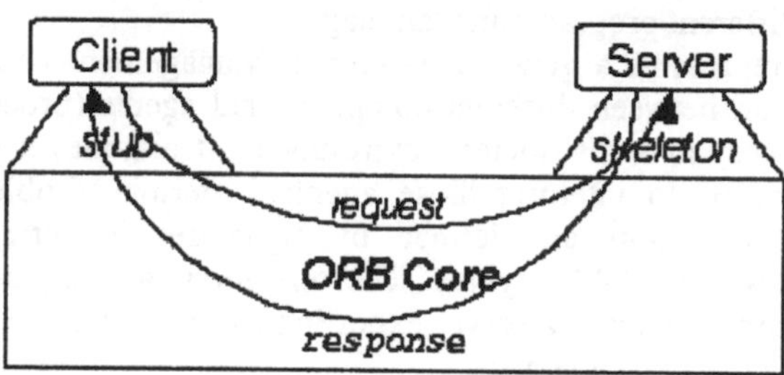

Figure 5.3 Generalized Object Request Broker Architecture

OMG specifies a Core Object Model that includes a small number of basic concepts such as objects, operations, types and type inheritance. A specific application domain can provide extensions to the Core Object Model. OMG also specifies a set of general-purpose Common Object Services, which are fundamental for developing CORBA-based applications. For example, it includes Persistent State Service (PSS, formerly, Persistent Object Service), which provides a uniform interface to manipulate persistent objects. The main task of the OMG PSS is storing/restoring objects on different *persistent storages*. The implementation details of the storage media (i.e., data format) should not influence the client application, i.e., storage implementation is transparent to the client. PSS can also co-operate with other OMG services such as transactions, concurrency, etc.

OMG architecture, as it was explained above, does not support LBC. At the time of the first releases of CORBA the mobile and wireless technologies were not well developed and widely accepted. As a result, there were no well-established standards in location-based protocols. Conceptually CORBA is flexible and extensible to utilize principles of LBC. For example, LBC and LBSs might be implemented as a part of Common Object Services. This approach has significant performance drawbacks. Another option would be to have a part of ORB Core to implement location-based protocols. However, as indicated by recent experiences implementing ORB, it is desirable to keep the ORB core as small as possible in order to be efficient.

5.3.2 Location Interoperability Protocols

In this subsection we discuss standards for location interoperability protocols that could be used to merge open distributed processing with LBC. They include *Location Interoperability Forum (LIF)* and *Wireless Application Protocol (WAP)*

Location Framework. These protocols can be used to implement invocation interoperability for LBC.

5.3.2.1 Location Interoperability Forum (LIF)

LIF (LIF, 2001) was established in 2000, more than ten years after the first CORBA specification had been released. LIF has produced a specification for a *Mobile Location Protocol (MLP)*, which is an application-level protocol for the positioning of mobile terminals. It is independent of the underlining network technology and of the positioning method. MLP serves as the interface between a Location Server and a Location-Based Client (see Figure 5.4). Typically, a client initiates the dialogue by sending a request to the location server and the server responds to the query. The LIF API is defined as a set of XML Document Type Definitions (DTDs).

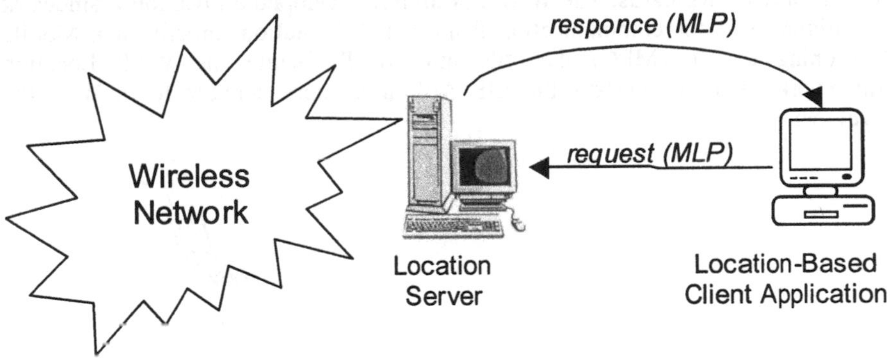

Figure 5.4 LIF Framework

MLP has a multiplayer structure. It includes a transport layer protocol separated from the XML content. On the lowest level, the transport protocol defines how XML content is transported. Possible MLP protocols include HTTP, Simple Object Access Protocol (SOAP), etc. The only currently defined MLP transport is HTTP. The next MLP level is the Element Layer that defines all common elements used by the services in the service layer. The Service Layer is the last layer that defines the actual services offered by the MLP framework.

The application requests a location using an HTTP POST message to the Location Server. The LIF supports the following location services.

- Standard Location Immediate Service (SLIS) is used when a single location response is required immediately.
- Emergency Location Immediate Service (ELIS) is used for querying the location of a mobile client that has initiated an emergency call. The response is required immediately.
- Standard Location Reporting Service (SLRS) is used when a Mobile Subscriber wants a client to receive its location.

- Emergency Location Reporting Service (ELRS) is used when the wireless network automatically initiates the positioning at an emergency call.
- Triggered Location Reporting Service (TLRS) is used when the location of the mobile subscribers should be reported at a specific time interval or on occurrence of a specific event.

An LIF user can request not to be positioned by setting the PFLAG_STATUS attribute. This request can be overridden only by an emergency request. The LEV_CONF (level of confidence) attribute allows application to set quality requirements on location information. It indicates the probability that the mobile host is actually located in the returned position.

5.3.2.2 *Wireless Application Protocol (WAP) Location Framework*

WAP is a communication protocol designed to access information from micro browsers running on small handheld devices. It is based on HTML, XML and TCP/IP Internet standards. The WAP Forum has developed a Location Framework (see Figure 5.5) for communication from a mobile station directly to a Mobile Positioning Centre (MPC) that fits into WAP. While the WAP Location Framework partially overlaps the LIF API, it is more suitable for a Web site developer.

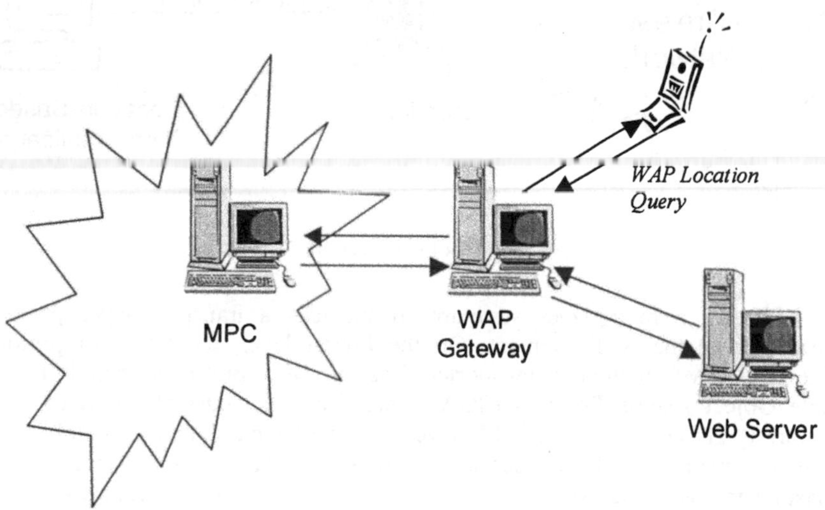

Figure 5.5 WAP Location Framework

The WAP Location Query Functionality provides three basic services:

- The Immediate Query Service (IQS) allows an application to query the location of a WAP client and get an immediate response.

- The Deferred Query Service (DQS) allows an application to receive the location of a WAP client on a continuous basis with deferred responses (e.g., every minute).
- The Location Attachment Service (LAS) allows an application to include the current location of the user in the location request and personalize the response with respect to the user location (e.g., finding the nearest ATM).

Both request and response are encapsulated in an XML message. However, the WAP client should not necessarily be a WAP browser. It can be a device with no user interface, such as a telematics device in a car. It is also possible to request the location of several clients at the same time, or to request different kinds of location information in the same query. The Quality of Position element allows applications to request information of a certain accuracy or maximum age.

5.3.3 Location Specification Languages

Standard location specification languages enable location-based data interoperability. Below we introduce *Geography Markup Language (GML)* and *Point of Interest Exchange (POIX) Language.*

5.3.3.1 Geography Markup Language

In 1999, the OGC (OpenGIS, 2002) specified the GML as an XML extension for encoding the transport and storage of geographic information, including both the geometry and properties of geographic features. GML serves to separate content from presentation in the realm of geography in the same way that XML does in the Web in general. The geographic data in GML is distinct from any graphic interpretation of that data. Keeping presentation separate from content allows GML data to be used for many different purposes on many different devices.

The design goals of GML includes:

- Establishing the foundation for Internet-based GIS in an incremental and modular fashion;
- Providing easy-to-understand encoding of spatial information and spatial relationships; and
- Providing a set of common geographic modelling objects to enable interoperability of independently-developed applications.

The OGC provides the following reasons for developing GML (OpenGIS, 2002):

- GML is securely based on the widely adopted public XML standard. This ensures that GML data will be viewable, editable and transformable by a wide variety of software tools.
- GML, as an extension of XML, is easy to transform. Programmers can perform data visualization, coordinate transformations, spatial queries, etc., using different programming languages.
- GML can be used as a major format of storing geographic information. XML-like linking capacity allows users to build complex and distributed geographic data sets. Because of XML's flexible architecture, the same data can be used

in different ways by different applications. Thus, GML is a public encoding standard for spatial information available for wide adoption.

- GML is based on a common model of geography (the OpenGIS Abstract Specification), which has been accepted by the vast majority of GIS vendors.
- Being based on XML, GML provides a natural way to verify integrity of geographic data.
- Being ASCII based, any GML document can be read and edited using a simple text editor.
- GML, together with non-spatial XML subsets, will enable integration of spatial and non-spatial data.
- GML is a natural way to package spatial data for LBSs involving mobile Internet devices.

GML is designed to support data interoperability through the provision of basic geometry tags, a common data model, and a mechanism for creating and sharing application schemas. An application schema defines the characteristics of a class of objects.

GML 2.0 defines three base schemas for encoding spatial information: Geometry schema, Feature schema and Xlink schema and provides three corresponding XML Schema documents. The Geometry schema (*geometry.xsd*) includes type definition for both abstract geometry elements, and concrete point, line and polygon elements, as well as complex type definitions for the underlying geometry types. The Feature schema (*feature.xsd*) makes the GML geometry constructs available in defining feature classes. The Xlink schema (*xlink.xsd*) provides the Xlink attributes to support linking functionality. These schema documents provide base types and structures, which may be used by an application schema.

A geographic feature is a named list of properties. Some or all of these properties may be geospatial, describing the position and shape of the feature. Each feature has a type (class) that prescribes the properties that a particular feature of this type is required to have. The properties are modeled as attributes of the feature class. An example of a GML feature is *Road* with such properties as *name* and *destination*.

Besides specific application-dependent properties, GML defines a set of property elements that corresponds to a small set of basic geometries. They include *pointProperty*, *lineStringProperty*, *polygonProperty*, *boundedBy*, etc. There are no restrictions in the type of geometry property a feature type may have.

The base GML schemas provide a set of foundation classes, from which an application schema can be constructed. As an example consider a feature type called *Person* with *lastName*, *firstName* and *age* properties, which can be specified in an XML schema as follows:

```
<element_name="Person", type:="ex:PersonType" />
<complexType name="PersonType" >
     <sequence>
          <element name="lastName" type="string" />
          <element name="firstName" type="string" />
          <element name="age" type="integer" />
```

```
    </sequence>
</complexType>
```

A single instance of this schema in XML might look as follows:

```
<Person>
      <lastName> Brown<\lastName>
      <firstName> John<\firstName>
      <age>35<\age>
</Person>
```

Using GML, we can associate location information with the above person as follows:

```
<Person>
      <lastName> Brown<\lastName>
      <firstName> John<\firstName>
      <age>35<\age>
      <gml:location>
        <gml:Point>
            <gml:coord><gml:X>0.5</gml:X><gml:Y>0.5</gm
            l:Y></gml:coord>
       </gml:Point>
      </gml:location>
</Person>
```

which is based on the following extended application schema:

```
<element_name="Person" type:="ex:PersonType"
        substitutionGroup="gml:_Feature" />
<complexType name="PersonType" >
  <complexContent>
     <extension base="gml:AbstractFeatureType">
      <sequence>
            <element name="lastName" type="string" />
            <element name="firstName" type="string" />
            <element name="age" type="integer" />
            <element ref="gml:location"/>
      </sequence>
    </extension>
  </complexContent>
</complexType>
```

In general, an application does not deal with the entire GML specification, which is quite complex. Instead, it may employ a subset of constructs corresponding to specific application requirements. This can be done using GML *profiles* (GML, 2001).

In summary, GML provides a standard way to encode, transport and store geographic information. It is anticipated that GML will increase the ability of sharing geographic information between different location-specific content providers.

5.3.3.2 Point of Interest Exchange Language

The *Point of Interest Exchange (POIX) Language Specification* defines a general-purpose specification language for describing location information using XML. Mobile device developers and LBS providers can use POIX specification for exchanging location data over the Internet by embedding it in HTML and XML documents. In contrast to GML, POIX is intended only to indicate a location of a mobile host and its structure is simple enough to be supported by portable terminals and car navigation systems.

The POIX specifies a set of DTDs of the basic elements that can be used to describe the location-related information (e.g., point, latitude, longitude, speed, direction, etc.). The following is a fragment of description of location-related information for a mobile host (POIX).

```
<?xml version="1.0" ?>
<!DOCTYPE poix PUBLIC "-//MOSTEC?POIX V2.0//EN"
"poix.dtd">
<poix version="2.0">
<format>
<datum>wgs84</datum>
<unit>dms</unit>
</format>
<poi>
<point>
<pos>
<lat> 35.41,28.7</lat>
<lon>139.45,02.4</lon>
<herror>30</herror>
</pos>
</point>
<move>
<method>car</method>
<speed>30</speed>
<dir>45</dir>
<locus>
<pos><lat>35.41,29.3</lat><lon>139.45,04.3</lon></pos>
<pos><lat>35.41,30.1</lat><lon>139.45,07.4</lon></pos>
<pos><lat>35.41,30.6</lat><lon>139.45,09.0</lon></pos>
</locus>
```

```
</move>
</poi>
</poix>
```

It should be noted that the W3Consortium (W3C) made current POIX specification available for discussion only. This indicates no endorsement of its content, nor that the W3C has had any editorial control in its preparation. It is too early to make any conclusions about the possibility of wide adoption of POIX for LBC.

5.4 LOCATION-BASED DATA MANAGEMENT

It is expected that a considerable part of computational load in the LBC environment will result from *Data Intensive Applications (DIAs)* that deal with relatively simple computations over large amounts of data. LBC environments with millions of interoperating DIAs that run simultaneously and concurrently access remote repositories of data will be quite common in near future. Efficient LBC would not be possible without efficient location-based data management.

Location-based data management is a relatively new research area that has been explored in the context of mobile wireless networks. In general, issues of efficient data management in mobile wireless networks remain an important challenge of current database research. For example, a common database technique consisting of the client sending a request to the server to initiate data transfers (*pull-based* approach) is not suitable for mobile LBC environments. LBC occurs in Asymmetric Communication Environments (*ACE*), where the communication capacity from server to client (*downstream capacity*) is much greater than communication capacity from client back to server (*upstream capacity*). Mobile clients can transmit only over a lower bandwidth link, while servers may have high bandwidth capacity. This requires using non-traditional data access methods, such as *broadcast disk* (Acharya *et al*, 1995), a *push-based* architecture where data is pushed from server to client. This architecture tries to utilize abundance in downstream communication capacity.

Accepting a push-based architecture for data access requires re-consideration of major issues of query processing. In particular, it introduces a new paradigm of queries over data streams where using traditional blocking operators during the query execution should be avoided. Query processing has to be more adaptive with respect to data delivery. Research on adaptive dataflow query processing partially addresses this problem (Hellerstein and Avnur, 2000). This research considers special query operators over data streams. For example, *eddy* is an operator that continuously reorders other operators in a query execution plan as it runs. The adaptive dataflow query processing is applicable when assumptions made at the time a query is submitted do not hold throughout the duration of query processing. The adaptive query processing system should function robustly in an unpredictable and constantly fluctuating environment, which makes it especially useful for wireless data networks.

The goal of research on *data recharging* (Cherniack *et al,* 2001) is to provide a service that supports recharging the mobile device-resident data cache from the Internet in a similar way as a device's battery is recharged from a fixed power grid.

This analogy is very preliminary since in contrast to power, data is much more personalized. A device should be associated with a user profile that allows searching the available data for data items relevant to the user. Data recharging is implemented as profile-based data dissemination and synchronization. Altinel and Franklin (2000) consider an efficient strategy to filter documents for the profile-based data dissemination. It uses a sophisticated index structure and a modified finite state machine approach to quickly locate and examine relevant profiles. This approach provides a highly efficient matching of XML documents to a large number of user profiles.

Recently, the concept of *dataspace* (Imielinski and Goel, 1999) and related concept of *device database* (Bonnet and Seshadri, 2000) were suggested to deal with information about physical world represented in highly distributed geographically based networks. In contrast to the traditional concept of database that stores information about remote objects locally, in a dataspace or device database, data is a part of the physical object and may be queried by reaching that object through the network. Characteristic routing (Navas and Wynblatt, 2001) and directed diffusion (Intanagonwiwat *et al*, 2000) are other advanced techniques that, in particular, can be used to retrieve location data from highly distributed wireless networks.

A data management issue in mobile wireless environments that attracted much attention was the tracking of mobile objects. Identifying the current location of a mobile device is important to both the delivering of data as well as to support LBSs. The issue of location data management in mobile wireless environments arises in the context of identifying current locations of mobile objects. Indeed, in wireless access networks, communicating with a mobile node and running mobile applications include an extra task of searching and updating their locations. With the increase of population of mobile users, the overhead of network traffic for the location processing becomes quite significant.

A major trade-off in implementation of location data management is the proper balance between the cost of location lookups and the cost of location updates with different levels of data availability, freshness and precision (Pitoura and Samaras, 2001). For example, the requirement that the current and precise location data is available at every node (i.e., every node maintains a local copy of the location database) makes location lookups trivial but implies heavy overheads from location updates. In contrast, we could consider another extreme when no location data is stored at any network node. In this case location lookups consist of an expensive global network search, while location updates are not required. Existing approaches to design of location databases in wireless networks, implement reasonable options in the above lookup/update trade-off.

The common implementation of location databases is a two-tier scheme that supports two kinds of location register: *Home Location Register (HLR)* and *Visitor Location Register (VLR)*. Every mobile user has a home location/zone with assigned HLR. Current user location is maintained in his/her HLR. In addition, each zone is also associated with a VLR to store information about all the users currently located in that zone, which is not their home location. The HLRs and VLRs should be consistently updated as users move. The user search procedure is based on querying the location registers (see more details about location registers in Chapter 4).

The above approach is not very flexible and has several disadvantages with respect to the lookup/update trade-off that are addressed in alternative implementation schemes, among them are hierarchical location (Wang, 1993) and regional matching (Awerbuch and Peleg, 1995). Different optimization techniques, such as proper database placement, caching, replication, etc. were suggested to improve performance of queries over location databases. An approach to use a centralized database to maintain information about moving objects is suggested in (Wolfson *et al*, 1998) where important deficiencies of existing database management systems, such as the inability to manage continuously changing data, are addressed. A good overview of this subject is in (Pitoura and Samaras, 2001).

Despite significant research activity in location-based data management, the potential of mobile devices as information assistants running data intensive applications is not utilized (Franklin, 2001). Frequent changes in the mobile environment require more adaptive methods to be utilized. We further discuss this issue in the next section.

5.5 ADAPTIVE LOCATION-BASED COMPUTING

As mentioned earlier, an important characteristic of LBC is a high level of interoperability between mobile applications and services. Meanwhile, some of the well-known problems for wireless mobile environments are expected to remain in the foreseeable future. These include wide disparity in the availability of remote services, limitations on local resources imposed by weight and size constraints, concern for battery power consumption and lowered trust and robustness resulting from exposure and motion, and unpredictable network behavior. The result is a challenge of *adaptive LBC*, which is based on interoperability between wireless and mobile applications that adjusts in real time to match varying capabilities and constraints of wireless mobile environments. An adaptive LBC environment is based on efficient self-monitoring, which impose significant requirements to the knowledge maintained by the LBC environment about itself (system metadata). In this section we introduce advanced *pervasive catalog infrastructure* to meet high metadata requirements of adaptive LBC and to support complex queries to monitor the LBC environment.

5.5.1 Motivating Example

Consider an example of mobile database queries. Assume that the fixed infrastructure includes several servers with electronic catalogs of products and prices maintained by major software vendors. A group of wireless mobile clients is interested in keeping track of price increases for products. To achieve this, in the absence of any broadcast push service, the clients submit database queries to the servers in the fixed infrastructure. As a consequence, multiple clients can download the same or overlapping data. Consider the query *find database-related products developed by Oracle and IBM with the price rage of less then $2000 for university customers*. We assume that the schemas of the corresponding Oracle and IBM data sources are already integrated (e.g., using wrapper/mediator technology (Zadorozhny, 2002)). A wireless client *C1* submits the following SQL expression:

(Select product_name, product_price from Oracle_server
where product_price < 2000 and license= "university")
union
(Select product_name, product_price from IBM_server
where product_price < 2000 and license= "university").

The alternative data sources that can be used in answering this query include the *Oracle_server*, the *IBM_server* and other peer wireless nodes. Let us assume two such mobile nodes *S1* and *S2* already downloaded relevant data. *It is important for the client to have accurate information about location of such peer nodes.* If they are in a reasonably close proximity the client can take advantage of their processing capabilities. Obviously, these data sources have different and varying capabilities. In contrast to the *Oracle_server* and the *IBM_server*, the two mobile data sources are not expected to support full SQL query processing functionality. For instance, let us assume that *S1* does not support SQL queries although it provides data caching services and has downloaded a significant part of the requested data, for example from *Oracle_server*. On the other hand, *S2* does support partial SQL query processing, but it has downloaded few of the requested data items.

A query plan will attempt to maximize the completeness of the result, taking into consideration the processing capabilities of *S1*, *S2* and the client. Assuming *C1* has basic SQL processing capabilities, a possible query plan would be to execute remotely *select* subqueries on *Oracle_server* and *IBM_server* and do the union on the wireless client *C*. However, in the case that one or both of the above servers are not available or overloaded, the other plans might be:

- If the bandwidth and latency are acceptable, download data from *S1* and execute query locally on *C1*.
- If *S2* has enough power, execute query locally on *S2* and send the result to *C1*.
- If the bandwidth and power are not enough to use any of sources *S1* and *S2* to execute the whole query, try to decompose the query so that subqueries can meet the resource requirements and capabilities of *S1* and *S2*. The result will be aggregated from the subquery results. Note that although *S1* and *S2* individually might not fully answer the example query, their aggregated service might.

5.5.2 Metadata Management for Adaptive Location-Based Computing

In the above example, the proper choice of a query plan depends on the current and accurate knowledge of data distribution, replication and caching, as well as statistics about the wireless network, such as bandwidth and power limits. To answer the above query, the system should not only realize what part of data is cached at *S1,* but also possibly tolerate a partial answer. Query planner and execution engine should be able to request the relevant statistics, which is a part of *application metadata*. Such meta-data queries may be quite complex.

In general, to provide a proper level of application interoperability under given conditions the mobile environment should efficiently monitor itself. It should support an infrastructure for executing complex monitoring queries, such as: *What nodes have enough resources (power, available memory, bandwidth, etc.) to implement certain levels of interoperability?* The information that supports the previous queries (e.g., information about power, available memory, bandwidth, etc.) represents *mobile network metadata*.

Maintaining a metadata repository for adaptive mobile environments in a way that location databases are maintained (Section 4) is not an appropriate solution. Issues of adaptive interoperability impose more significant demand on the system metadata and the algorithms that use that knowledge (Franklin, 2001). First, location databases are typically used to evaluate simple queries such as *given an object name, find the object location*. In principle it is possible to use location databases to execute more complex location-aware queries, like finding services and points of interests based on their location attributes. However, efficiency of such queries in a wireless environment with millions of mobile devices may not be acceptable. Second, visible location updates (e.g., due to crossing the cell boundary) are not very frequent compared to significant fluctuations in the mobile environment that should be reflected in the wireless network metadata for interoperable applications. In addition to the traffic related to tracking mobile users, an interoperable wireless access network should carry an extra traffic with metadata that supports implementation of specific interoperability strategies (e.g., availability of services, device capabilities, user preferences, local network conditions, and access bandwidth).

The availability, freshness and precision requirements for that metadata are different from the requirements for location data: the frequency of updates is much higher, the imprecision is less tolerable, and the availability requirements are stricter. As a result, we are not able to re-use directly the approaches for implementing location databases to provide metadata support for interoperability in wireless networks. Thus, one of the major goals of adaptive LBC is to devise an infrastructure that utilizes the flexibility and performance advantages of location databases while providing the components that implement the extra needed functionality for interoperable mobile services.

5.5.3 Pervasive Catalog Infrastructure

To meet the requirements of freshness, precision and availability of metadata in the interoperable wireless environments, as well as to support complex monitoring queries, we discuss the concept of *pervasive catalog infrastructure*. The pervasive catalog, together with an advanced query processing component, form a core of fully supported adaptive interoperable LBC. The concept of pervasive catalog subsumes the concept of location database and provides the additional functionality to meet the metadata requirements of interoperable wireless applications. The design and implementation of the pervasive catalog utilize efficient methods of distributed data and metadata management. It should be noted that the concept of pervasive data management also subsumes issues of data management in wireless and mobile environments. The pervasive catalog infrastructure extends them in a wider context of wireless and mobile application interoperability. It will also provide a uniform basis to integrate networked data management with service location and interaction protocols supported by the network layer and mobile IP.

In general, the existing approaches to distributed data management can be grouped into one of the following two categories (Bonnet and Seshadri, 2000). The *warehousing approach* assumes that the raw data is transferred from wireless devices to a central database. The raw data is structured in the appropriate data model and integrated in the centralized database. With a *distributed query processing approach* (e.g., device databases (Bonnet and Seshadri, 2000), dataspaces (Imielinski and Goel, 1999) information is stored on the network nodes and some queries, or parts of queries, can be executed on the devices.

Existing techniques to implement distributed metadata management can also be broadly classified into two categories: *Naming services with fixed structure* in which the discovery of data/services/operations is based on the traditional client-server architecture (CORBA) and *Peer-to-Peer (P2P)* in which the discovery of data/services/operations is dynamically provided by distributed communities of hosts (Pieper and Munafo, 2000), (P2P). As opposed to naming services with a central registry, in P2P information about searches is passed around in a relay, from one host to the next. The P2P approach is more suitable in frequently changing distributed environments, such as wireless networks. Metadata about current host communities is more accurate in dynamic P2P systems compared to naming services with a fixed structure. The disadvantage of the P2P approach is that the network may be over-flooded with the P2P traffic in the case of large host communities and restricted network resources.

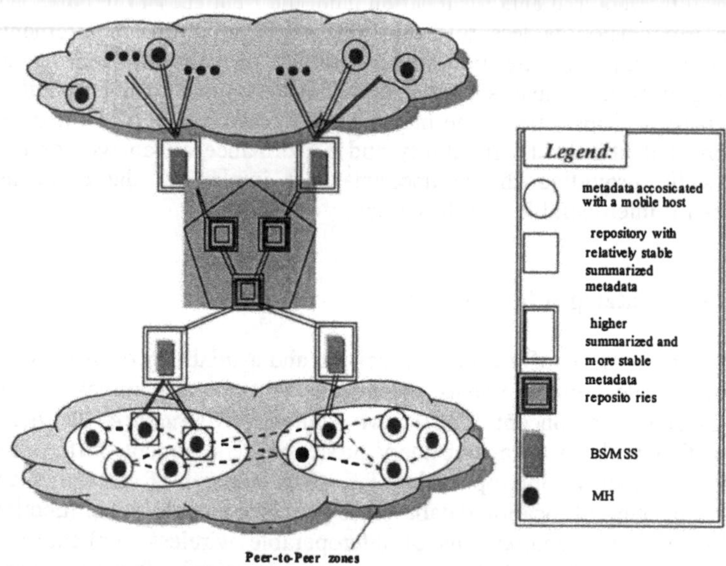

Figure 5.6 Pervasive Catalog Infrastructure

In (Chrysanthis and Zadorozhny, 2002) a hybrid approach for the pervasive catalog with *adaptive dynamic distribution granularity* (see Figure 5.6) is proposed. This approach represents a compromise between the fixed structure client/server and P2P approaches, and combines advantages of the warehouse and distributed query processing techniques. Under this approach every node in the wireless network is associated with most current and accurate node-specific metadata (e.g., percentage of CPU and battery utilization and other information about capabilities of the device). Some of the nodes can host a more summarized local database that stores and monitors certain metadata about a subnet of devices. A cost-based decision support system (on-line optimizer) is responsible for the choice of a monitored subnet and the content of corresponding local databases. The local databases are arranged in a monitoring hierarchy. As the wireless environment changes, the on-line optimizer periodically relocates the local databases and re-arranges the monitoring hierarchy.

Thus, the pervasive catalog is implemented as a monitoring hierarchy of meta-repositories dynamically distributed over the wireless network. The decisions about the relocation of local databases and re-arrangement of the monitoring hierarchy are based on the metadata stored in the pervasive catalog. To break this loop, the on-line optimizer uses a set of basic heuristics (bootstrapping base) to build an initial catalog hierarchy. The heuristics in the bootstrapping base should take into account initial approximations of processing and communication capabilities of the wireless devices to minimize meta-data traffic, as well as energy consumption while providing a certain level of data accuracy and freshness. The on-line optimizer can be tuned for either maximum lookup or update performance. In this way it is possible to implement different trade-offs between data availability, freshness, precision and overall system efficiency.

The pervasive catalog technology allows for adjusting of the trade-off between the efficiency of data access and data freshness, and introduces a natural metadata filtering and summarization scheme. The efficient filtering is important because the amount of metadata collected from constantly changing communities of millions of mobile devices can grow to an unmanageable size quite fast.

5.5.4 Querying Pervasive Catalog

One of the tasks of the pervasive catalog infrastructure is to provide access cost distributions to data sources that can be used by tools such as WebPT (Gruzer *et al*, 2000) as part of query planning. This also includes providing answers to network monitoring queries like *what nodes have enough resources (power, available memory, bandwidth, etc.) to implement certain level of interoperability?* The catalog should also support accurate network condition report generation, such as current and expected link utilization percentage, current and expected throughput of each device (node), etc.

The catalog queries would be submitted to the highly distributed system of meta-repositories that establish the pervasive catalog infrastructure. The meta-repositories represent dynamic data sources with possibly overlapping information and different levels of summarization. Issues of querying highly distributed data sources were addressed in (Zadorozhny *et al*, 2002): in particular, a two-phased approach to query planning and optimization in wide area environments is proposed. The first phase utilizes a cost-based pre-optimizer to choose among alternative data sources. In the second phase, an optimizer produces a good query execution plan.

The pre-optimizer constructs a pre-plan, which identifies relevant data sources and their query processing capabilities that can be used to answer a query. The pre-plan is a higher-level abstraction that circumscribes and describes a space of query execution plans. Each query execution plan in the pre-plan space is a physical plan for evaluating the query that has a specific cost. Consider the motivating example of wireless database queries from Section 5.1. The query pre-optimizer will build a pre-plan including all alternative data sources that can be used to provide information relevant for the query, i.e., *Oracle_server*, *IBM_server* and two peer wireless nodes *S1* and *S2* that already downloaded relevant data.

To make a proper choice, the query planner and optimizer should send a *catalog query* for the relevant statistics. Since the pervasive catalog is implemented as a highly distributed dynamic system of meta-repositories (i.e., data sources with possibly overlapping information and different levels of summarization), the catalog query itself should be carefully planned and optimized using the same two-phase approach. While planning the catalog query, the optimizer should consider the trade-off between the accuracy and availability of the wireless metadata. For example, the first, most general request would be sent to a highly available and summarized meta-repository on a powerful network server to find out what would be a set of candidate alternative data sources for a given application query. The next, more refined request could be sent to more accurate/current and less summarized meta-repositories closely monitoring the relevant data sources, up to sending a request for the most current metadata to specific sources. The decision on sending more refined catalog requests should be made on the results of the previous general catalog requests.

Since planning the catalog queries is based on the metadata stored in the pervasive catalog itself, the catalog should provide an appropriate bootstrapping base, a set of heuristics, to avoid infinite recursion in catalog queries. As in the case of building the catalog hierarchy, the heuristics in the bootstrapping base should take into account initial approximations of processing and communication capabilities of the wireless devices to minimize data traffic, as well as resource consumption while providing acceptable level of data accuracy and freshness.

5.6 LOCATION-BASED ROUTING AS ADAPTIVE LBC

In this section we consider how location information can be used by an adaptive LBC infrastructure, using the example of location-based routing. A routing method, which is not location-aware, is commonly based on a flooding technique assuming that the route request message potentially reaches every host in the network. This may result in overloading the wireless network with the route discovery messages and have a negative impact on the network performance and scalability of the routing protocol. Our discussion will be focused mainly on location-based routing in mobile ad hoc networks. We elaborate on some location-aware routing protocols and compare them with the simple flooding procedure that ignores information about the location of mobile hosts. We show how LBC enables scalable and efficient algorithms for discovering routes between mobile hosts in an ad hoc network. We also briefly outline location-based routing in the Metricom packet radio system (Metricom).

Mobile ad-hoc networks consist of wireless mobile hosts that communicate with each other in the absence of fixed infrastructure. Routes between two hosts in an ad-hoc network may consist of hops through other hosts in the network. Host mobility results in frequent unpredictable topology changes. The task of finding and maintaining routes in mobile ad-hoc networks is non-trivial and represents an important area of LBC. We discuss how efficient routing protocols can benefit from location information.

Without location information, routing in ad-hoc networks is commonly based on *flooding technique* (Ko and Vaidya, 1998). The flooding technique works as follows. Consider the example of database query from section 5.1. To use processing capabilities of the host *S1*, host *C1* needs to find a route to *S1*. Thus, *C1* broadcasts a route request to all its neighbors. Two mobile hosts are neighbors if they can communicate with each other directly over a wireless link. Any mobile host *X* receiving a route request compares the desired destination with its own identifier. If there is a match, the host *X* does not broadcast the route request anymore. Otherwise, *X* broadcasts the routing request to its neighbors.

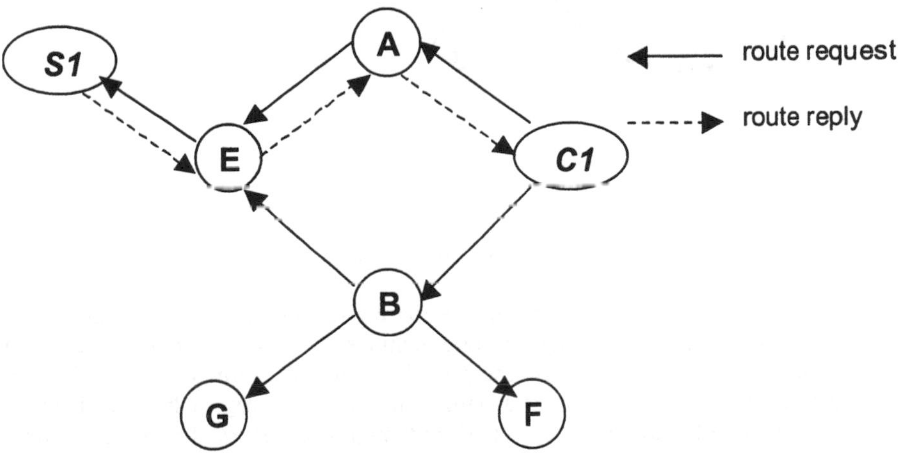

Figure 5.7 Routing via Flooding

Figure 5.7 illustrates the flooding technique. As above, the host *C1* needs a route to the host *S1*. To do that, *C1* broadcasts a route request to its neighbors: hosts *A* and *B*. When *A* and *B* receive the route request they broadcast it to all their neighbors (except *C1*). Thus, host *E* receives the route request from both *A* and *B*. Those requests are identical, and *E* drops one of them. Finally, *E* broadcasts the route request to all its neighbors (except *A* and *B*). The request reaches the destination *S1*. The destination host responds by sending a route reply to the sender.

If either the destination host does not receive a route request message, or the sender host does not receive the route reply message, the sender needs to reinitialize the above *route discovery* process (e.g., using a predefined timeout).

With the flooding technique the route request message potentially reaches every host in the ad hoc network. This is a significant drawback that may result in overloading the network with the route discovery messages. It may have a negative impact on the network performance and scalability of the routing protocol. To solve this problem, an adaptive LBC infrastructure could utilize the location information. As an example, we consider a basic Location-Aided Routing (LAR) protocol (Ko and Vaidya, 1998) and outline other more advanced location-based routing techniques. LAR protocol assumes that every mobile host knows its physical location (e.g., from the GPS).

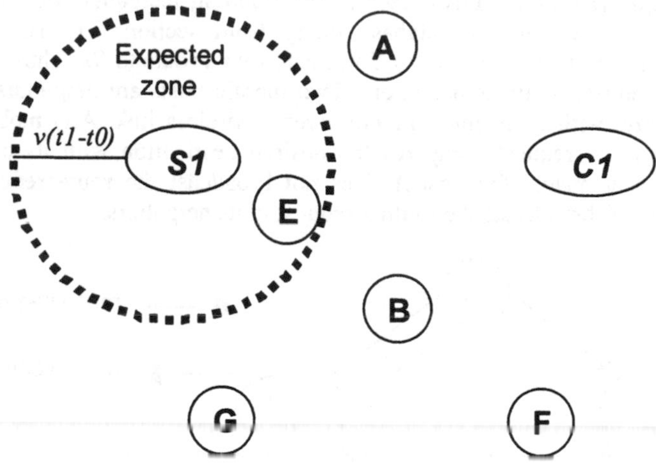

Figure 5.8 Expected Zone

According to the LAR algorithms the routing request is forwarded only to the hosts in a *request zone*. The request zone includes the *expected zone* (Figure 5.8) and some other regions to increase the probability that the route request will reach the destination host. In Figure 5.9 a path from *C1* to *S1* may include *B*. Thus, *B* should be included in the request zone, even though *B* is outside the expected zone. There are different schemes to determine request zones (Basagni *et al*, 1998).

Other location-based protocols improve the above algorithm utilizing additional location-related information. For example, DREAM (Distance Routing Effect Algorithm for Mobility) protocol (Basagni *et al*, 1998) considers the *distance effect* and *mobility rate* for mobile hosts. Indeed, the greater the distance separating two hosts, the slower they appear to be moving with respect to each other. Thus, nodes that are far apart need to update each other's locations less frequently than nodes staying closer. Concerning the mobility rate, the faster a node moves, the more often it must communicate its location. This allows each node to optimize its dissemination frequency transmitting location information only when needed and without sacrificing accuracy.

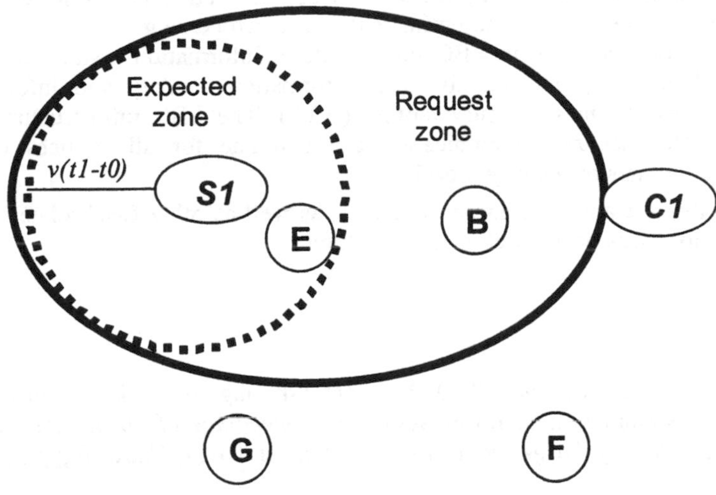

Figure 5.9 Request Zone

An example of industrial network infrastructure that uses location-based routing is the Metricom packet radio system (Metricom). The Metricom network infrastructure includes fixed base stations (pole top radios) whose precise location is determined using GPS at the time of installation. The pole top radios route packets over a wireless link towards or from the nearest wired access point after which the packet is routed through the Internet or to the mobile host. The routing is performed geographically on the basis on the latitude and longitude of the pole top radios with respect to the final destination.

5.7 CONCLUDING REMARKS

Location-Based Computing is growing in importance. It is finding increasing use in day-by-day business. With the explosive growth of the wireless Web, successful deployment of LBC infrastructure becomes a critical prerequisite of the future computing environments.

It is not enough to say that LBC has been enabled by advances in such diverse technologies as wireless telecommunication, GIS, open distributed processing and distributed data management. In fact, LBC integrates those areas in a new powerful paradigm and establishes a visible milestone in the progress of computing. At the same time it also reveals new challenging research directions. Below we enumerate just some of them:

- Location-based semantic interoperability. Since the physical world becomes a context of LBC, it should be described in a semantically rich and non-ambiguous way. The problem of context integration becomes especially hard for LBC.

- Location-based query processing in distributed databases. Both query optimizer and execution engine should be able take advantage of location data and current query context to optimize the query processing.
- Information assurance in LBC infrastructure. Information assurance can be defined as the provisions for both information security and information availability via network survivability (WIA). The LBC infrastructure must ensure that secure communication can continue for all services despite attacks, failures or security breaches.

We expect that the above research directions, as well as other LBC-related issues will be a major focus of research in the near future.

REFERENCES

Altinel, M. and Franklin M., 2000, Efficient Filtering of XML Documents for Selective Dissemination of Information. In *Proceedings of 26 th International Conference on Very Large Data Bases*, September 10-14, Cairo, Egypt, pp. 53-64.

Acharya, S., Alonso, R., Franklin, M. and Zdonik S., 1995, Broadcast Disks: Data Management for Asymmetric Communications Environments. In *Proceedings of the ACM SIGMOD International Conference on Management of Data*, San Jose, California, May 22-25, pp. 199-210.

Awerbuch, B. and Peleg, D., 1995, Online Tracking of Mobile Users. *Journal of the ACM*, 42(5), pp. 1021-1058.

Basagni, S., Chlamtac, I., Syrotiuk, V. and Woodward, B., 1998, A Distance Routing Effect Algorithm for Mobility (DREAM). In *Proceedings of the Fourth Annual ACM/IEEE International Conference on Mobile Computing and Networking*, October 25-30, 1998, Dallas, pp. 76-84.

Bonnet, P. and Seshadri., P., 2000, Device Database Systems. In *Proceedings of the 16th International Conference on Data Engineering*, 28 February 28 – March 3, San Diego, California, pp. 194.

Burkhardt, J., Henn, H., Hepper, S., Rindtorff, K. and Schack, T., 2002, Pervasive Computing. Technology and Architecture of Mobile Internet Applications, (Edinburgh: Addison-Wesley).

Cherniack, M., Franklin, M. and Zdonik, S., 2001, Expressing User Profiles for Data Recharging. *IEEE Personal Communications: Special Issue on Pervasive Computing,* August, pp. 6-13.

CORBA (The OMG's Common Object Request Broker Architecture). Online. Available HTTP: <http://www.corba.com> (accessed 7 April, 2003).

Chrysanthis, P. and Zadorozhny, V., 2002, From Location Databases to Pervasive Catalog. In *Proceedings of DEXA 5th International Workshop on Mobility in Databases and Distributed Systems*, September 5, Aix-en-Provence, France, pp. 739-746.

Franklin, M., 2001, Challenges in Ubiquitous Data Management. In *Informatics: 10 Years Back, 10 Years Ahead*, R. Wilhiem (Ed.), LNCS N 2000, Springer-Verlag, pp. 24-33.

Imielinski, T. and Goel, S., 1999, DataSpace - Querying and Monitoring Deeply Networked Collections in Physical Space. In *Proceedings of the ACM*

International Workshop on Data Engineering for Wireless and Mobile Access, August 20, Seattle, WA, pp. 44-51.

Intanagonwiwat, C., Govindan, R. and Estrin, D., 2000, Directed Diffusion: A Scalable and Robust Communication Paradigm for Sensor Networks. In *Proceedings of the Sixth Annual International Conference on Mobile Computing and Networking*, August 6-11, Boston, MA, pp. 56-67.

GML (Geography Markup Language 2.0. OpenGIS Implementation Specification.) 2001, Online. Available HTTP: <http://opengis.net/gml/01-029/GML2.html> (accessed 7 April, 2003).

Pieper, J. and Munafo, R., 2000, Gnut Manual. Online. Available HTTP: <http://carnagepro.com/pub/Docs/Gnut/gnut.html#toc1> (accessed 7 April, 2003)

Gruser, J.R., Raschid, L., Zadorozhny, V. and Zhan T., 2000, Learning Response Time for WebSources using Query Feedback and Application in Query Optimisation. *The International Journal on Very Large Databases*, 9(1), pp. 18-37.

Hellerstein, J. M. and Avnur, R., 2000, Eddies: Continuously Adaptive Query Processing. In *Proceedings of the ACM SIGMOD International Conference on Management of Data*, May 16-18, 2000, Dallas, Texas, pp. 261-272

Hjelm, J., 2002, Creating Location Services for the Wireless Web. Wiley Computer Publishing.

Ko, Y., and Vaidya, N., 1998, Location-Aided Routing (LAR) in Mobile Ad Hoc Networks. In *Proceedings of the Fourth Annual ACM/IEEE International Conference on Mobile Computing and Networking*, October 25-30, Dallas, Texas, pp. 66-75.

LIF (Location Interoperability Forum. Mobile Location Protocol Specification). 2001, Online. Available HTTP: <http://www.openmobilealliance.org/lif/> (accessed 7 April, 2003).

Metricom (Metricom Corporation Home Page). Online. Available HTTP: <http://www.metricom-corp.com> (accessed 7 April, 2003).

P2P (Stretching the fabric of the net: examining the present and potential of peer-to-peer technologies.) Online. Available HTTP: <http://www.siia.net/sharedcontent/piracy/pubs/peer1101.pdf > (accessed 7 April, 2003).

MSDN (MSDN Data Access Technology). Online. Available HTTP: <http://www.microsoft.com/data> (accessed 7 April, 2003).

Navas, J.C. and Wynblatt, M., 2001, The Network is the Database: Data Management for Highly Distributed Systems. *Proceedings of the ACM SIGMOD International Conference on Management of Data*, Santa Barbara, California, May 21-24, pp. 544 – 551.

OpenGIS (OpenGIS Reference Model)., 2002, Online. Available HTTP: <http://www.opengis.org/info/orm/> (accessed 7 April, 2003).

Pitoura, E. and Samaras, G., 2001, Locating Objects in Mobile Computing. *IEEE Transactions on Knowledge and Data Engineering*, 13 (4), pp. 571-592.

POIX (Point of Interest exchange Language Specification. W3C Note.) Online. Available HTTP: <http://www.w3.org/TR/poix> (accessed 7 April, 2003).

Sophica (Mobile Web Services Interoperability Test Bed)., 2001, Online. Available HTTP: <http://www.commerce.net/projects/ngi/grants/2001/sophica.html> (accessed 7 April, 2003).

Walker, J. (Ed.)., 2002, Advances in Mobile Information Systems. Artech House Publishers.

WAP (Wireless Application Protocol Forum Releases. <http://www.wapforum.org/what/technical.htm>.

WIA (Wireless Information Assurance Architecture Group.) Online. Available HTTP: <http://www.sis.pitt.edu/~wireless/wia/index.html> (accessed 7 April, 2003).

Wang, J.Z., 1993, A Fully Distributed Location Registration Strategy for Universal Personal Communication Systems. *IEEE Journal on Selected Areas in Communications*, 11(6), pp. 850-860.

Wolfson, O., Chamberlain, S., Dao, S., Jiang, L. and Mendez, G., 1998, Cost and Imprecision in Modelling the Position of Moving Objects. In *Proceedings of the 14th International Conference on Data Engineering*, February 23-27, Orlando, Florida, pp. 588-596.

Zadorozhny, V., Bright, L., Vidal, M.E., Raschid, L. and Urhan, T., 2002, Efficient Evaluation of Queries in a Mediator for WebSources. In *Proceedings of the ACM SIGMOD International Conference on Management of Data*, Madison, Wisconsin, June 3-6, pp. 85-96.

CHAPTER SIX

Location-Based Services

Xavier R. Lopez

6.1 INTRODUCTION

Location-Based Services (LBSs) consist of a broad range of services that incorporate location information with contextual data to provide a value-added experience to users on the Web or wireless devices. In contrast to the passive fixed Internet, users in the mobile environment are demanding personalized, localized, and timely access to content and real-time services. Targeted data, combined with location determination technology, is essential to create personalized value to an end-user's mobile experience. This allows wireless carriers and portals to significantly increase the value of services to subscribers while opening up new revenue opportunities.

Through new applications, mobile offerings can be personalized to a user's lifestyle, tastes and preferences and can be immediately synchronized with other portable devices. The variety and breadth of applications and services are large, from pure content and advertising, to emergency 911, navigational services, fleet and asset management, logistics, and location-sensitive billing. The high level of interest in these services coupled with corresponding technology developments have spurred the emergence of the LBS industry and created a multifaceted assortment of players, service concepts, and business models. This chapter is designed to introduce the major concepts in this rapidly evolving technology and services market.

6.2 TYPES OF LOCATION-BASED SERVICES

The separation of services from the telecommunications networks will lead to the establishment of an open, Internet Protocol (IP) based service environment. In the initial phase it will primarily be digital content services that are implemented in this IP environment. However, it will not be long before technologies such as voice over IP will enable multimedia messaging services and real-time call control services to be implemented in an open, IP-based service environment. The first generation LBSs can roughly be divided into six categories:

- **Safety Services:** End-user assistance services, such as Enhanced 911 (E-911), are low usage services designed to provide the end-user assistance in the case of an emergency. These type of services can expect to gain a high market acceptance due to the general concern of the public for the personal security. With a push from the United States Federal Communications Commission (FCC) E-911 mandate and new location solutions, wireless carriers will be able to route an emergency call based on the caller's location and the Public Safety Answering Point (PSAP) jurisdictional boundary, determining the nearest emergency center and drastically reduce response time (FCC, 1996).

- **Information Services:** These services comprise a vast area of applications. Some of these include traffic information, navigation assistance, yellow pages, travel/tourism services, etc. Users will come to expect voice-enabled driving directions and walking directions, as well as information services, whereby requested information is delivered in various ways, such as Wireless Application Protocol (WAP), by a Short Message Service (SMS) message, by Interactive Voice Response (IVR), Multimedia Mark-up Language (MML), or by a call center operator.
- **Enterprise Services:** These services include vehicle tracking, logistic systems, fleet management, workforce management, and "people finding." Today, many of these services are offered by legacy, mobile data systems. However, with the growing availability of broadband wireless capability, it is likely that many of these services will be merged into the digital wireless networks. It is the enterprise applications where the deployment of mobile LBSs is taking hold first.
- **Consumer Portal Services:** As consumer technology platforms and wireless carrier infrastructures are upgraded to support ubiquitous, accurate location information, consumers will begin to access navigational services, such as driving directions. Location-aware services will enable the delivery of "local" news, weather, and traffic information determined by location of devices – all provided through an icon-based user interface.
- **Telematics Services**: Telematics is most often used to describe vehicle navigation systems, such as OnStar, where drivers and passengers employ Global Positioning System (GPS) technology to obtain directions, track their location, and obtain assistance when a vehicle is involved in an accident. In-car systems, however, are car or machine centric as opposed to hand-held mobile devices, which are user centric by nature. Unlike static CD-ROM-based in-car navigation systems, online mobile systems provide accessibility to up-to-date, time sensitive information and databases.
- **Triggered Location Services**: As carriers form partnerships with location-based application providers and develop direct content relationships with businesses, they will be able to trigger services as consumers or corporate clients enter predetermined areas. Example triggered services include: location-sensitive advertising, location-sensitive billing, and location-sensitive logistics.

6.3 WHAT IS UNIQUE ABOUT LOCATION-BASED SERVICES?

There are important similarities and differences between LBS technology and geographic information systems (GISs). For beginners, much of the underlying mapping, spatial indexing, spatial operators, geocoding and routing technology that is used to deliver LBSs originate from the GIS industry. However, what makes LBS technology different is that it is a service deployed on a foundation of IT and wireless technology. The value chain of a GIS is generally limited to the providers of a desktop or client-server solution, whereas the value chain of LBSs includes many players ranging from hardware and software vendors, content and online service providers, wireless network and infrastructure providers, wireless handset vendors, and branded portal sites. Another major difference is that LBSs impose significant technology and service capabilities that exceed the general requirements of static GIS uses, namely:

- **High Performance**: Delivers sub-second queries required for Internet and wireless
- **Scalable**: Supports thousands of concurrent users and terabytes of data
- **Reliable**: Capable of delivering up to 99.9999 up-time
- **Current**: Supports the delivery of real-time, dynamic information
- **Mobile**: Available from any device (wireless and wired) and from any location
- **Open**: Supports common standards and protocols, HTTP, WAP, Wireless Markup Language (WML), Extensible Markup Language (XML) and Multimedia Markup Language (MML)
- **Secure**: Leverages the underlying database locking and security services
- **Interoperable**: Integrated with e-Business applications (Customer Relationship Management, Billing, Personalization) and wireless positioning gateways

There are also significant performance and scalability requirements that further differentiate GIS solutions and LBS solutions. One might consider the delivery of wireless LBSs as service infrastructure – unlike the delivery of other utilities services. Online content services generally require large data servers, large enterprise hardware offerings, and significant mid-tier cached application servers that allow the service to scale and perform. LBSs routinely publish personalized content to tens of thousands of users on an hourly and daily basis. Contrast this with GISs, where a handful of users perform relatively complex spatial queries on desktops of client-server systems.

Interoperability is a fundamental requirement of any LBS. Unlike GISs that manage all vector and attribute data locally using proprietary data structures and data models, LBSs rely on Standard Query Lanaguage (SQL) to access locally hosted content, and XML and Simple Object Access Protocol (SOAP) interfaces to incorporate syndicated online content. Building upon standard query languages, interfaces, encodings, and protocols, it becomes possible to chain basic LBS functions (e.g, geocoding, mapping, routing, real-time traffic) for the creation and delivery of a complex end-user service. Take for example the delivery of a wireless location-enhanced dining service. A wireless restaurant finding service requires a provider to fuse handset position acquisition services with a yellow page proximity search. These inputs can also be referenced with a personalized profile of dining preferences. This example can be further enhanced with syndicated access to an online restaurant menu and reservation service which simplifies the decision making process for the hungry diner. While this example of a "location-enhanced" dining guide is consumer oriented, one can envision other types of business-to-business applications in domains like logistics, customer care, field sales, real estate and marketing, and electronic marketplaces.

Customers also want the provision of LBSs to be automatic – they want carriers and wireless portals to take care of integrating a variety of Internet and enterprise information services with a customer's preferences, enabling a user to focus on informed decision making. For example, a real-time telematics service may access multiple syndicated data services (traffic, business directories, driving directions) from multiple sources – and integrate the information in a meaningful manner for the customer. A customer checking on the availability of a hotel in a given city might access geocoding services that identify his/her location and the nearest hotels, and would cull data from real-time travel services to check availability and book a room, and from a driving directions service to route him/her to the hotel.

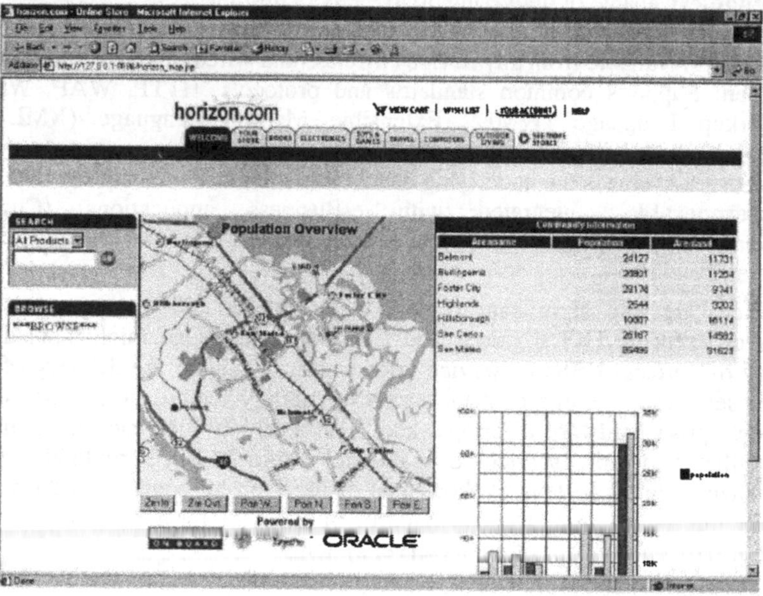

Figure 6.1 CRM Application Incorporating Location Information

6.3.1 Integration With e-Business Solutions

A unique capability of LBSs is the ease to which they can be integrated with or incorporated into corporate e-Business applications like Customer Relationship Management (CRM), Enterprise Resource Planning (ERP), and Business Intelligence (BI). Organizations have demonstrated that effectively managing location can provide strategic differentiation when managing customer information and efficiently managing corporate assets, both critically important in business. LBSs make use of spatial information and the functions that operate on this information, thus enabling the incorporation of "location awareness" and "location sensitivity" into an organization's e-Business applications, field operations, Web offerings and more. Wireless carriers for example, now recognize that they will compete on the basis of how effectively they can integrate their CRM and field service operations with those of customers and suppliers to create a positive business experience.

By integrating enterprise information with location-enhanced customer information, carriers obtain comprehensive business intelligence, and value builds exponentially. Mobile operators become better positioned to use real customer information to determine wireless service expansion, improve service delivery, and determine load demands. On the customer end, by automating information integration and interpretation, the customer is able to deal with a much richer set of location-enhanced information for better decision-making (see Figure 6.1). With the introduction of event-driven e-Business, wireless carriers and portals can send fresh information as it becomes available or as users roam into a new location, rather than waiting for customers to check in with the service. Customers, mobile operators, and partners can react immediately to the changed location of a handset user by delivering personalized services for his/her new roaming region.

6.4 ENABLING TECHNOLOGIES

The performance and capability requirements expected for wireless LBSs can easily approach that of a top Internet portal – millions of queries on a daily basis, hundreds of concurrent transactions, and millisecond query response times. Thus, a typical LBS must support all the unique CPU-intensive location queries, and provide scalability, storage, and interoperability.

6.4.1 Spatial Data Management

A robust LBS platform should be carefully designed to meet the unique performance, scalability, and flexibility requirements of wireless and Internet portals. For example, an LBS service is generally built around a robust database and middleware technology stack that supports multiple internally and externally hosted applications (see Figure 6.2). This Internet platform must also interoperate with the wireless network and with a variety of client devices. A spatially-enabled database serves as a foundation for deploying Internet and wireless LBS. It provides data management for location information such as road networks, wireless service boundaries, and geocoded customer addresses. It enhances the development and deployment of LBS applications by allowing users to easily incorporate location information directly in their applications and services. Spatial databases by vendors such as Oracle provide spatial object type storage, SQL access, spatial operations, fast R-tree indexing, and projection and coordinate transformation support. These databases can also perform location queries on geocoded yellow page databases, find the nearest hotels, restaurants, and gas stations. Spatial databases, in combination with specialized geocoding, routing, and mapping tools enable wireless carriers, portals and e-Business applications with the capability to incorporate location into their services.

Platforms for LBSs are designed to meet the unique performance and scalability requirements of wireless and Internet portals. In Figure 6.2, business and location content is stored in local database and spatial database servers. The middle-tier applications server delivers critical load balancing, caching, messaging, and security (see details below). In addition, specialized LBS tools like routing, geocoding, and map rendering can run inside the application server's Java 2 Platform, Enterprise Edition (J2EE) container for fast performance. The wireless extension to the application server handles content transformation (using XML) to various wireless devices, while it also incorporates the syndication of LBS Web Services that may be hosted externally. The middle-tier server must also support the various positioning gateways from vendors like Ericsson, Nokia, and Qualcomm, preferably via the Location Interoperability Forum's (LIF) XML positioning interfaces.

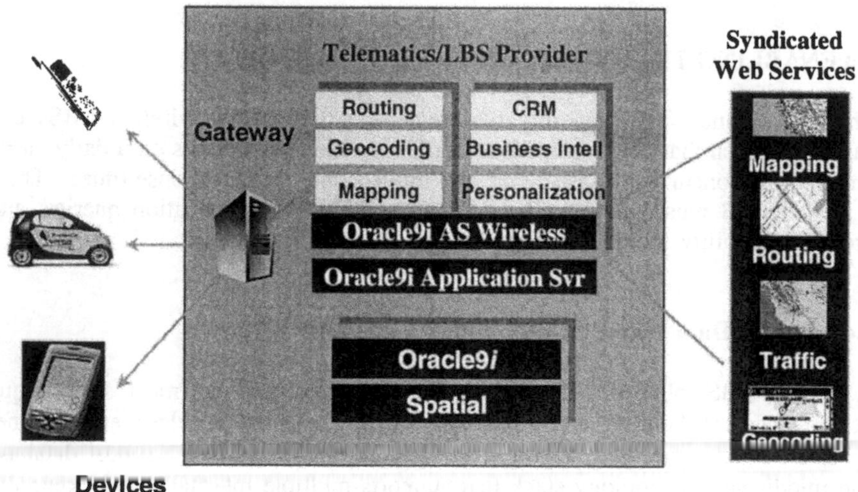

Figure 6.2 A LBS Architecture Platform

Routing engines, mapping, and geocoding can be run in the middle tier or the database. Results of LBS queries could be delivered to mobile devices in the form of raster, vector, text, or text to voice objects. Locally hosted business directory and customer information pages can be managed in a local database which allows the results of powerful server side location queries (e.g., within distance, nearest neighbor, route buffering) to be fused with other types of map content or externally syndicated web services. By intelligently leveraging database and mid-tier processes, developers can increase performance, optimize processing, and minimize the amount of data transmitted between the server and application tier. On the client side, web browsers, Personal Digital Assistants (PDA), and wireless handsets can now readily handle the new types of LBS content, such as raster, vector, Scalar Vector Graphics (SVG), Geographic Markup Language (GML), and Macromedia Flash, which also provide enhanced graphic interface, query and analysis capabilities.

6.4.2 Mobile Middleware

The core technology for Internet and mobile solutions is an application server. Application servers provide built-in features like portal software, wireless and voice, Web page caching, powerful business intelligence features, complete integration, and more, pre-integrated in a single product. For example, the Oracle9i Application Server (9iAS) supports all major J2EE, Web services and XML industry standards, and its open and integration-ready architecture ensures that Web applications can integrate with standard IT environments (see Figure 6.3). Application servers provide the critical scalability, reliability and security necessary to keep critical LBS applications up and running (Lopez, 2000).

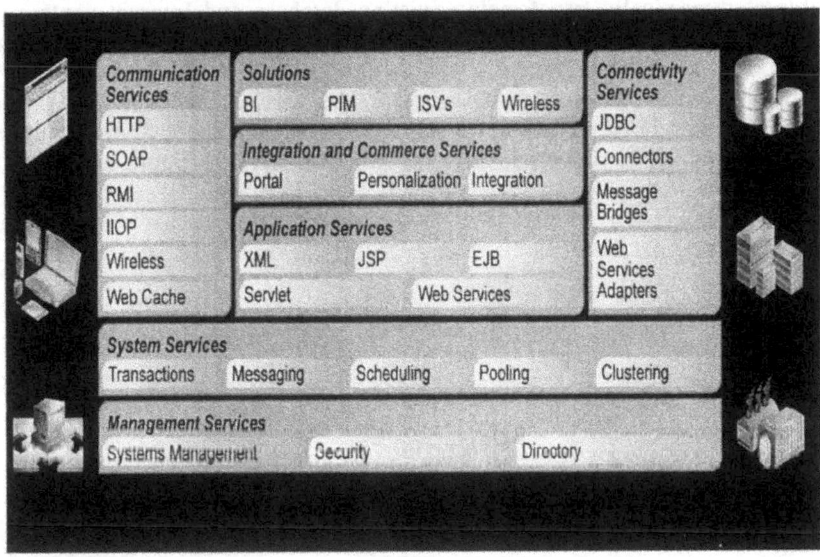

Figure 6.3 Core Application Server Capabilities

Wireless application servers also allow end-users of the portal to customize and tailor their wireless browsing experience to their personal requirements. In addition to sophisticated, personalized content transformation, mobile middleware should also provide a portal interface for the user. An important feature of mobile middleware is personalization, which allows users to define, organize and personalize the services that they want to access from the mobile device. This personalization capability should enable users to easily define the class of LBSs requested and how to trigger them (query, event driven). Other key features of mobile middleware include:

- Separates application logic from service access by always calling Java or XML interfaces;

- Supports well-defined XML interfaces for transfer of yellow pages, geocoding, mapping, driving; directions, and real-time traffic information; with ability to define XML services;
- Repurposes available online services in a variety of wireless and e-Business applications;
- Makes development of value-added applications easier and faster by brokering services from variety of online and wireless service providers; and
- Enables robust and dynamic improvement of services without affecting existing applications.

Mobile operators, content providers, or wireless Internet Service Providers (ISPs) create custom portal sites that utilize all kinds of content, from existing web pages to custom Java applications to all new XML-based applications. Wireless-enabled application servers from vendors like BEA Systems, IBM and Oracle are capable of dynamically transforming existing database and Internet content to a generic XML format, and then generating any device specific output desired. Applications can use any content available on the Web, in the database, or a file system. In the case of Oracle9*i*AS Wireless, the wireless application server can either accept data in MobileXML (Oracle's device-independent XML) or use one of its many adapters to convert content into MobileXML (see Figure 6.4).

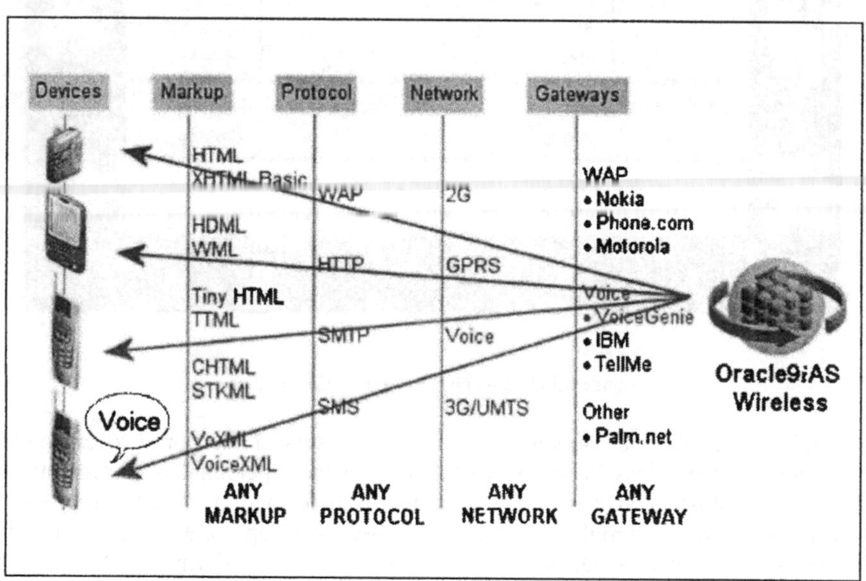

Figure 6.4 Wireless Application Server Capabilities

Opportunities for reusing code and content are considerable, and the advantages for development time and cost are apparent. Oracle9*i*AS Wireless Transformers then convert the MobileXML into the markup language required by each mobile device (WML, TinyHTML, c-HTML, VoiceXML, etc.). Obviously, devices vary in their ability to display certain content in a reasonable manner, or even to store it in memory. However, MobileXML is flexible enough to enable different input and output options depending on a specific device's capability. The application server exploits the maximum hardware capability of the device to present information. Most importantly, applications that work on today's devices will continue to work without limitation with tomorrow's more advanced devices and markup languages. This approach makes the creation of services very simple. Once a web service is created, it is immediately available on all mobile devices.

In short, the application server tier provides a simple yet flexible framework that enables wireless and Internet services to build value-added services. This platform enables rapid deployment of new services, ultimately increasing the quality and reducing the development cycle of Internet applications.

Real-time, transaction-based LBSs have specific feature and performance requirements listed in Table 6.1.

Table 6.1 Feature and Performance Requirements for LBSs

Feature Requirements	Performance Requirements
Address verification and matching Map rendering Yellow page directory query Driving directions Personalization by location Proximity analysis Standardized location interfaces Personal/in-car navigation Voice (VoXML) capability XML integration Web Services Directories Multi-lingual Lightweight Directory Access Protocol (LDAP) support	Scalable architecture Gigabytes to terabytes of data Multiple CPU processing DBMS table partitioning Distributed computing Native spatial data management Online services interoperability Millisecond location query Million + daily queries 25,000+ user sessions per hour Portal caching.

6.4.3 Open Interface Specifications

Location-awareness for mobile applications is incomplete without specialized services, such as geocoding, reverse geocoding, driving directions (routing), yellow pages, white pages, maps, weather forecasts, traffic reports, and demographic information. The OpenGIS Consortium (OGC) has defined a suite of Open Location Services (OpenLS) interface specifications that enables developers to incorporate remotely hosted/syndicated LBSs using HTTP protocols (OpenGIS, 2002; Bishr, 2002). This enables developers of LBSs to easily ingest different sources of location services providers worldwide using a single, consistent XML or SOAP interface. For companies that do not have the resources or data to locally host their LBSs, OpenLS interfaces enable them to syndicate LBS data streams from third-party providers. This gives the developer the choice of selecting those services to be hosted locally and those that are accessed remotely. It also allows them to register redundant services in case one fails.

Underlying services supported through these interfaces include:

- **Geocoding**: Geocoding determines the longitude and latitude coordinates of an address and is the most fundamental service, because it is used directly or indirectly by the other LBSs.
- **Reverse Geocoding**: This feature returns address information associated with a given longitude/latitude. This information could include postal code, city, and street intersections.
- **Routing**: Commonly known as *driving directions*, provides turn-by-turn driving instructions based on an address of origin and destination. Routing engines might also provide maps of each turn and of the complete route or provide a solution to more complex problems such as "traveling salesman."
- **Mapping**: Mapping enables users with capable devices to visualize rendered map data that is generated from locally hosted map server or an externally syndicated data.
- **Find Nearest**: Given an address or location, this service is able to return nearby geographic features, such as restaurants, health centers, or gas stations, ranked by distance.
- **Real-time Traffic**: Provides traffic reports from Traffic Data Sources. LBS developers can incorporate this traffic on real-time basis or cache it in their server for integration with routing instructions.
- **Directory Services**: A directory service helps identify one or more businesses falling within a given geographical region and consisting of a business name or a category. Directory services are generally hosted locally and use spatial database technology to perform large numbers of spatial queries (e.g., find nearest, within distance)

This list of LBS interfaces is by no means comprehensive. These core interfaces do, however, serve as building blocks for the delivery of more complex services that fuse location with other types of static and real-time content.

6.4.4 Network-Based Service Environment

The deployment of wireless- (and Web-) based LBSs relies on a common standards-based network infrastructure. The telecom market is currently experiencing a transition in the delivery of services from proprietary and network closed implementations to an open, IP-based service environment. An operator's IP-based infrastructure is becoming the strategic service environment where new classes of data and LBS services are hosted, integrated and delivered as new services.

The IP-based service environment, for example a positioning application, will inter-work with the telecom network and gateways that provide access to information that resides in the network. The positioning service will be an important component in this open, IP-based service environment. The bundling of positioning information with LBSs, personalization, security, and messaging will be key for any operator when offering service packages to the subscribers.

The LBS response/request flow is generally carried out within a wireless carrier network. Components include: the mobile phone, the wireless network, the positioning server, various gateway servers, the geospatial server, and finally the LBS application (see Figure 6.5). While this flow is by no means complete, it does highlight the important components of most LBS implementations. The mobile phone provides a keypad for query and either a numeric or graphical interface for display. The wireless network securely handles the voice or text communication. The mobile positioning server, usually embedded in the wireless carrier's infrastructure, calculates the position of the device using one or more positioning approaches. The gateways handle the smooth transformation of content and protocols between the wireless and IP-based networks. The geospatial server is a database managing (i.e., storage, query, index) of spatial and attribute content (maps, road networks, directory information). Finally, the LBS servers (routing, geocoding, mapping) are embedded into an application server infrastructure.

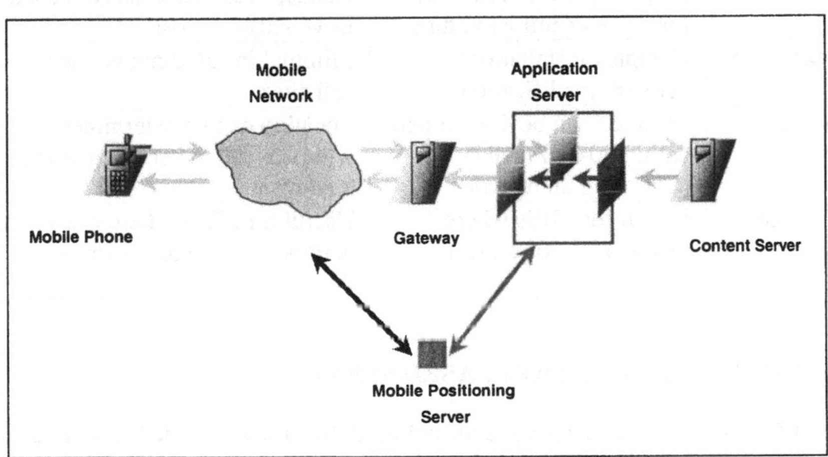

Figure 6.5 Positioning Request Flow

6.4.5 Positioning Equipment

The various wireless location measurement technologies fall into two broad categories: network-based and handset-based solutions (Campbell, 2001). Readers interested more on these approaches are referred to Chapter 3.

Network-based solutions rely on base station radios to triangulate the position of a roaming mobile device, either with received radio signals or transmitted synchronization pulses. The principal advantage of such an approach is that it enables every user to access LBS without the need to upgrade the handsets.

Handset-based solutions or terminal-based solutions, on the other hand, are systems that incorporate the measuring and processing of the location information within the handset. GPS is the principal technology and has been further enhanced by the development of Assisted-GPS (A-GPS), which allows faster and more accurate service (Moeglein, 2001). Handset-based solutions can be complementary to base station triangulation methods, because they address areas where triangulation methods fall short. For example, stand-alone GPS may perform very well in suburban and rural areas, where the view of the sky is unobstructed. However, GPS signals are easily obstructed by structures in urban areas. GPS is expected to perform better in dense urban areas by combining observations from GPS satellites and the network infrastructure. The capabilities of both network and handset-based positioning are compared in Table 6.2. Implementation experience is suggesting that the major carriers are demonstrating an initial preference for handset solutions that result in customers upgrading their terminals.

Table 6.2 Comparison of Alternative Positioning Determining Options (Source: www.snaptrack.com)

	Network Positioning	Handset Positioning
Cost	Large capital investments for network infrastructure	Handset-based and server-based; Low marginal cost
Installation	Complex installation; network modification	Minimal installation; No new cell sites
Reception	Location can be determined within 50-200 meters depending on network	Location can be determined within 5-100 meters depending on device
Capability	Useful for LBSs where precision is not critical	Useful for LBSs where precision is important: dispatch, driving directions, billing

6.5 MARKET FOR LOCATION-BASED SERVICES

The LBS market is developing and being defined and fielded under trying economic conditions. The market is also directly dependant on a wireless industry that is suffering under the burden of 3G spectrum acquisition costs. Because of this, initial predictions for market growth and schedules for LBSs rollout and corresponding market penetration have not been consistent with initial forecasts. In fact, analysts' estimates today are much more conservative than they were in mid-2000, both in terms of the rate of growth and in terms of the services that will be available (BWCS, 2002). Today's estimates are far more pragmatic reflecting the realism that comes from a reduced capital base. Moreover, the years 2000-2002 witnessed the consolidation of many players in the LBS industry. The burst of the "Dot-Com" bubble eliminated many of the smaller, non-viable technology players and resulted in mergers and acquisitions of many others. Acquisitions of MapQuest by AOL, SnapTrack by Qualcomm, and Signalsoft by OpenWave, are examples of the industry restructuring currently underway.

According to the Strategis Group (2001), a leading telecommunications research firm, the market for wireless LBSs is expected to generate $4 billion a year in annual service revenue for carriers. Many vendors (Nokia, Ericsson, and Motorola) are already in the process of developing such services for carriers like Verizon, Vodaphone, Sprint, and AT&T. Despite the slowdown in the wireless industry, analysts are still predicting rapid take-up of this technology by 2005 and will remain strong through 2010. This industry can expect a major boost in the near future as service providers undertake significant expenditures for hardware, software, content, and location servers and enhanced networks. Soon, the number of web-enabled mobile handsets will surpass the number of desktops. To consumers, this will bring widespread availability of powerful location and mapping capability into the hands of every user with a Web browser or web-enabled wireless handset.

How did the LBSs market emerge? Surprisingly, the United States Federal Government regulation has been an important factor in the early development of LBSs. The FCC introduced regulations requiring mobile service operators to provide comprehensive location information for E-911 situations. By October 2001, wireless service providers in the United States were required to provide the location of mobile handset users within approximately 125 meters, for 67 percent of the time. However, most carriers requested waivers and extensions of time. Nevertheless, some progress has been made, and the business opportunities for LBSs are still there, even if they have been postponed.

In September 2001, Sprint PCS became the first carrier to sell a GPS-enabled mobile phone, known as the SPH-N300; Samsung developed the phone and Qualcomm developed the GPS chip set. Since then, Samsung has begun producing similar phones for Verizon. Both Sprint and Verizon are in the process of equipping their nation-wide networks to support this new positioning capability.

Much of the same infrastructure needed to support mandated E-911 services can also be used for chargeable LBSs like car navigation, traffic information, travel information, and yellow page directory access. Moreover, as wireless and wired Internet infrastructures converge, service providers will make LBSs a mainstay of their offerings.

6.5.1 Location-Based Service Market Players

The LBS marketplace is both complex and still evolving. There are a variety of players vying for market leadership in various parts of the LBS value chain. For simplicity, the following categorizations of key LBS players are used:

- **Infrastructure Software Providers**: BEA, IBM, Lucent, Microsoft, Motorola, Nokia, Oracle and Sun provide much of the core software infrastructure (e.g., databases, application servers, enterprise applications, and positioning servers) necessary for wireless and Web service delivery.
- **Network and Handset Positioning Vendors**: Alcatel, Cambridge Positioning Systems, Ericsson, Lucent, Motorola, Nokia, Qualcomm, Siemens provide positioning technology as part of the network infrastructure. In addition, handset providers are beginning to embed A-GPS positioning technology into a new generation of handsets.
- **Specialty Tools and LBS Platform Vendors**: Autodesk, Environmental Systems Research Institute (ESRI), Ionic, Intelliwhere, MapInfo, Telcontar,

and others, provide specialized mapping, geocoding, and routing technologies for LBS deployments.

- **LBS Services**: These specialized LBS providers offer re-branded services like real-time traffic (SmartTraveler, TrafficMaster), mapping, driving directions and yellow pages (MapQuest, Vicinity, Webraska, Yahoo), geocoding and gazetteers (Whereonearth), and satellite imagery services (GlobeXplorer).
- **Content Providers**: Geographic Data Technologies, Navigation Technologies, and Tele Atlas. Meanwhile, major business listing companies providers like Acxiom, Claritas, InfoUSA, and Polk are important sources of demographic and business directories. Finally, local, state and federal government agencies are an important source of both raw and value-added content mapping and directories information.
- **Wireless and Web Portals**: Portals come in various flavors: (1) Wireless carriers such as British Telecom, Vodaphone, Verizon, AT&T, France Telecom, Telefonica, and Deutsche Telekom; (2) Branded Internet portals like AOL and Yahoo; (3) Wireless Application Service Providers (ASPs), like InfoSpace; and (4) Telematics portals, like the General Motors' OnStar service.

As the LBS market achieves its potential over the next five years, we can expect market consolidation and the emergence of clear market leaders in each of these domains.

6.6 IMPORTANCE OF ARCHITECTURE AND STANDARDS

To fully realize the benefits of LBSs, the design of system architecture becomes paramount. As with the deployment of mission critical data warehouses, CRM, and e-commerce solutions, wireless LBSs require a solid architecture that is extensible, secure and standards based. This generally means a consolidation of database, application servers, and tools for reasons of performance, scalability, and cost-effectiveness. LBS will also take advantage of performance-enhancing IT features like caching, parallelism, partitioning, and high availability, all of which are delivered by leading database and application vendors. For flexibility and robustness, these systems are now leveraging the power and flexibility of Java across all tiers of solutions architecture (Lopez, 2001).

The delivery of location applications introduces challenging technical issues as the system architecture (client, middleware, data server) grows in complexity and as the number of system dependencies increases. Dependencies exist between software and operating system compatibility, system and network interoperability, and network capability. As LBSs incorporate locally hosted and externally syndicated content services, dependencies spill into the inter-organizational realm. These system dependencies can make or break a system. And although they cannot be completely avoided, if these dependencies are not managed carefully, they can decrease reliability and performance, as well as increase the lifetime cost of a system.

An important way to minimize the problem of co-dependencies across technology layers and systems is to use standards-based technology. Open standards permit interoperability and minimize system degradation caused by versioning conflicts. Standards also minimize vendor lock-in dependencies by moving away from proprietary technology, application programming interfaces (APIs), and unique data formats. Industry consortia and standards bodies like the World Wide Web Consortium (W3C), International Organization for Standardization (ISO), Open Mobile Alliance (OMA), and OpenGIS are playing a significant role in addressing LBS interoperability. The extraordinary success of the Web is, after all, a direct result of the rapid adoption of W3C protocols (HTTP) and markup languages (HTML, XML). The richness of available content and services provide an opportunity to link proximity searches to a whole new category of services. This is why open-standards-based technology and adherence to standards is critical – it enables wireless and Internet portals to quickly extend their services and to interoperate with other systems and on-line services.

A number of standards efforts are currently underway that will steer the evolution of LBSs. These include the OpenGIS Consortium, the OMA, the Internet Engineering Task Force (IETF), and the WAP forum, among others. Of particular note is the OpenGIS, which recently launched an industry-led interoperability program for LBSs. The OpenGIS is working with the OMA, IETF and WAP forum to demonstrate the delivery of end-to-end LBSs – from handset, through the wireless carrier infrastructure, hosted LBS applications, links with Internet content providers, and back to the handset. While the LIF and IETF are focusing their interoperability work on positioning, the OpenGIS is advancing standards for key LBS interfaces like geocoding, web mapping, driving directions, yellow page search, and real-time traffic acquisition.

GML is a related XML encoding specification that is likely to impact the evolution of LBSs and GISs in the coming years. The significant trend here is a move away from proprietary GIS and mapping formats and APIs that have prevented the diffusion of spatial technology in the past.

6.6.1 Java and Location-Based Services

Recently, there has been an unprecedented level of acceptance of Java as an emerging standard for the deployment of LBSs (Niedzwiadek, 2002). Developers and end users alike recognize the simplicity and power of Java applets, servlets, and beans for the delivery of LBSs. Java for spatial and location applications, or Java location services as it is now being referred to, delivers some unique capabilities such as:

- **Power**: Java is a modern, object-oriented language complete with single-state inheritance and multiple interfaces. This combination has been found to be both powerful and efficient, and is a good complement to standard programming models.
- **Simplicity**: Java retains most of the power of C++, but with far less complexity. It is designed for automatic storage management, which is an enormous simplification since the programmer no longer needs to deal with pointer arithmetic. Java is essentially C++ done right.
- **Familiarity**: Java borrows heavily from the syntax and semantics of the ubiquitous C language. While a relatively new language, Java is familiar to a large and rapidly expanding population of developers because of its widespread acceptance and its derivation from C.

- **Efficiency**: Java's design enables highly efficient interpretation and compilation. Its virtual machine-based organization defines a highly compact set of byte codes that can be efficiently transported in the Internet/Intranet environment.
- **Portability**: To maximize its intrinsic portability, Java has been carefully specified with a standard, platform-neutral format at both the source and binary levels. Furthermore, Java defines both a language and a set of standard class libraries (packages) to ensure that real-world applications may be easily constructed to run on any Java virtual machine in any environment.
- **Safety**: While powerful, Java is fundamentally safer than low-level languages such as C or C++. Because Java has uniform reference semantics and automatic storage management, it completely avoids various storage management and pointer errors. All Java operations are type-safe, and the language provides a sophisticated lexical exception mechanism by which developers can produce robust code.

6.7 EXAMPLE LOCATION-BASED SERVICES: J-PHONE J-NAVI (JAPAN)

In May 2000, J-Phone, Japan's second largest wireless telecom, launched their J-Navi service. J-Navi delivers all the standard types of LBS that one expects from a wireless carrier (geocoding, driving directions, yellow page information). In addition, J-Navi rich graphic handsets are also capable of displaying the results of location queries using color maps resulting in the world's first operational graphical map delivery to mobile phones.

J-Phone is Japans #3 wireless service provider. The J-Phone J-Navi location applications were written in Java and run on a spatial database. Java Server scripts running in the database and the mid-tier provides lightweight and scalable geocoding, map rendering, and location capability (Figure 6.6). This particular deployment runs nearly all of its LBS functions directly from a spatial database and is able to achieve scalability requirements of 30,000 user sessions per hour. The result is the ability to deliver over 1 million color vector and raster maps per day to a new class of General Packet Radio Service (GPRS) and Universal Mobile Telecommunications System (UMTS) enabled multimedia handsets. The average query processing is less than 200ms and average download time is two seconds.

The J-Phone deployment, in combination with partner technologies and services leveraged performance-enhancing features like caching, parallelism, partitioning, and high availability. This is particularly relevant to wireless location-based applications where new application components may need to be created and enhanced regularly to differentiate service offerings. The two other leading Japanese carriers NTT DoCoMo (#1) and KDDI (#2) also use Oracle database technology and application server infrastructure for the deployment of their advanced LBS content services.

Figure 6.6 J-Phone Handset and J-Navi Map (Images Courtesy of J-Phone)

6.8 CONCLUSIONS

The recent convergence of network computing and wireless telecommunications with spatial technologies is giving rise to a new class of location-based applications and services. Internet and wireless service providers are beginning to deliver web mapping, street routing, and electronic mobile yellow services to both web and wireless handsets. The broader Internet and wireless industry has now realized the importance of incorporating location information into existing software solutions. LBSs provide a means to deploy horizontal services that can easily be embedded in existing tools and applications or as a stand-alone point-based solution. In delivering LBSs, IT vendors are developing germane software and hardware products that enable the delivery of location-based voice and data services to wireless network operators. These services unlock the central element of mobile telephone networks, the location of their users. The location of the wireless user is central to the delivery of a whole new class of services by wireless operators. In short, LBSs will become a key enabler of mobile commerce, which is the extension of Internet-based electronic commerce, to mobile phones.

LBSs imply a broad range of concepts, technologies and potential applications and services. In fact, the LBS value chain includes the development and delivery of wireless handsets, wireless network positioning infrastructure, positioning servers, application servers, database technology, LBS specific tools (e.g., mapping, routing, and geocoding engines). LBSs also include content and mechanisms to aggregate and publish that content in the form of personalized information services to users. These mission-critical LBSs all require a solid architecture that is extensible, secure and standards-based. This has led to the centralization of services for reasons of performance, scalability, and cost-effectiveness. The successful adoption of LBSs will rely on flexible and secure approaches that can be readily scaled-up without being locked into a particular vendors solution. The technology infrastructure now exists for application developers to build a "best of class" technology platform that is open, scalable, secure, manageable, and standards-based.

REFERENCES

Bishr, Y., 2002, *OGC's Open Location Services Initiative & Location Interoperability Forum: Putting them Together*, Image Matters LLC, Leesburg, VA. Online. Available HTTP: <http://www.jlocationservices.com/EducationalResources/EducationalMain.htm> (accessed 7 January, 2003).

BWCS, 2002, *Mobile Location Based Services: Where's the Revenue?*", Herefordshire, UK. Online. Available HTTP: <http://www.bwcs.com> (accessed 7 January, 2003).

FCC, 1996, *FCC Docket No. 96-264: Revision of the Commission's Rules to Ensure Compatibility with Enhanced 911 Emergency Calling Systems*. Federal Communications Commission, Washington DC. Online. Available HTTP: <http://www.fcc.gov/Bureaus/Wireless/Orders/1996/fcc96264.txt> (accessed 7 January, 2003).

IDC, 2001, *Spatial Information Management: 2001–2005 Trends and Analysis*, International Data Corp., Framingham, MA.

Lopez, X., 2000, *Deploying Location-Based Services with Oracle9i Application Server*, Oracle Corporation, Redwood Shores, CA. Online. Available HTTP: <http://technet.oracle.com/products/spatial> (accessed 7 January, 2003).

Lopez, X., 2001, Deploying Location-Based Services with Oracle9i AS Wireless Edition and Oracle Spatial. *Oracle Magazine*, **15**, pp. 32-36.

Niedzwiadek, H., 2002, *Tutorial: Java Location Services*. Image Matters LLC, Leesburg, VA. Online. Available HTTP: <http://www.jlocationservices.com/EducationalResources/EducationalMain.htm> (accessed 7 January, 2003).

OpenGIS, 2002, *Open Location Services (OpenLS) Interface Specification*, Wayland, MA. Online. Available HTTP: <http://www.openls.org> (accessed 7 January, 2003).

Moeglein, M., 2001, *An Introduction to SnapTrack Server-Aided GPS Technology*, SnapTrack, Campbell, CA. Online. Available HTTP: <http://www.snaptrack.com/AtWork/ion.pdf> (accessed 7 January, 2003).

Strategis Group, 2001, *US Telematics Marketplace: 2002*, Strategis Group, Washington D.C.

CHAPTER SEVEN

Wearable Tele-Informatic Systems for Personal Imaging

Steve Mann

7.1 INTRODUCTION

Clynes defined Cybernetic organisms (also known as "cyborgs," "borgs," and somewhat as "posthumans") by way of a synergy between human and machine such that operation of the machine does not require conscious thought or effort on the part of the human (Clynes and Kline, 1960). The theory of Humanistic Intelligence (HI) makes this concept more precise, and focuses on machines of an informatic nature (Mann, 1988a). HI is defined as intelligence that arises from the human being (being) in the feedback loop of a computational process in which the human and computer are inextricably intertwined. This inextricability usually requires the existence of some form of body-borne computer. When a body-borne computer functions in a successful embodiment of HI, the computer uses the human's mind and body as one of its peripherals, just as the human uses the computer as a peripheral. This reciprocal relationship, where each uses the other in its feedback loop, is necessary for a successful implementation of HI. This theory is in sharp contrast to many goals of Artificial Intelligence (AI) where the computer replaces or emulates human intelligence.

Early cyborg communities of the late 1970s and early 1980s were constructed to explore the creation of visual art within a computer-mediated reality. Then with the advent of the World Wide Web in the 1990s, cyborg logs (glogs, short for cyborglogs) became shared spaces. Such logfiles resulted in Wearable Wireless Webcam (a WWW readable cyborg logfile of daily activities), and more recently, various others have started keeping personal daily logfiles with portable devices. Such logfiles were later referred to as Web Logs (weblogs, or "blogs" for short).

The main difference between the recent weblogs and the earlier cyborglogs is that blogs often originate from a desktop computer, wheras glogs can originate while walking around, often without any conscious thought and effort, as stream of (de)consciousness glogging.

Not all glogs need be webcast, e.g. a glogger Personal Safety Device (PSD) is to the individual person as the "black box" flight recorder is to an airplane. Thus a glog may simply be a personal digital diary that functions as, for example, a visual memory prosthetic (Mann, 1996b) (e.g., to help an Alzheimer's patient remember names and faces).

Webcasting of glogs is also quite useful, for example, to assist the visually impaired by seeking "Seeing Eye People" volunteers, and there are numerous other useful applications of glogs in the context of a Telegeoinformatic space.

Creation of glogs often involves the use of portable cameras, and as these devices get easier and easier to use, glogs can grow with little or no effort.

Telegeoinformatic cyborg communities take many of the concepts of the internet beyond the confines of the desktop. Wearable Computer Mediated Reality (EyeTap devices, digital eyeglasses, etc.) also blurs the boundary between cyberspace and the real world. But the most profound effect, is probably that of decentralized personhood made possible through ambiguous and fragmented identity collectives. Telegeoinformatic cyborg communities also capture the ideas of inverse surveillance (sousveillance, from French "sous" meaning "from below" and "veiller," meaning "to watch").

In this context, the body-borne computer is known as an "Architecture of One" (i.e. a "building" made for a single occupant), yet it provides a shell of community around that individual.

The ambiguous identity of the cyborg body has transformed the world from the modernist ideal of universally agreed-upon global objective reality, to the postmodernist era of fragmented indeterminate subjective collective individualism. But its weakness is on its reliance upon centralized wireless infrastructure that suggests it may give way to a post-cyborg (pastmodernist) model of authoritarian, dictated, and centralized control. Thus the future may very well rest upon the development of independent indestructible wireless peer-to-peer networks that have the unstoppable nature promised by the early Internet.

7.2 HUMANISTIC INTELLIGENCE AS A BASIS FOR INTELLIGENT IMAGE PROCESSING

Personal Imaging is an integrated personal technologies, personal communications, and mobile multimedia tele-informatic methodology well suited to adaptive mobile computing. In particular, Personal Imaging devices are visual information devices with an "always ready" usage model, and comprise a device or devices that are typically carried or worn so that they are always with the user (Mann, 1977b).

An important theoretical development in the field of Personal Imaging is that of HI.

HI is a new information processing framework in which the processing apparatus is inextricably intertwined with the natural capabilities of our human body and intelligence. Rather than trying to emulate human intelligence, HI recognizes that the human brain is perhaps the best neural network of its kind, and that there are many new signal processing applications, within the domain of personal imaging, that can make use of this excellent, but often overlooked processor, that we already have attached to our bodies. Devices that embody HI are worn (or carried) continuously during all facets of ordinary day-to-day living, so that, through long-term adaptation, they begin to function as a true extension of the mind and body.

7.3 HUMANISTIC INTELLIGENCE

HI is a new form of "intelligence" whose goal is to not only work in extremely close synergy with the human user, rather than as a separate entity, but more importantly, to arise, in part, because of the very **existence** of the human user (Mann, 1998a). This close synergy is achieved through an intelligent user-interface to signal processing hardware that is both in *close physical proximity* to the user, and is *constant.*

There are two kinds of constancy; one is called *operational constancy*, and the other is called *interactional constancy* (Mann, 1998a). Operational constancy also refers to an always ready-to-run condition, in the sense that although the apparatus may have power saving ("sleep") modes, it is never completely "dead" or shut down or in a temporary inoperable state that would require noticable time from which to be "awakened."

The other kind of constancy, called interactional constancy, refers to a constancy of user-interface. It is the constancy of user-interface that separates systems embodying a *personal imaging* architecture from other *personal* devices such as pocket calculators, Personal Digital Assistants (PDAs), and other *imaging* devices such as handheld video cameras and the like. Note that interactional constancy does not necessarily mean that the user is always interacting with the device, but it does mean that the device is always ready to be interacted with.

For example a handheld calculator left turned on but carried in a shirt pocket lacks interactional constancy, since it is not always ready to be interacted with (e.g., there is a noticeable delay in taking it out of the pocket and getting ready to interact with it). Similarly, a handheld camera that is either left turned on or is designed such that it responds instantly, still lacks interactional constancy because it takes time to bring the viewfinder up to the eye in order to look through it. In order for it to have interactional constancy, it would need to always be held up to the eye, even when not in use. Only if one were to walk around holding the camera viewfinder up to the eye during every waking moment, could we say it has true interactional constancy at all times.

By interactionally constant, what is meant is that the inputs and outputs of the device are always potentially active. Interactionally constant implies operationally constant, but operationally constant does not necessarily imply interactionally constant. The above examples of a pocket calculator, worn in a shirt pocket, and left on all the time, or of a handheld camera even if turned on all the time, were said to lack interactionally constancy because they could not be used in this state (e.g. one still has to pull the calculator out of the pocket or hold the camera viewfinder up to the eye to see the display, enter numbers, or compose a picture). A wristwatch is a borderline case; although it operates constantly in order to continue to keep proper time, and it is wearable, one must make some degree of conscious effort to orient it within one's field of vision in order to interact with it.

HI attempts to both build upon, as well as re-contextualize, concepts in *intelligent signal processing* (Haykin, 1994)(Haykin, 1992), and related concepts, such as AI (Marvin Minsky, 1960). HI also suggests a new goal for signal processing hardware, that is, to directly assist, in a truly personal way, rather than replace or emulate human intelligence.

HI is a means of re-situating computing in a different context, namely the truly personal space of the user. The idea here is to move the tools of computing, computationally mediated visual communication, and imaging technologies, directly onto the body, giving rise to not only a new genre of truly personal image computing, but to some new capabilities and affordances arising from direct physical contact between the computational imaging apparatus and the human mind and body. This gives rise to new forms of computing, such as visual associative memory in which the apparatus might, for example, play previously recorded video back into the wearer's eyeglass mounted display, in the manner of a *visual memory prosthetic* (Mann, 1996b).

7.4 'WEARCOMP' AS A MEANS OF REALIZING HUMANISTIC INTELLIGENCE

WearComp (Mann, 1997b) is now proposed as an apparatus upon which a practical realization of HI can be built, as well as a research tool for new studies in intelligent image processing. It is a device that computationally modifies our visual perception of reality, and thus sets forth a new framework for wearable computer vision systems.

7.4.1 Basic Principles of WearComp as a Tele-Informatic Device

WearComp will now be defined in terms of its three basic modes of operation.

Operational Modes of WearComp

The three operational modes in this new interaction between human and computer, as illustrated in Fig 7.1 are:

- Constancy: The computer runs continuously, and is "always ready" to interact with the user. Unlike a handheld device, laptop computer, or PDA, it does not need to be opened up and turned on prior to use. The signal flow from human to computer, and computer to human, depicted in Fig 7.1(a) runs continuously to provide a constant user-interface.
- Augmentation: Traditional computing paradigms are based on the notion that computing is the primary task. WearComp, however, is based on the notion that computing is NOT the primary task. The assumption of WearComp is that the user will be doing something else at the same time as doing the computing. Thus the computer should serve to augment the intellect, or augment the senses. The signal flow between human and computer, in the augmentational mode of operation, is depicted in Fig 7.1(b).
- Mediation: Unlike handheld devices, laptop computers, and PDAs, WearComp can encapsulate the user (Fig 7.1(c)). It doesn't necessarily need to completely enclose us, but the basic concept of mediation allows for whatever degree of encapsulation might be desired, since it affords us

the possibility of a greater degree of encapsulation than traditional

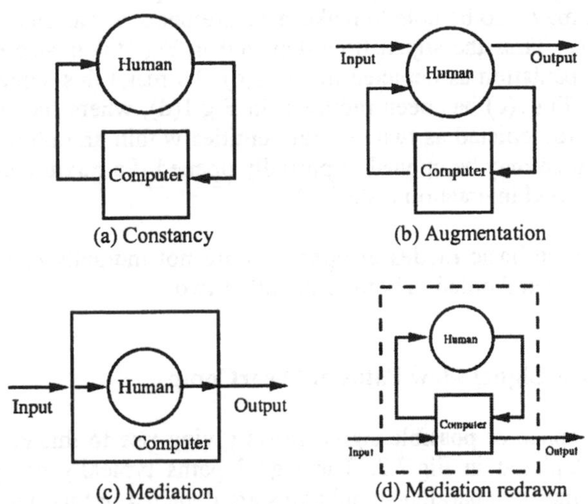

Figure 7.1 The Three Basic Operational Modes of WearComp

portable computers. Moreover, there are two aspects to this encapsulation, one or both of which may be implemented in varying degrees, as desired:

- Solitude: The ability of WearComp to mediate our perception can allow it to function as an information filter, and allow us to block out material we might not wish to experience, whether it be offensive advertising, or simply a desire to replace existing media with different media. In less extreme manifestations, it may simply allow us to alter aspects of our perception of reality in a moderate way rather than completely blocking out certain material.
- Privacy: Mediation allows us to block or modify information leaving our encapsulated space. In the same way that ordinary clothing prevents others from seeing our naked bodies, WearComp may, for example, serve as an intermediary for interacting with untrusted systems, such as third party implementations of digital anonymous cash, or other electronic transactions with untrusted parties. Moreover, the close synergy between the human and computers makes it harder to attack directly, e.g. as one might look over a person's shoulder while they are typing, or hide a video camera in the ceiling above their keyboard.[1]

[1] For the purposes of this chapter, privacy is not so much the absolute blocking or concealment of personal information, but it is the ability to control or modulate this outbound information channel. Thus, for example, one may wish certain people, such as members of one's immediate family, to have

Because of its ability to encapsulate us, e.g. in embodiments of WearComp that are actually articles of clothing in direct contact with our flesh, it may also be able to make measurements of various physiological quantities. Thus the signal flow depicted in Fig 1(a) is also enhanced by the encapsulation as depicted in Fig 1(c). To make this signal flow more explicit, Fig 1(c) has been redrawn, in Fig 1(d), where the computer and human are depicted as two separate entities within an optional protective shell, which may be opened or partially opened if a mixture of augmented and mediated interaction is desired.

Note that these three basic modes of operation are not mutually exclusive in the sense that the first is embodied in both of the other two.

7.4.2 The Six Basic Signal Flow Paths of WearComp

Collectively, the space of possible signal flows giving rise to this entire space of possibilities, is depicted in Fig 7.2. The signal paths typically comprise vector quantities. Thus multiple parallel signal paths are depicted in this figure to remind the reader of this vector nature of the signals.

These signal flow paths each define one of the basic underlying principles of WearComp, and are each described, in what follows, from the human's point of view. Implicit in these six properties is that the computer system is also operationally constant and personal (inextricably intertwined with the user). The six basic properties are:

1. **UNMONOPOLIZING of the user's attention**: it does not necessarily cut one off from the outside world like a virtual reality game or the like does. One can attend to other matters while using the apparatus. It is built with the assumption that computing will be a secondary activity, rather than a primary focus of attention. In fact, ideally, it will provide enhanced sensory capabilities. It may, however, facilitate mediation (augmenting, altering, or deliberately diminishing) these sensory capabilities.
2. **UNRESTRICTIVE to the user**: ambulatory, mobile, roving, one can do other things while using it, e.g. one can type while jogging, running down stairs, etc.
3. **OBSERVABLE by the user**: It can get the user's attention continuously if the user wants it to. The output medium is constantly perceptible by the wearer. It is sufficient that it be almost-always-observable, within reasonable limitations such as the fact that a camera viewfinder or computer screen is not visible during the blinking of the eyes.

---

greater access to personal information than the general public. Such a family-area-network may be implemented with an appropriate access control list and a cryptographic communications protocol.

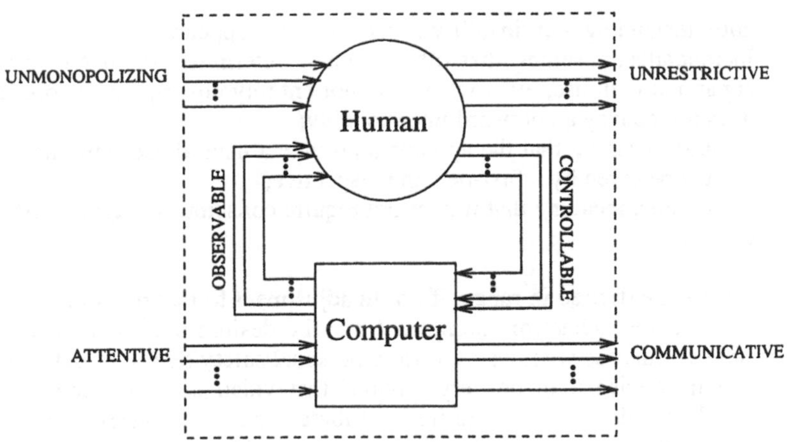

Figure 7.2 The six signal flow paths for the new mode of human-computer interaction provided by WearComp.

4. **CONTROLLABLE by the user** (Responsive): The user can take control of it at any time the user wishes. Even in automated processes the user should be able to manually override the automation to break open the control loop and become part of the loop at any time desired. Examples of this controllability might include a "Halt" button the user can invoke as an application mindlessly opens all 50 documents that were highlighted when the user accidentally pressed "Enter."
5. **ATTENTIVE to the environment**: Environmentally aware, multimodal, multisensory. (As a result, this ultimately gives the user increased situational awareness.)
6. **COMMUNICATIVE to others**: WearComp can be used as a communications medium when the user wishes. Expressive: WearComp allows the wearer to be expressive through the medium, whether as a direct communications medium to others, or as means of assisting the user in the production of expressive or communicative media.

7.5 WHERE ON THE BODY SHOULD A VISUAL TELE-INFORMATIC DEVICE BE PLACED?

Although the final conclusion is that both the tele-informatic sensory and display devices should be placed, effectively, right within the eye itself, various other possibilities are considered and explained first, in terms of various desirable properties, such that the apparatus be:

- Covert, e.g. it does not have an unusual appearance that may otherwise cause objections or ostracization owing to an unusual physical appearance, it being known, for example, that blind or visually challenged persons are very concerned about their physical appearance notwithstanding their inability to see their own appearance;

- Incidentalist, meaning that others cannot determine whether or not the apparatus is in use, even when it is not entirely covert, e.g. its operation does not convey an outward intentionality;
- Natural, meaning that the apparatus provide a natural user interface, such as may be given by a first-person perspective;
- Cybernetic, meaning that it does not require conscious thought or effort to operate.

These attributes are desired in range, if not in adjustment to that point of the range of operational modes. Thus for example, it may be desired that the apparatus be highly visible at times as when using it for a personal safety device to deter crime, wherein one may wish it to be very obvious that video is being recorded and transmitted. Thus ideally these desired attributes are affordances rather than constraints. Therefore, for example, the apparatus is ideally covert but with an additional means of making it obvious *when desired*. Such an additional means may include a display viewable by others, or a blinking red light indicating transmission of video data. Thus such a system would ideally be operable over a wide range of obviousness levels, over a wide range of incidentalism levels, etc.

7.6 TELEPOINTER: WEARABLE HANDS-FREE COMPLETELY SELF CONTAINED VISUAL AUGMENTED REALITY WITHOUT HEADWEAR AND WITHOUT ANY INFRASTRUCTURAL RELIANCE

Telepointer is a wearable hands-free, headwear-free device that allows the wearer to experience a visual collaborative telepresence, with text, graphics, and a shared cursor, displayed directly on real world objects. A person wears the device clipped onto a tie, which sends motion pictures to a video projector at a base (home) where another person can see everything the wearer sees. When the person at the base points a laser pointer at the projected image of the wearer's site, the wearer's device points a laser at the same thing the wearer is looking at. It is completely portable and can be used almost anywhere since it does not rely on infrastructure. It is operated through a Reality User Interfaces (RUI) that allows the person at the base to have **direct** interaction with the real world of the wearer, establishing a kind of computing that is completely free of metaphors, in the sense that a laser at the base controls the wearable laser servo controls in the wearable device.

7.6.1 No Need for Headwear or Eyewear if Only Augmenting

Using a Reality Mediator (to be described in a later section) to do only augmented reality (which is a special case of mediated reality) is overkill. Therefore, if all that is desired is augmented reality (e.g. if no Diminished Reality or Altered/Mediated Reality is needed), the telepointer is proposed as a Direct User Interface.

The wearable portion of the apparatus, denoted WEAR STATION in Fig 7.3, contains a camera, denoted WEAR CAM, which sends pictures perhaps thousands of miles away, to the other portion of the apparatus, denoted BASE STATION, where the motion picture is stabilized by a mathematical algorithm called VideoOrbits (running on a base station computer denoted BASE COMP) and then shown by a projector, denoted PROJ., at the BASE STATION. Rays of light denoted PROJ. LIGHT reach a beamsplitter, denoted B.B.S., in the apparatus of the BASE STATION, and are partially reflected; some projected rays are considered wasted light and denoted PROJ. WASTE. Some of the light from the projector will also pass through beamsplitter B.B.S., and emerge as light rays denoted BASE LIGHT. The projected image thus appears upon a wall or other projection surface denoted as SCREEN. A person at the BASE STATION can point to projected images of any of the SUBJECT MATTER, by simply pointing a laser pointer at the SCREEN where images of the SUBJECT MATTER appear. A camera at the base station, denoted as BASE CAM provides an image of the SCREEN to the base station computer (denoted BASE COMP), by way of the beamsplitter B.B.S. The BASE CAM is usually equipped with a filter, denoted FILT., which is a narrowband bandpass filter having a passband to pass light from the laser pointer being used. Thus the BASE CAM will capture an image primarily of the laser dot on the SCREEN, and especially since a laser pointer is typically quite bright compared to a projector, the image captured by BASE CAM can be very easily made, by an appropriate exposure setting of the BASE CAM, to be black everywhere except for a small point of light from which it can be determined where the laser pointer is pointing.

The BASE CAM transmits a signal back to the WEAR COMP which controls a device called an AREMAC[2], after destabilizing the coordinates (to match the more jerky coordinate system of the WEAR CAM). SUBJECT MATTER within the field of illumination of the AREMAC scatters light from the AREMAC, so that the output of AREMAC is visible to the person wearing the WEAR STATION. A beamsplitter, denoted W.B.S., of the WEAR STATION, diverts some light from SUBJECT MATTER to the wearable camera, WEAR CAM, while allowing SUBJECT MATTER to also be illuminated by the AREMAC.

This shared telepresence facilitates collaboration, which is especially effective when combined with the voice communications capability afforded by the use of a wearable hands-free voice communications link used together with the telepointer apparatus. (Typically the WEAR STATION provides a common data communications link having voice, video, and data communications routed through the WEAR COMP.)

Fig 7.4 illustrates how the telepointer works to use a laser pointer (e.g. in the living room) to control an AREMAC (wearable computer controlled laser in the grocery store). For simplicity, Fig 7.4 corresponds to only the portion of the signal flow path shown in bold lines of Fig 7.3.

[2] An AREMAC is to a projector as a camera is to a scanner. The AREMAC directs light at 3D objects.

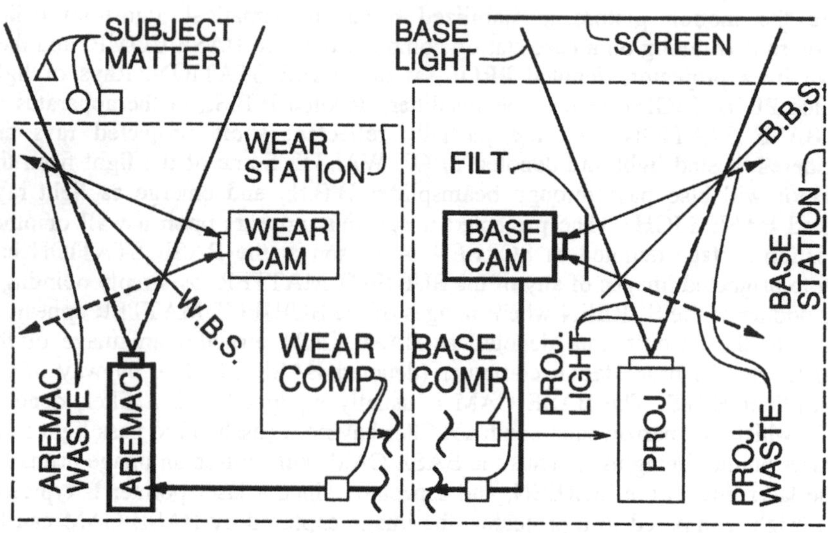

Figure 7.3 Telepointer system for collaborative visual telepresence without the need for eyewear or headwear or infrastructural support: The wearable apparatus is depicted on the left, whereas the remote site is depicted on the right.

SUBJECT MATTER in front of the wearer of the WEAR STATION is transmitted and displayed as PICTURED SUBJECT MATTER on the projection SCREEN. The SCREEN is updated, typically, as a live video image in a graphical browser such as glynx, while the WEAR STATION transmits live video of the SUBJECT MATTER.

One or more persons at the BASE STATION are sitting at a desk, or on a sofa, watching the large projection SCREEN, and pointing at this large projection SCREEN using a laser pointer. The laser pointer makes, upon the SCREEN, a bright red dot, designated in the figure as BASE POINT.

The BASE CAM, denoted in this figure as SCREEN CAMERA, is connected to a vision processor (denoted VIS. PROC.) of the BASE COMP, which simply determines the coordinates of the brightest point in the image seen by the SCREEN CAMERA. The SCREEN CAMERA does not need to be a high quality camera since it will only be used to see where the laser pointer is pointing. A cheap black and white camera will therefore suffice for this purpose.

Selection of the brightest pixel will tell us the coordinates, but a better estimate can be made by using the vision processor to determine the coordinates of a bright red blob, BASE POINT, to sub-pixel accuracy. This helps reduce the resolution needed, so that smaller images can be used, and therefore cheaper processing hardware and a lower resolution camera can be used for the SCREEN CAMERA.

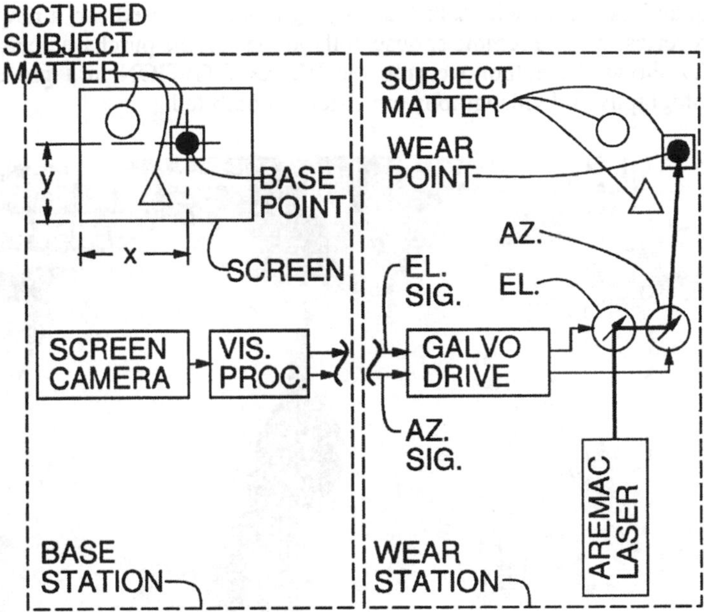

Figure 7.4 Details of the telepointer (TM) AREMAC and its operation: For simplicity, the livingroom or manager's office is depicted on the left, where she can point at the screen with a laser pointer.

These coordinates are sent as signals denoted EL. SIG. and AZ. SIG. and are received at the WEAR STATION and are fed to a galvo drive mechanism (servo) which controls two galvos. Coordinate signal AZ. SIG. drives azimuthal galvo AZ. Coordinate signal EL. SIG. drives elevational galvo EL. These galvos are calibrated by the unit denoted as GALVO DRIVE in the figure. As a result, the AREMAC LASER is directed to form a red dot, denoted WEAR POINT, on the object that the person at the BASE STATION is pointing at from her living room or office.

The AREMAC LASER together with the GALVO DRIVE and galvos EL and AZ comprise the device called an AREMAC which is generally concealed in a brooch (broche, broach) pinned onto a shirt, or exists in the form of a tie clip device for attachment to a necktie, or is built into a necklace. The author generally wears this device on a necktie. The AREMAC and WEAR CAM must be registered, mounted together (e.g. on the same tie clip), and properly calibrated. The AREMAC and WEAR CAM are typically housed in a hemispherical dome where the two are combined by way of beamsplitter W.B.S.

7.6.2 Computer Mediated Collaborative Living (CMCL)

While much has been written about Computer Supported Collaborative Work (CSCW), there is more to life than work, and more to living than pleasing one's employer. The apparatus of the invention can be incorporated into ordinary day-to-day living, and used for such "tasks" as buying a house, a used car, a new sofa, or buying groceries, while a remote spouse collaborates on the purchase decision.

Fig 7.5 shows the author wearing the WEAR STATION in a grocery store where photography and videography are strictly prohibited.

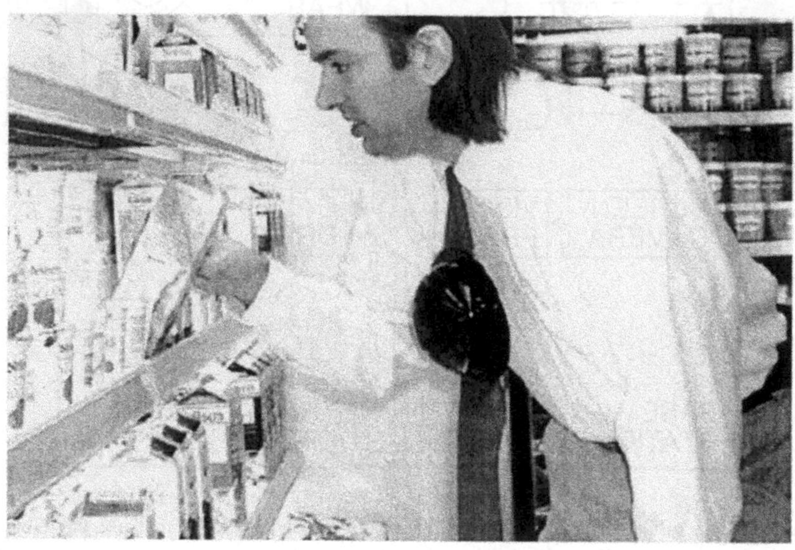

Figure 7.5 Wearable portion of apparatus, as worn by author: The necktie-mounted visual augmented reality system requires no headwear or eyewear. The apparatus is concealed in a smoked plexiglass dome of wine-dark opacity.

Fig 7.6 shows a close-up view of the necktie clip portion of the apparatus. The use of domes as housings is ideal because it matches the decor of nearly any department store, gambling casino, or other place where one might use such an apparatus. Thus the apparatus is a true détournement that re-situates the everyday familiar ceiling dome, in a different context (namely upon the shopper's body rather than the shopkeeper's ceiling), thus expanding our thinking on the otherwise one-sided nature of surveillance. Indeed such an apparatus gives rise to a phenomenon referred to as sousveillance, from "sous" (French for below), and "veiller" (to watch, from vigil). Sousveillance thus refers to an "inverse surveillance" made possible by personal Telegeoinformatics wear (as presented in the author's Keynote Address at the Virtual Reality session of Geographies, Genders and Genres, University of Toronto, 2002).

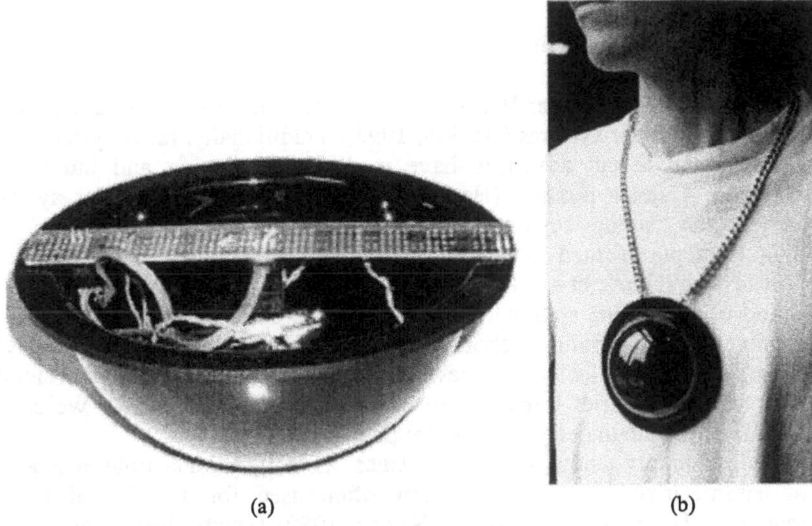

(a) (b)

Figure 7.6 Necktie clip portion: (a) The necktie mounted visual augmented reality system. (b) Necklace version as an alternate solution for those who do not wear neckties.

7.7 PORTABLE PERSONAL PULSE DOPPLER RADAR VISION SYSTEM

Telepointer, the necktie worn dome ("tiedome") of the previous section bears a great similarity to radar, and how radar in general works. In many ways, the telepointer tiedome is quite similar to the radomes used for radar antennas. The telepointer was a front-facing two-way imaging apparatus. We now consider a backward-facing imaging apparatus built into a dome that is worn on the back.

Time-Frequency and q-chirplet based signal processing is applied to data from a small portable battery operated pulse Doppler radar vision system designed and built by the author. The radar system and computer are housed in a miniature radome backpack together with video cameras operating in various spectral bands, to be backward-looking, like an eye in the back of the head. Therefore all the ground clutter is moving away from the radar when the user walks forward, and is easy to ignore, since the radar has separate in-phase and quadrature channels that allow it to distinguish between negative and positive Doppler. A small portable battery powered computer built into the miniature radome allows the entire system to be operated while attached to the user's body, The fundamental hypothesis upon which the system operates is that actions such as an attack or pickpocket by someone sneaking up behind the user, or an automobile on a collision course from behind the user, are governed by accelerational intentionality. Intentionality can change abruptly and gives rise to application of roughly constant force against constant mass. Thus the physical dynamics of most situations lead to piecewise uniform acceleration, for which the Doppler returns are piecewise quadratic chirps. These q-chirps are observable as peaks in the q-chirplet transform.

7.7.1 Radar Vision: Background, Previous Work

Haykin coined the term "Radar Vision" in the context of applying methodology of machine vision to radar systems (Haykin, 1992). Traditionally, radar systems were not coherent, but recent advances have made the designing and building of coherent radar systems possible (Haykin et al., 1991). Coherent radar systems, especially when having separate in-phase and quadrature components (e.g. providing a complex-valued output) are particularly well suited to Doppler radar signal processing (Currie et al., 1990). Time–Frequency analysis makes an implicit assumption of short time stationarity, which, in the context of Doppler radar, is isomorphic to an assumption of short time constant velocity. Thus the underlying assumption is that the velocity is piecewise constant. This assumption is preferable to simply taking a Fourier transform over the entire data record, but we can do better by modeling the underlying physical phenomena.

Instead of simply using sines and cosines, as in traditional Fourier analysis, sets of parameterized functions are now often used for signal analysis and representation. The wavelet transform (Strang, 1989) (Daubechies, 1990) is one such example having parameters of time and scale. The chirplet transform (Mann and Haykin, 1991) (Mihovilovic and Bracewell, 1992) (Baraniuk and Jones, 1993) (Chen, 1995) has recently emerged as a new kind of signal representation. Chirplets include sets of parameterized signals having polynomial phase (e.g. piecewise cubic, piecewise quadratic, etc.) (Mann and Haykin, 1991), sinusoidally varying phase, and projectively varying periodicity. Each kind of chirplet is optimized for a particular problem. For example, warbling chirplets (W - *chirplets*), also known as warblets (Mann and Haykin, 1991) were designed for processing Doppler returns from floating iceberg fragments which bob around in a sinusoidal manner.

Of all the different kinds of chirplets, it will be argued that the q-chirplets (quadratic phase chirplets) are the most well suited to processing of Doppler returns from land-based radar where accelerational intentionality is assumed. Q-chirplets are based on q-chirps (also called "linear FM"), $\exp(2\pi i(a + bt + ct^2)$ with phase a, frequency b, and chirpiness c. The Gaussian *q-chirplet*, $\psi_{t_0, b, c, \sigma} = 1/\sqrt{2\pi\sigma}\exp(2\pi i(a + bt_c + ct_c^2) - 1/2(t_c/\sigma)^2)$ is a common form of q-chirplet (Mann and Haykin, 1991), where $t_c = t - t_0$ is a movable time axis. There are four meaningful parameters, phase a being of lesser interest when looking at the magnitude of:

$$\langle \psi_{t_0, b, c, \sigma} \mid z(t) \rangle \tag{7.1}$$

which is the q-chirplet transform of signal $z(t)$ taken with a Gaussian window. Q-chirplets are also related to the fractional Fourier transform (Ozaktas and Onural, 1994).

7.7.2 Apparatus, Method, and Experiments

Variations of the apparatus to be described were originally designed and built by the author for assisting the blind. However the apparatus has many uses beyond use by the blind or visually challenged. For example, we are all blind to objects and hazards that are behind us, since we only have eyes in the forward looking portion of our heads.

A key assumption is that objects in front of us deserve our undivided attention, whereas objects behind us only require attention at certain times when there is a threat. Thus an important aspect of the apparatus is an intelligent rearview system that alerts us when there is danger lurking behind us, but otherwise does not distract us from what is in front of us. Unlike a rearview mirror on a helmet (or a miniature rearview camera with eyeglass based display), the radar vision system is an intelligent system that provides us with timely information only when needed, so that we do not suffer from information overload.

Rearview Clutter is Negative Doppler

A key inventive step is the use of a rearview radar system, so that ground clutter is moving away from the radar when the user is going forward. This rearview configuration comprises a backpack in which the radome is behind the user, and facing backwards.

This experimental apparatus was designed and built by the author in the mid-1980s, from low power components for portable battery powered operation. A variation of the apparatus, having several sensing instruments, including radar, and camera systems operating in various spectral bands, including infrared, is shown in Fig. 7.7. Note that the museum artifact pictured in this figure was a very crude early embodiment of the system. The author has since designed and built many newer systems that are now so small that they are almost completely invisible.

On the Physical Rationale for the q-Chirplet

The apparatus is meant to detect persons such as stalkers, attackers, assailants, or pickpockets sneaking up behind the user, or to detect hazardous situations, such as arising from drunk drivers or other vehicular traffic.

It is assumed that attackers, assailants, pickpockets, etc., as well as ordinary pedestrian, bicycle, and vehicular traffic are governed by a principle of accelerational intentionality. The principle of accelerational intentionality means that an individual attacker (or a vehicle driven by an individual person) is governed by a fixed degree of acceleration that is changed instantaneously and held roughly constant over a certain time interval. For example, an assailant is capable of a certain degree of exertion defined by the person's degree of fitness which is unlikely to change over the short time period of an attack. The instant the attacker spots a wallet in a victim's back pocket, the attacker may accelerate by applying a

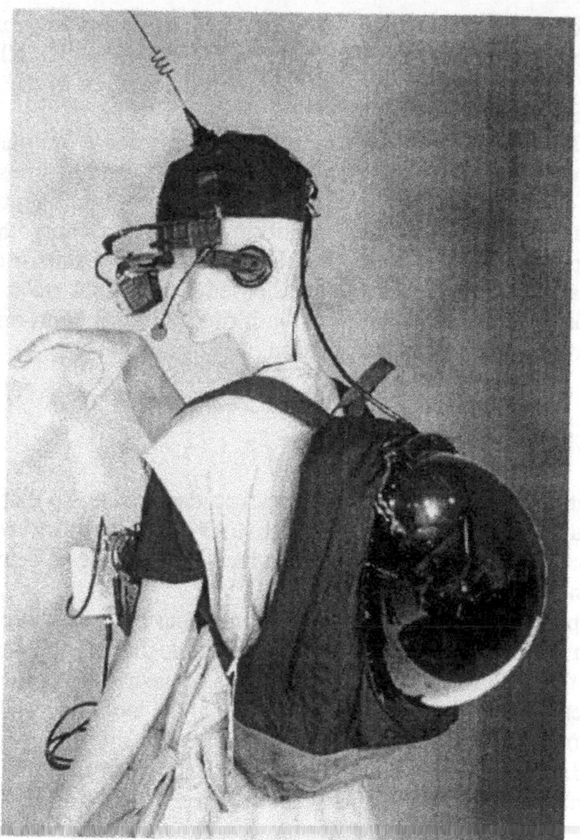

Figure 7.7 Early Personal Safety Device (PSD) with Radar Vision system designed and built by the author, as pictured on exhibit at an American museum (LVAC, 1997). The system contains several sensing instruments, including radar, and camera systems operating in various spectral bands, including infrared.

roughly constant force (defined by his fixed degree of physical fitness) against the constant mass of the attacker's own body. This gives rise to uniform acceleration which shows up as a straight line in the Time Frequency distribution.

Some examples following the principle of accelerational intentionality are illustrated in Fig. 7.8. These radar returns are processed using the Chirplet Transform (Mann, 2001), a mathematical signal representation that compares a signal with a family of variously parameterized windowed chirp functions.

7.8 WHEN BOTH THE CAMERA AND DISPLAY ARE HEADWORD: PERSONAL IMAGING AND MEDIATED REALITY

When both the image acquisition and image display embody a headworn first-person perspective (e.g. computer takes input from a headworn camera and

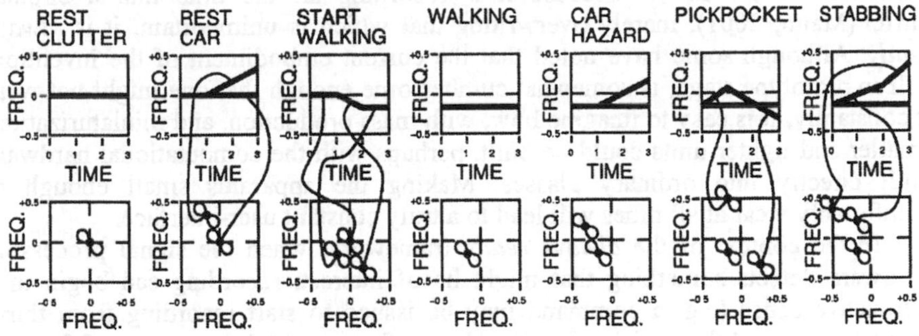

Figure 7.8 Seven examples illustrating the principle of accelerational intentionality, with Time Freq distribution shown at top, and corresponding chirplet transform Freq Freq distribution below.

provides output to a headworn display), a new and useful kind of experience results, beyond merely augmenting the real world with a virtual world.

Images shot from a first person perspective give rise to some very new research directions, most notably, new forms of image processing that take into account the fact that smaller lighter cameras can track and orient themselves much faster, relative to the effort needed to move one's entire body. The mass (and general dynamics) of the camera is becoming small as compared to that of the body, so the ultimate limiting factor becomes the physical constraints of moving the body itself. Accordingly, the laws of projective geometry have greater influence, and in fact, newer methods of image processing emerge.

7.8.1 Some Simple Illustrative Examples

Always-Ready Cyborg Logs ("glogs")

Current day commercial personal electronics devices we often carry are just useful enough for us to tolerate, but not good enough to significantly simplify our lives. For example, when we are on vacation, our camcorder and photographic camera require enough attention that we often either miss the pictures we want, or we become so involved in the process of video or photography that we fail to really experience the immediate present environment (Norman, 1992).

One ultimate goal of the proposed apparatus and methodology is to "learn" what is visually important to the wearer, and function as a fully automatic camera that takes pictures without the need for conscious thought or effort from the wearer. In this way, it might summarize a day's activities, and then automatically generate a gallery exhibition by transmitting desired images to the World Wide Web, or to specific friends and relatives who might be interested in the highlights of one's travel. The proposed apparatus, a miniature eyeglass-based imaging system, does not encumber the wearer with equipment to carry, or with the need to remember to use it, yet because it is recording all the time into a circular buffer (Mann, 2001), merely overwriting that which is unimportant, it is *always ready*. Although some have noted that the current embodiment of the invention, still in prototype stage, is somewhat cumbersome enough that one might not wear it constantly, it is easy to imagine how, with mass production, and miniaturization, smaller and lighter units could be built, perhaps with the computational hardware built directly into ordinary glasses. Making the apparatus small enough to comfortably wear at all times will lead to a truly constant user-interface.

In the context of the *always ready* framework, when the signal processing hardware detects something that might be of interest, recording can begin in a retroactive sense (e.g. a command may be issued to start recording from thirty seconds ago), and the decision can later be confirmed with human input. Of course this apparatus raises some important privacy questions discussed previously, and also addressed elsewhere in the literature (Mann, 1996a) (Mann, 1997a).

The system might use the inputs from the biosensors on the body, as a multidimensional feature vector with which to classify content as important or unimportant. For example, it might automatically record a baby's first steps, as the parent's eyeglasses and clothing-based intelligent signal processor make an inference based on the thrill of the experience. It is often moments like these that we fail to capture on film: by the time we find the camera and load it with film, the moment has passed us by.

Personal Safety Device for Reducing Crime

A simple example of where it would be desirable that the device operate by itself, without conscious thought or effort, is in an extreme situation such as might happen if the wearer were attacked by a robber wielding a shotgun, and demanding cash.

In this kind of situation, it is desirable that the apparatus would function autonomously, without conscious effort from the wearer, even though the wearer might be aware of the signal processing activities of the measuring (sensory) apparatus he or she is wearing.

As a simplified example of how the processing might be done, we know that the wearer's heart rate, averaged over a sufficient time window, would likely increase dramatically[3] with no corresponding increase in footstep rate (in fact footsteps would probably slow at the request of the gunman). The computer would then make an inference from the data, and predict a high visual saliency. (If we simply take heart rate divided by footstep rate, we can get a first-order approximation of the visual saliency index.) A high visual saliency would trigger

[3]Perhaps it may stop, or "skip a beat" at first, but over time, on average, in the time following the event, experience tells us that our hearts beat faster when frightened.

recording from the wearer's camera at maximal frame rate, and also send these images together with appropriate messages to friends and relatives who would look at the images to determine whether it was a false alarm or real danger.

Such a system is, in effect, using the wearer's brain as part of its processing pipeline, because it is the wearer who sees the shotgun, and not the WearComp apparatus (e.g. a much harder problem would have been to build an intelligent machine vision system to process the video from the camera and determine that a crime was being committed). Thus HI (intelligent signal processing arising, in part, because of the very existence of the human user) has solved a problem that would not be possible using machine-only intelligence.

Furthermore, this example introduces the concept of 'collective connected HI,' because the signal processing systems also rely on those friends and relatives to look at the imagery that is wirelessly send from the eyeglass-mounted video camera and make a decision as to whether it is a false alarm or real attack. Thus the concept of HI has become blurred across geographical boundaries, and between more than one human and more than one computer.

7.8.2 Deconfigured Eyes: The Invention of the Reality Mediator

One of the best ways to bring the human into the loop is through the concept of Mediated Reality.

Virtual Reality (Artaud, 1958) (VR) allows us to experience a new visual world, but deprives us of the ability to see the actual world in which we live. Indeed, many VR game spaces are enclosed with railings or the like so that players will not fall down, since they have replaced their reality with a new space, and are therefore blind to the real world.

Augmented reality attempts to bring together the real and virtual. The general spirit and intent of Augmented Reality (AR) is to *add* virtual objects to the real world. A typical AR apparatus might consist of a video display with partially transparent visor, upon which computer-generated information is *overlaid* over the view of the real world.

In this section, Mediated Reality (MR) is proposed. We will also see how MR forms a basis for Personal Imaging, and a possible new genre of documentary video, electronic news gathering, and the like.

MR differs from typical AR in two respects:

1. The general spirit of MR, like typical AR, includes *adding* virtual objects, but also includes the desire to *take away*, *alter*, or more generally to visually 'mediate' real objects. Thus MR affords the apparatus the ability to augment, diminish, or otherwise alter our perception of reality.

2. Typically, an AR apparatus is tethered to a computer workstation which is connected to an AC outlet, or constrains the user to some other specific site (such as a workcell, helicopter cockpit, or the like). What is proposed (and reduced to practice), in this chapter, is a system which facilitates the augmenting, diminishing, or altering of

the visual perception of reality in the context of ordinary day-to-day living.

MR uses a body-worn apparatus where both the *real* and *virtual* objects are placed on an equal footing, in the sense that both are presented together via a synthetic medium (e.g. a video display).

Successful implementations have been realized by *viewing* the real world using a head-mounted display (HMD) fitted with video cameras, body-worn processing, and/or bidirectional wireless communications to one or more remote computers, or supercomputing facilities. This portability enabled various forms of the apparatus to be tested extensively in everyday circumstances, such as while riding the bus, shopping, banking, and various other day-to-day interactions.

The proposed approach shows promise in applications where it is desired to have the ability to deconfigure reality. For example, color may be deliberately diminished or completely removed from the real world at certain times when it is desired to highlight parts of a virtual world with graphic objects having unique colors. The fact that vision may be *completely* deconfigured also suggests utility to the visually handicapped.

7.8.3 Personal Cyborg Logs ("glogs") as a Tool for Photojournalists and Reporters

In the early and mid-1990s the author experimented with personal imaging as a means of creating personal diary (logfiles), personal documentary, etc., and in sharing this personal documentary video on the World Wide Web, in the form of *wearable wireless webcam* (Jones, 1995).

Such Cyborg Logs (cyborglogs, or "glogs" for short) are a predecessor of the presently popular Web Logs (weblogs, or "blogs" for short). Of course a glog does not necessarily need to be a blog, since it can be a personal diary that is never sent to the Web, but often the web is used as the medium for glogging, as in Wearable Wireless Webcam (Mann and Niedzviecki, 2001) (See Fig 7.9.) Additionally, cybor**glogs** can be used for evidence gathering and personal safety (cyborg law, or "glaw" for short).

An example of an early glog from 1995 is illustrated in Figure 7.10. This shows how Computer Supported Collaborative Photojournalism (CSCP) emerged from *wearable wireless webcam*. In actual practice, multiple images are "stitched together" to make a picture good enough for a full-page newspaper-size photograph despite the fact that each of the images has relatively low resolution.

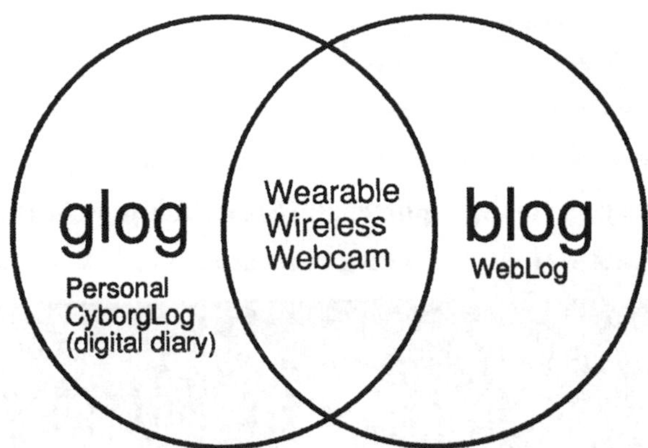

Figure 7.9 Relationship between cyborglogs and weblogs: Wearable Wireless Webcam, invented, designed, built, and worn by the author, ran continuously from 1994 to 1996, and was perhaps the world's first example of a portable weblog using wireless technology.

7.9 PERSONAL IMAGING FOR LOCATION-BASED SERVICES

A visual tele-informatic device opens many new possibilities for location-based interfaces. Visually context-aware computing thus becomes possible. Informatic content may be shared geographically, or associated with a particular name, face, billboard, building, or scene. This is archived using the VideoOrbits freewear, available from http://comparametric.sourceforge.net and maintained by the author together with two of the author's students: Corey Manders and James Fung. These programs allow pictures (comprised of pixels) to be converted to lictures (comprised of lixels) and built into location-based environment maps. One interesting result from this research is the VideoOrbits head tracker.

7.9.1 VideoOrbits Head Tracker

The VideoOrbits algorithm is a mathematical formulation that gives rise to a new method of head-tracking based on the use of a video camera (Mann, 1997b). The VideoOrbits algorithm performs head-tracking, visually, based on a natural environment, and works without the need for object recognition. Instead it is based on algebraic projective geometry, and a featureless means of estimating the change in spatial coordinates arising from movement of the wearer's head, as illustrated in Figure 7.11.

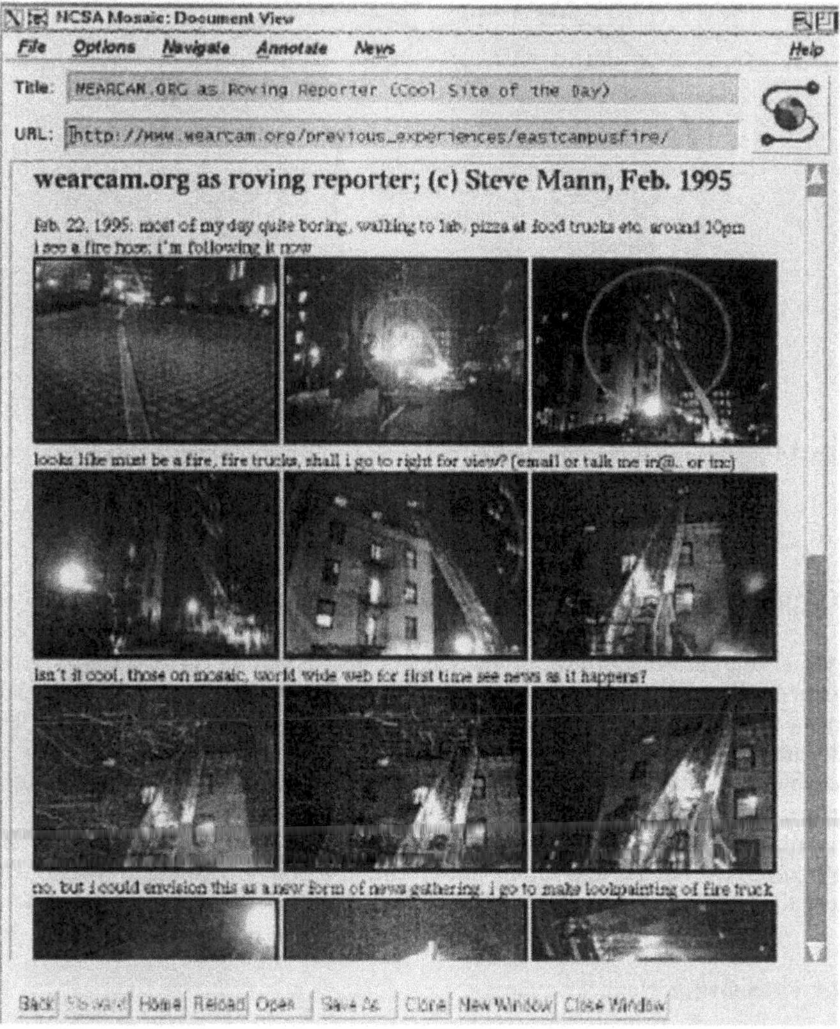

Figure 7.10 CyborgLog ("glog") serendipitously arising from Computer Mediated Collaborative Photojournalism (CMCP). Author encountered an event serendipitously through ordinary everyday activity.

Figure 7.12 shows some frames from a typical location-based image sequence. Figure 7.13 shows the same frames brought into the coordinate system of frame (c), that is, the middle frame was chosen as the *reference frame*.

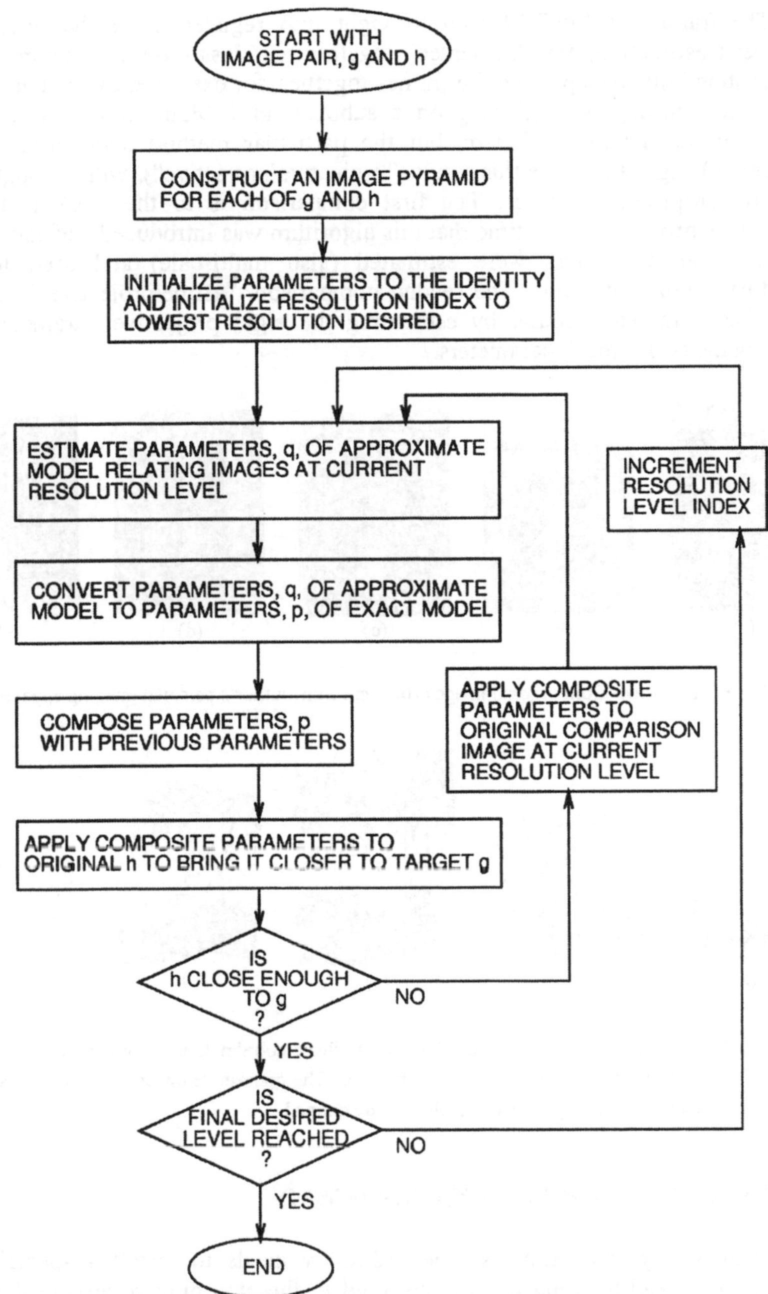

Figure 7.11 The 'VideoOrbits' head-tracking algorithm: The new head-tracking algorithm requires no special devices installed in the environment.

The frames of Fig 7.12 were brought into register using the differential parameter estimation, and "cemented" together seamlessly on a common canvas. "Cementing" involves piecing the frames together, for example, by median, mean, or trimmed mean, or combining on a subpixel grid (Mann and Picard, 1994). (Trimmed mean was used here, but the particular method made little visible difference.) Fig 7.14 shows this result ("projective/projective"), with a comparison to two non-projective cases. The first comparison is to the work that other researchers proposed at the time that this algorithm was introduced, "affine/affine" where affine parameters were estimated (also multiscale) and used for the coordinate transformation. The second comparison, "affine/projective," uses the six affine parameters found by estimating the eight projective parameters and ignoring the two "chirp" parameters.

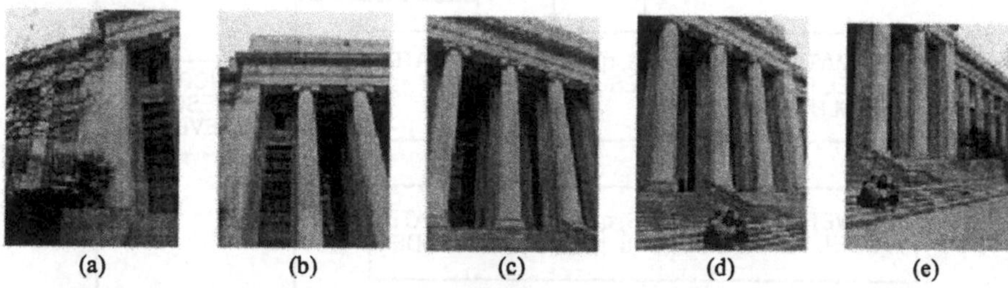

(a) (b) (c) (d) (e)

Figure 7.12 Frames from original image orbit, sent from author's personal imaging apparatus.

(a) (b) (c) (d) (e)

Figure 7.13 Frames from original images video orbit after a coordinate transformation to move them along the orbit to the reference frame (c). The coordinate-transformed images are alike excerpt for the region over which they are defined.

7.10 REALITY WINDOW MANAGER (RWM)

In virtual reality environments, one ordinarily needs to install a special head tracker, with head-tracking devices installed in the environment around the user. However, this installation assumes a cooperative environment.

Figure 7.14 Frames of Fig 7.13 "cemented" together on a single image "canvas," with comparison of affine and projective models.

In certain situations, the environment may be noncooperative. Extreme examples of noncooperative environments might include criminal organizations, gambling casinos, and department stores owned and operated by crooks. In a department store owned by a criminal organizations, or in any other establishment that might in fact be a money laundering front, it is doubtful that the organization will welcome the use of a camera, let alone provide infrastructure to make it work better.

Accordingly, for day-to-day use in ordinary environments like department stores, banks, and out on the street, a visual prosthetic must rely on itself, and not entirely on the environment, to function well.

The VideoOrbits system provides a featureless environment (object) tracker in the scene, so that as one looks around, the apparatus has a sense of where one is looking. This means that one does not need a special separate head-tracking device that might otherwise require a cooperative environment.

7.10.1 A Simple Example of RWM

Window managers often provide the ability to have a large virtual screen, and to move around within that virtual screen. When the VideoOrbits head tracker is used, the mediation zone of a reality mediator can become a viewport into this virtual screen (Fig 7.15).

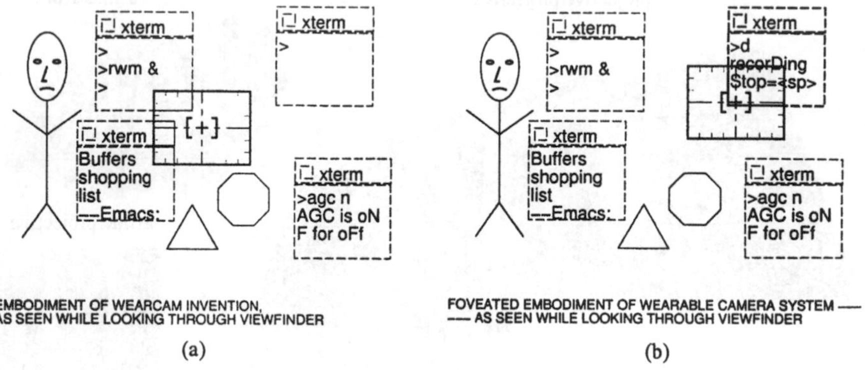

Figure 7.15 Reality Window Manager (RWM): The viewport is defined by an EyeTap or Laser EyeTap device, or the like, and serves as a view into the real world. This viewport is denoted by a reticle, graticule, with crosshairs. It is the mediation zone over which the visual perception of reality can be altered.

7.10.2 The Wearable Face Recognizer as an Example of a Reality User Interface

Once we have an RWM running, it can also be used for a variety of other purposes, such as providing a constantly running automatic background process such as face recognition in which virtual name tags exist in the wearer's own visual world. These virtual name tags appear as illusory planar patches arising through the process of marking a reference frame (Mann and Picard, 1995) with text or simple graphics, where it is noted that by calculating and matching homographies of the plane using VideoOrbits (as described earlier in this chapter), an illusory rigid planar patch appears to hover upon objects in the real-world, giving rise to a form of computer-mediated collaboration (Mann, 1997b).

This collaborative capability suggests an application of HI to the visually challenged, or those with a visual memory disability (Mann, 1996b). In this application, a computer program, or remote expert (be it human or machine) may assist in way finding, or by providing a photographic/videographic memory, such as the ability to never forget a face. (see Fig 7.16.). In Fig 7.16, note that the tracking remains at or near subpixel accuracy even when the subject matter begins to move out of the mediation zone, and in fact even after the subject matter has left the mediation zone. This is owing to the fact that the tracking is featureless and global.

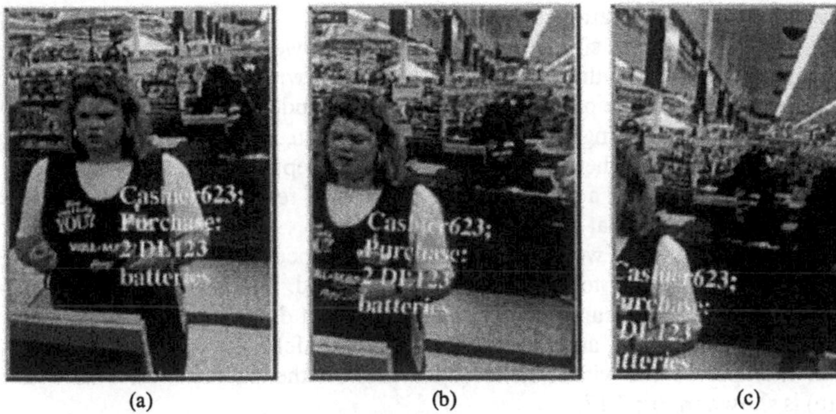

(a) (b) (c)

Figure 7.16 Mediated reality as a photographic/videographic memory prosthesis: RWM provides a window into reality, in which the virtual window tracks the real scene to within subpixel accuracy. (a) Wearable face-recognizer with virtual "name tag" (and grocery list) appears to stay attached to the cashier, (b) even when the cashier is no longer within the field of view of the *mediation zone* (c).

With RWM running on such a reality-mediator, one might leave a grocery list on the refrigerator that would be destined for a particular individual (not everyone wearing the special glasses would see the message — only those wearing the glasses and on the list of intended recipients).

If desired, the message may be sent right away, and "live" dormant on the recipient's WearComp until an image of the desired object "wakes up" the message.

Such "instant messages" include a virtual "Post-It" note (to use Feiner's "Post-It" metaphor), in addition to the name. This note provides additional information, associated with (bound to) the face of the cashier. The additional information pertains to the purchase. These notes may be notes the wearer gives to himself/herself, or they may be notes the wearer sends to other people.

7.11 PERSONAL TELEGEOINFORMATICS: BLOCKING SPAM WITH A PHOTONIC FILTER

This section deals primarily with solitude, defined as the freedom from violation by an inbound channel controlled by remote entities. Solitude, in this context, is distinct from privacy which is defined, for the purposes of this chapter, as the freedom from violation by an outbound channel controlled by remote entities.

While much has been written and proposed in the way of legislation and other societal efforts at protecting privacy and solitude, the purpose of this chapter is to concentrate on a personal approach at the level of the point of contact between the individual and his or her environment. A similar personal approach to privacy issues has already been published in the literature (Mann, 1998b). The main thesis of the argument in (Mann, 1998b) and (Mann, 1998a) is that of personal empowerment through wearable cybernetics and HI.

This section applies a similar philosophical framework to the issue of solitude protection. In particular, the use of MR, together with the VideoOrbits-based RWM, is suggested for the protection of personal solitude.

WearComp, functioning as a Reality Mediator can, in addition to augmenting reality, also diminish or otherwise alter the visual perception of reality.

Why would one want a diminished perception of reality? Why would anyone buy a pair of sunglasses that made one see worse?

An example of why we might want to experience a **diminished** reality, is when driving and trying to concentrate on the road. Sunglasses that not only diminish the glare of the sun's rays, but also filter out distracting billboards could help us see the road better, and therefore drive more safely.

An example of the visual filter (operating on the author's view in Times Square) is shown in Fig 7.17.

Thanks to the visual filter, the spam (unwanted advertising material) gets filtered out of the wearer's view. The advertisements, signs, or billboards are still visible, but they appear as windows, containing alternate material, such as email, or further messages from friends and relatives. This personalized world welcomes the wearer's own choices, not the world that is thrust upon us by advertisers and solitude thieves.

As an example of how MR can prevent theft of solitude, for the spam shown in Fig 7.17, light entering the front of the apparatus, which is depicted in the form of an image sequence in Fig 7.17, is used to form a photoquantigraphic image composite. This photoquantigraphic representation captures the essence of the gaze pattern of the wearer of the apparatus, and thus represents a trace of what the wearer would normally see if it were not for the apparatus.

Within a mathematical representation, the WearComp apparatus deletes or replaces the unwanted material with more acceptable (non-spam) material (Fig 7.18). In this case, the non-spam substitute comprises various windows from the wearer's own information space. From this revised mathematical representation of the wearer's gaze pattern, a new sequence of frames for the new spam-free image sequence is rendered as shown in Fig 7.18. This is what the wearer sees. Note that only the frames of the sequence containing spam are modified.

By wearing the apparatus that contains a visual filter, it is thus possible to filter out offensive advertising and turn billboards into useful cyberspace.

7.12 CONCLUSION

Telegeoinformatic cyborg communities provide a new and useful communications modality. But going beyond the obvious utility of Telegeoinformatic systems and communities, there is a new **decon**textualization of **decon**structionism (decon^2, or "decon squared"). The cyborg age is largely the postmodern age in which we live, yet the centralized nature of current wireless services suggests a vulnerability that might embrace the postcyborg age (pastmodernism). So what's next? The answer may be distributed Telegeoinformatic collectives having no reliance on centralized infrastructure. Such decon ("decon cubed", or "deconism" as it is becoming known in the arts community) may hold the ultimate answer.

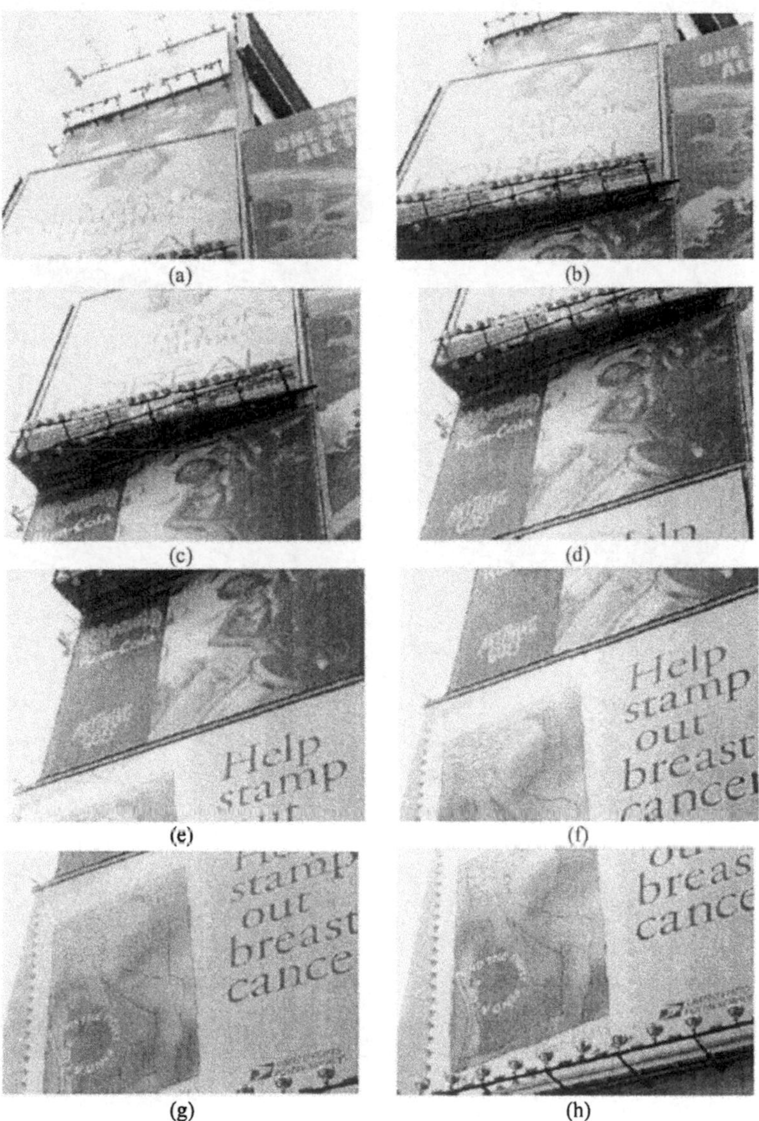

Figure 7.17 Successive video frames of Times Square view (frames 142-149)

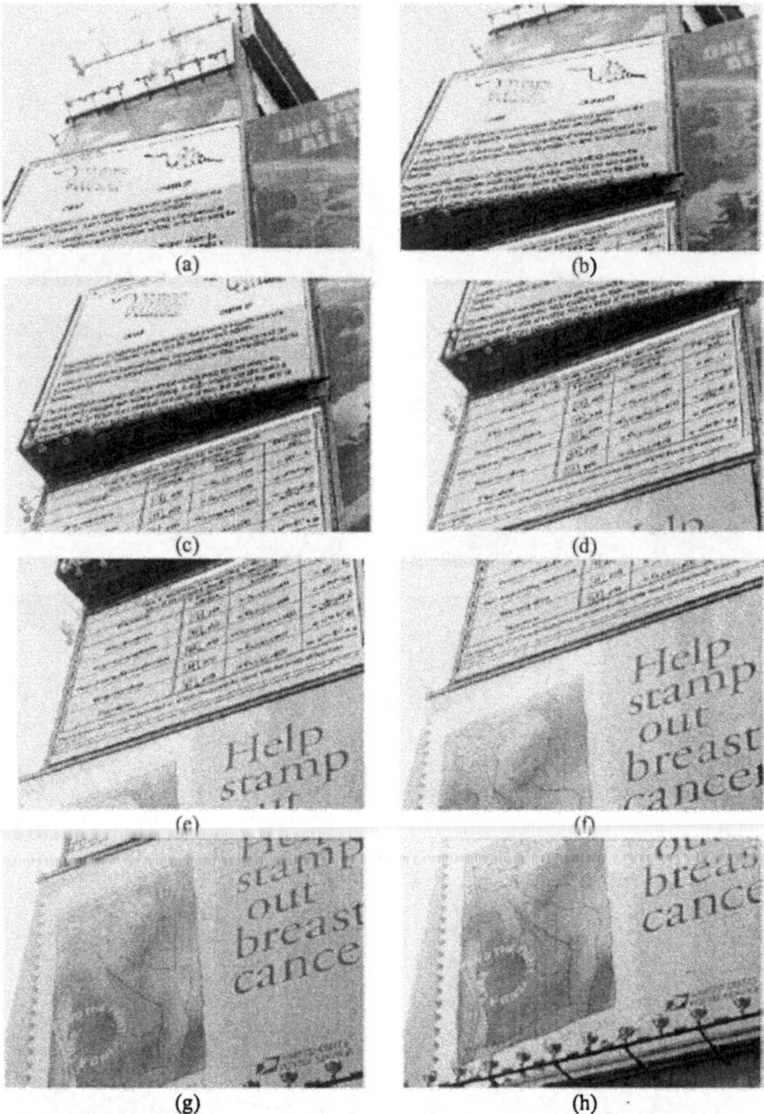

Figure 7.18 Filtered video frames of Times Square view (frames 142-149 filtered): Note the absence of spam within those regions of images where spam was originally present.

A small portable device, owned, operated, and controlled by the individual may well create the metaphysics and epistemology of choice necessary for an existential self determination and mastery over one's own destiny. Likewise, the proliferation of new "Wi Fi" unlicensed unregulated wireless could also bring about the New Telegeoinformatic Age, wherein billboards and other spaces get reclaimed as truly "public space." Eric McLuhan's idea of "electric crowds" may well come to fruition in the context of détournement of what was once an erosion of public space, that might then become the new electric commons.

With Reality Mediators, Telegeoinformatics cyborglogs allow communities to share their own interpretation of visual reality. Such systems may therefore hold the promise that the Web and the Internet were supposed to deliver.

REFERENCES

Artaud, A., 1958, The theater and its double. first published reference to "la realite virtuell" ("virtual reality").

Baraniuk, R. and Jones, D., December 1993, Shear madness: New orthonormal bases and frames using chirp functions. *Trans. Signal Processing, vol. 41*. Special Issue on Wavelets in Signal Processing.

Chen, S. S. and Donoho D. L., M. A. S., Atomic decomposition by basis pursuit. *<http://www- stat.stanfovd.edu/~donoho/Reports/1995/30401.pdf>*. pp. 1-29.

Clynes, M. and Kline, N., September 1960, Cyborgs and space. *Astronautics*, 14(9): pp.26-27,and pp.74-75.

Currie, B. W., Haykin, S., and Krasnor, C.,1990, Time-varying spectra for dual-polarized radar returns from targest in an ocean environment. *In IEEE Conference Proceedings RADAR0*, pp. 365-369, Arlington, Virginia. IEEE Aerospace and Electronics Systems Society.

Daubechies, I., 1990, The wavelet transform, time- frequency localization and signal analysis. *IEEE Trans. on Inf. Theory*, 36(5), pp. 961-1005.

Haykin, S., 1992, Radar vision. Second international Specialist Seminar on Parallel Digital Processors, Portugal.

Haykin., S., 1994, *Neural Networks: A comprehensive foundation*. McMillan, New York.Haykin, S., Krasnor, C., Nohara, T. J., Currie, B. W., and Hamburger, D. (1991). A coherent dual-polarized radar for studying the ocean environment. *IEEE Transactions on Geosciences and Remote Sensing*, 29(1), pp. 189-191.

Jones, S. P., 1995, Turning the tables on video surveillance. Technical report, THE BOSTON HERALD Boston, Massachusetts.

Mann, S., 1996a, "Smart clothing": Wearable multimedia and "personal imaging" to restore the balance between people and their intelligent environments. pp. 163-174, Boston, MA. Proceedings, ACM Multimedia 96; http://wearcam.org/acm-mmg6.htm.

Mann, S., 1996b, Wearable, tetherless computer- mediated reality: Wearcam as a wearable face-recognizer and other applications for the disabled. TR 361, M.I.T. Media Lab Perceptual Computing Section; Also appears in AAAI Fall Symposium on Developing Assistive Technology for People with Disabilities 9-11 November 1996, MIT; http://wearcam.org/vmp.htm, Cambridge, Massachusetts.

Mann, S., 1997a, Humanistic intelligence. *Proceedings of Ars Electronica*, pages 217-231. Invited plenary lecture, Sep. 10, http://wearcam.org/ars/ http://www.aec.at/Eeshfactor, Republished in: Timothy Druckrey (ed.), Ars Electronica: Facing the Future, A Survey of Two Decades, MIT Press, pp. 420-427.

Mann, S., 1997b, Wearable computing: A first step toward personal imaging. IEEE

Computer; *http://wearcam.org/ieeecomputer.htm*, 30(2), pp.25-32.

Mann, S., 1998a, Humanistic intelligence/humanistic computing: 'wearcomp' as a new framework for intelligent signal processing. *Proceedings of the IEEE*, 86(11), pp. 2123-2151+cover. http://wearcm.org/procieee.htm.

Mann, S., 1998b, Reflectionism and diffusionism. *Leonardo*, *http://weavcam.org/leonavdo/index.htm*, 31(2): pp. 93-102.

Mann, S., 2001, *Intelligent Image Processing*. John Wiley & Sons. ISBN: 0-471-40637-6.

Mann, S. and Haykin, S., 1991, The chirplet transform: A generalization of Gabor's logon transform. *Vision Interface* '91 pp. 205-212. ISSN 0843-803X.

Mann, S. and Niedzviecki, H., 2001, *Cyborg: Digital Destiny and Human Possibility in the Age of the Wearable Computer*. Randomhouse (Doubleday). ISBN:0-385-65825-7.

Mann, S. and Picard, R. W., 1994, Virtual bellows: constructing high-quality images from video. In *Proceedings of the IEEE first international conference on image processing*, pp. 363-367, Austin, Texas.

Mann, S. and Picard, R. W., 1995, Video orbits of the projective group; a simple approach to featureless estimation of parameters. TR 338, Massachusetts institute of Technology, Cambridge, Massachusetts. Also appears in IEEE Trans. Image Proc., Sept 1997, Vol. 6 No. 9, pp. 1281-1295.

Marvin Minsky, P. L. E., 1996 (paper in IRE 1960), Steps toward artificial intelligence. In *Great papers on computer science*, West Publishing Company, Minneapolis/St. Paul.

Mihovilovic, D. and Bracewell, R., November, 1992, Whistler analysis in the time-frequency plane using chirplets. *Journal of Geophysical Research*, 97(A11): pp. 17199-17204.

Norman, D., 1992, *Turn signals are the facial expressions of automobiles*. Addison Wesley

Özaktas, H., Barshan, B., Mendlovic, D. and Onural, L., 1994. Convolution, filtering, and multiplexing in fractional fourier domains and their relation to chirp and wavelet transforms. *JOSA A*, Volume 11, pp. 547-559.

Strang, G., 1989, Wavelets and dilation equations: A brief introduction. SIAM Review, 31(4), pp. 614-627.

CHAPTER EIGHT

Mobile Augmented Reality

Tobias H. Höllerer and Steven K. Feiner

8.1 INTRODUCTION

As computers increase in power and decrease in size, new mobile, wearable, and pervasive computing applications are rapidly becoming feasible, providing people access to online resources always and everywhere. This new flexibility makes possible new kind of applications that exploit the person's surrounding context. Augmented reality (AR) presents a particularly powerful user interface (UI) to context-aware computing environments. AR systems integrate virtual information into a person's physical environment so that he or she will perceive that information as existing in their surroundings. Mobile augmented reality systems (MARS) provide this service without constraining the individual's whereabouts to a specially equipped area. Ideally, they work virtually anywhere, adding a palpable layer of information to any environment whenever desired. By doing so, they hold the potential to revolutionize the way in which information is presented to people. Computer-presented material is directly integrated with the real world surrounding the freely roaming person, who can interact with it to display related information, to pose and resolve queries, and to collaborate with other people. The world becomes the user interface.

This chapter provides a detailed introduction to mobile AR technology with in-depth reviews of important topics, such as wearable display and computing hardware, tracking, registration, user interaction, heterogeneous UIs, collaboration, and UI management for situated computing. As part of this introduction, we define what we mean by augmented reality, give a brief overview of the history of the field in general, and review some important mobile AR system considerations. Section 8.2 discusses the potential and possibilities of MARS technology, with a detailed overview of prototype application areas, and reviews the challenges that impede immediate widespread commercial adoption. In Section 8.3 we take a closer look at the requirements and specific components of MARS, before examining UI concepts in Section 8.4. We conclude the chapter with an outlook on research directions.

8.1.1 Definition

A*ugmented reality* is related to the concept of *virtual reality* (VR). VR attempts to create an artificial world that a person can experience and explore interactively, predominantly through his or her sense of vision, but also via audio, tactile, and other forms of feedback. AR also brings about an interactive experience, but aims to supplement the real world, rather than creating an entirely artificial environment. The physical objects in the individual's surroundings become the backdrop and target items for computer-generated annotations. Different researchers subscribe to narrower or wider definitions of exactly what constitutes AR. While the research community largely agrees on most of the elements of AR systems, helped along by the exchange and discussions at several international conferences in the field, there are still small differences in opinion and nomenclature. For the purpose of this chapter, we follow the definitions of Azuma (1997) and Azuma and colleagues (2001). We will define an AR system as one that combines real and computer-generated information in a real environment, interactively and in real time, and aligns virtual objects with physical ones. At the same time, AR is a subfield of the broader concept of *mixed reality* (MR) (Drascic and Milgram, 1996), which also includes simulations predominantly taking place in the virtual domain and not in the real world.

Mobile AR applies this concept in truly mobile settings, that is, away from the carefully conditioned environments of research laboratories and special-purpose work areas. Quite a few technologies, introduced in earlier chapters, must be combined to make this possible: global tracking technologies (Chapter 3), wireless communication (Chapter 4), location-based computing (LBC) and services (LBS) (Chapters 5 and 6), and wearable computing (Chapter 7).

After giving a brief historical overview of AR systems in the next subsection, we will take a look at the components needed to create a mobile AR experience. While AR can potentially supplement the physical environment with information perceptible by all human senses, visual and auditory overlays are currently the most commonly applied augmentations. In the case of visual AR, computer-generated graphics are spatially registered with, and overlaid on, real objects, using the display and tracking technologies that we describe in this chapter.

8.1.2 Historical Overview

While the term *augmented reality* was coined in the early 1990s, the first fully functional AR system dates back to the late 1960s, when Ivan Sutherland and colleagues (1968) built a mechanically tracked 3D see-through head-worn display, through which the wearer could see computer-generated information mixed with physical objects, such as signs on a laboratory wall. For the next few decades much research was done on getting computers to generate graphical information, and the emerging field of *interactive computer graphics* began to flourish. Photorealistic computer-generated images became an area of research in the late 1970s, and progress in tracking technology furthered the hopes to create the ultimate simulation machine. The field of VR began to emerge. Science fiction literature, in particular the early 1980s movement of *cyberpunk*, created visions of man-machine symbiosis. The entertainment industry jumped in with movies such as the *Terminator* series, which presented a specific rendition of what computer-annotated vision could look like. During the 1970s and 80s, AR was a research topic at some institutions, such as the U.S. Air Force's Armstrong Laboratory, the NASA Ames Research Center, the Massachusetts Institute of Technology, and the University of North Carolina at Chapel Hill. As part of the US Air Force *Super Cockpit* project,

Tom Furness developed a high-resolution heads-up overlay display for fighter pilots, supported by 3D sound (Furness, 1986).

(a) (b)

Figure 8.1 Mobile AR Restaurant Guide. (a) User with MARS backpack, looking at a restaurant. (b) Annotated view of restaurant, imaged through the head-worn display

It was not until the early 1990s, with research at the Boeing Corporation, that the notion of overlaying computer graphics on top of the real world received its current name. Caudell and Mizell (1992) worked at Boeing on simplifying the process of conveying wiring instructions for aircraft assembly to construction workers, and they referred to their proposed solution of overlaying computer-presented material on top of the real world as *augmented reality*. Even though this application was conceived with the goal of mobility in mind, true mobile graphical AR was out of reach for the available technology until a few years later. Also during the early 1990s, Loomis and colleagues (1993) at the University of California, Santa Barbara, developed a GPS-based outdoor system, presenting navigational assistance to the visually impaired with spatial audio overlays.

Since about the mid-1990s computing and tracking devices have become sufficiently powerful, and at the same time small enough, to support registered computer-generated graphical overlays in a dynamic mobile setting. The Columbia *Touring Machine* (Feiner *et al.*, 1997) is an early prototype of an outdoor MARS that presents 3D graphical tour guide information to campus visitors, registered with the buildings and artifacts the visitor sees. Figure 8.1 shows a more recent version of the Touring Machine, annotating restaurants in the Columbia University neighborhood.

In 1979, a mobile audio display changed the way people listened to music: the Sony Walkman. It was one of the three most successful consumer products of the 1980s, with the other two being roller skates and digital watches (another kind of mobile display). This commercial success paved the way for other mobile devices, among them personal digital organizers. The original Sony Walkman weighed in at

390g, not counting batteries and audiotape. Today, many MP3 players weigh less than 40 grams, including batteries.

Wearable computing (Mann, 1997; Starner *et al.*, 1997a), took off in the 1990s, when personal computers were becoming small enough to be carried or worn at all times. The earliest wearable system was a special purpose analog computer for predicting the outcome of gambling events, built in 1961 (Thorp, 1998). On the commercial front, palmtop computers embody the trend towards miniaturization. They date back to the Psion I organizer from 1984 and later became commonplace with the introduction of the Apple Newton MessagePad in 1993 and the Palm Pilot in 1996. Since the mid-1990s, wearable computing has received ever-increasing commercial backing, and the miniaturization and more cost-effective production of mobile computing equipment resulted in several companies now offering commercial wearable computing products (e.g., Xybernaut, Charmed Technology, ViA, Antelope Technologies).

In terms of the technologies necessary for a mobile AR experience, we will look briefly at the historical developments in the fields of *tracking and registration*, *wireless networking*, *display technology*, and *interaction technology* in Section 8.3. Now that the technological cornerstones of mobile AR have been placed, it might seem that it is purely a matter of improving the necessary components, putting it all together, and making the end result as reliable as possible. However, there are more challenges lying ahead, and, after giving an overview of the necessary components of a functional mobile AR system in the following subsection, we will come back to these challenges in Section 8.2.2.

8.1.3 Mobile AR Systems

Revisiting our definition of AR, we can identify the components needed for MARS. To begin with, one needs a *computational platform* that can generate and manage the virtual material to be layered on top of the physical environment, process the tracker information, and control the AR display(s).

Next, one needs *displays* to present the virtual material in the context of the physical world. In the case of augmenting the visual sense, these can be head-worn displays, mobile hand-held displays, or displays integrated into the physical world. Other senses (hearing, touch, or smell) can also be potentially augmented. Spatialized audio, in particular, is often used to convey localized information, either complementing or completely substituting for visual elements (Sawhney and Schmandt, 1998).

Registration must also be addressed: aligning the virtual elements with the physical objects they annotate. For visual and auditory registration, this can be done by *tracking* the position and orientation of the user's head and relating that measurement to a model of the environment and/or by making the computer "see" and potentially interpret the environment by means of cameras and computer vision.

Wearable input and interaction technologies enable a mobile person to work with the augmented world (e.g., to make selections or access and visualize databases containing relevant material) and to further augment the world around them. They also make it possible for an individual to communicate and collaborate with other MARS users.

Wireless networking is needed to communicate with other people and computers while on the run. Dynamic and flexible mobile AR will rely on up-to-the-second information that cannot possibly be stored on the computing device before application run-time. For example, this would make it possible to report train or bus delays and traffic conditions to the busy commuter.

This brings us to another item in the list of requirements for MARS: *data storage and access technology*. If a MARS is to provide information about a roaming individual's current environment, it needs to get the data about that environment from somewhere. Data repositories must provide information suited for the roaming individual's current context. Data and service discovery, management, and access pose several research questions that are being examined by researchers in the database, middleware, and context-based services communities. From the user's point of view, the important question is how to get to the most relevant information with the least effort and how to minimize information overload.

AR is an immensely promising and increasingly feasible UI technology, but current systems are still mostly research prototypes. There is no obvious overall best solution to many of the challenging areas in the field of AR, such as tracking and display technologies. New research results constantly open up new avenues of exploration. A developer, planning to deploy AR technology for a specific task, has to make design decisions that optimize AR performance for a given application, based on careful task analysis. In some cases, it might be possible to come up with specialized AR solutions that will work well in constrained areas with special purpose hardware support. If, on the other hand, the task analysis for the AR scenario reveals that the end-user of the system is to be supported in a wide variety of locations and situations, possibly including outdoor activities, then we enter the realm of true mobile AR.

8.2 MARS: PROMISES, APPLICATIONS, AND CHALLENGES

The previous chapters on LBC and LBSs have shown how location-aware technology opens up new possibilities in the way we interact with computers, gather information, find our way in unfamiliar environments, and do business. AR can provide a powerful UI for this type of computing; one might even say, the ultimate interface: to interact directly with the world around us — the world becomes the user interface.

Mobile AR is particularly applicable whenever people require informational support for a task while needing to stay focused on that task. It has the potential to allow people to interact with computer-supported information (which might come from databases or as a live feed from a remote expert), without getting distracted from the real world around them. This is a very important feature for the mobile worker, or for anybody who needs or wants to use their hands, and some of their attention, for something other than controlling a computer. The next subsection gives a few examples of such occupations and summarizes application areas for which mobile AR prototypes have been tested.

8.2.1 Applications

A high percentage of conference submissions on AR come from industrial research labs or are joint work between universities and industry. Many of the early AR publications are "application papers," describing applications of the new technology in varied fields. In the rest of this section, we give an overview of potential uses for mobile AR systems.

Assembly and construction. Over a nine-year period, researchers at Boeing built several iterations of prototypes for AR-supported assembly of electrical wire bundles for aircraft (Mizell, 2001). AR overlaid schematic diagrams, as well as accompanying documentation, directly onto the wooden boards on which the cables are routed, bundled, and sleeved. The computer led (and could potentially talk) assembly workers through the wiring process. Since the resulting wire bundles were long enough to extend through considerable portions of an aircraft, stationary AR solutions were not sufficient, and the project became an exercise in making mobile AR work for a specific application scenario.

(a) Installing a spaceframe strut

(b) View through the head-worn display

(c) Overview visualization of the tracked scene.

Figure 8.2 AR for Construction

"Augmented Reality for Construction" (Feiner *et al.*, 1999), is an indoor prototype for the construction of spaceframe structures. As illustrated in Figure 8.2 (a–b), a construction worker would see and hear through their head-worn displays where the next structural element is to be installed. The construction worker scans the designated element with a tracked barcode reader before and after installation to verify that the right piece gets installed in the right place. The possibility of generating virtual overview renderings of the entire construction scene, as indicated in a small-scale example via live-updated networked graphics in one of the prototype's demonstrations (see Figure 8.2 (c)), is a side benefit of tracking each individual worker and following their actions, and could prove immensely useful for the management of large complex construction projects. Such construction tasks would ultimately take place in the outdoors.

Maintenance and inspection. Apart from assembly and construction, inspection and maintenance are other areas in manufacturing that may benefit greatly from applying MARS technologies. Sato and colleagues (1999) propose a prototype AR system for inspection of electronic parts within the boundaries of a wide-area manufacturing plant. Their MARS backpack is tracked with a purely inertia-based (gyroscope orientation tracker plus acceleration sensor) tracking system, and calibration has to be frequently adjusted by hand. Zhang and colleagues (2001) suggest the use of visual coded markers for large industrial environments. Klinker and colleagues (2001) present the system architecture of a MARS prototype for use in maintenance for nuclear power plants. AR is well suited for situations that require "x-ray vision," an ability to see through solid structures. Using direct overlays of hidden infrastructure, AR can assist maintenance workers who are trying to locate a broken cable connection within the walls of a building, or the location of a leaking pipe beneath a road's surface. The exact position may have been detected automatically (e.g., by installed sensors), in which case direct visualization of the problem area via AR could be the fastest way to direct the worker's attention to the right area. Alternatively, AR may be used as a supporting tool for determining the problem in the first place, in that it could instantaneously and directly visualize any data the worker might gather, probing the environment with various sensors.

Navigation and path finding. Considering some other outdoor-oriented uses for mobile AR, an important application area for wearable systems is their use as navigational aids. Wearable computers can greatly assist blind users (Loomis *et al.*, 1993; Petrie *et al.*, 1996) via audio and tactile feedback. If auditory information relating to real-world waypoints and features of the environment is presented to the position-tracked wearer of the system via spatialized stereo audio, this clearly matches our definition of an AR system from Section 8.1.1. Visual AR can aid navigation by directly pointing out locations in the user's field of view, by means of directional annotations, such as arrows and trails to follow along (see Figure 8.3), or by pointing out occluded infrastructure, either directly by visually encoded overlays (Furmanski *et al.*, 2002), or indirectly via 2D or 3D maps that are dynamically tailored to the situation's needs and presented to the user (Figures 8.3 (a), 8.10, and 8.11).

Tourism. Taking navigational UIs one step further by including information about objects in a mobile person's environment that might be of interest to a traveler, leads naturally to applications for tourism (Feiner *et al.*, 1997; Cheverst *et al.*, 2000). In this case, AR is not only used to find destinations, but also to display background information. For example, instead of looking up a description and historic account of a famous cathedral in a guide book, (or even on a wirelessly connected palm-sized computer in front of the site, AR can make the air around the church come alive with information: 3D models of related art or architecture, the life and work of the architect, or architectural changes over the centuries can be documented in situ with overlays. The possibilities are endless and only limited by the amount and type of information available to the AR-enabled individual and the capabilities of the AR device the individual is wearing.

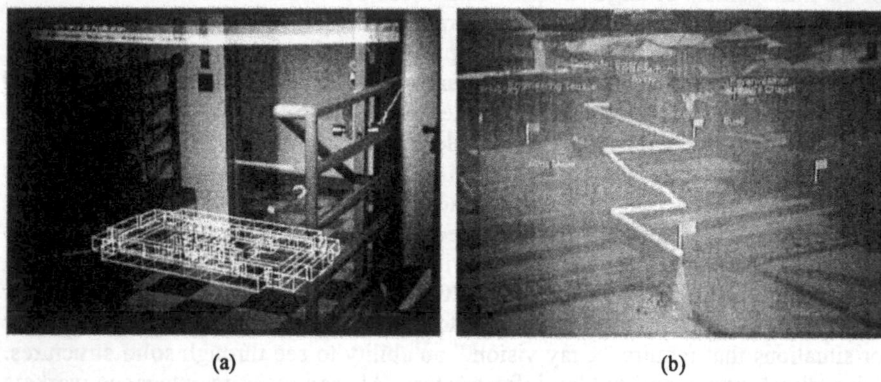

(a) (b)

Figure 8.3 Navigational AR Interfaces, Imaged through Head-Worn Displays. (a) Indoor guidance using overview visualization and arrows. (b) Virtual trails and flags outdoors (viewed from the roof of a building)

Figure 8.1 shows an example of a mobile AR restaurant guide developed at Columbia University (Feiner, 2002). This prototype MARS provides an interface to a database of the restaurants in Morningside Heights, New York City. Information about restaurants is provided either via an overview 3D map, so that the user can be guided to a specific place of his or her choice, or as direct annotations of the actual restaurant locations themselves. Having selected an establishment, the user can bring up a popup window with further information on it: a brief description, address and phone number, an image of the interior and, accessible at a mouse click: the menu and, if available, reviews of the restaurant, and its external web page. We will discuss related information display issues in Section 8.4.1.

Geographical field work. Field workers in geography and regional sciences could use AR techniques to collect, compare, and update survey data and statistics in the field (Nusser *et al.*, 2003). By assisting data collection and display, an AR system could enable discovery of patterns in the field, not just the laboratory. Instant verification and comparison of information with data on file would be possible.

Journalism. Journalism is another area in which MARS could have a major impact. Pavlik (2001) discusses the use of wireless technology for the mobile journalist, who covers and documents a developing news story on the run. MARS

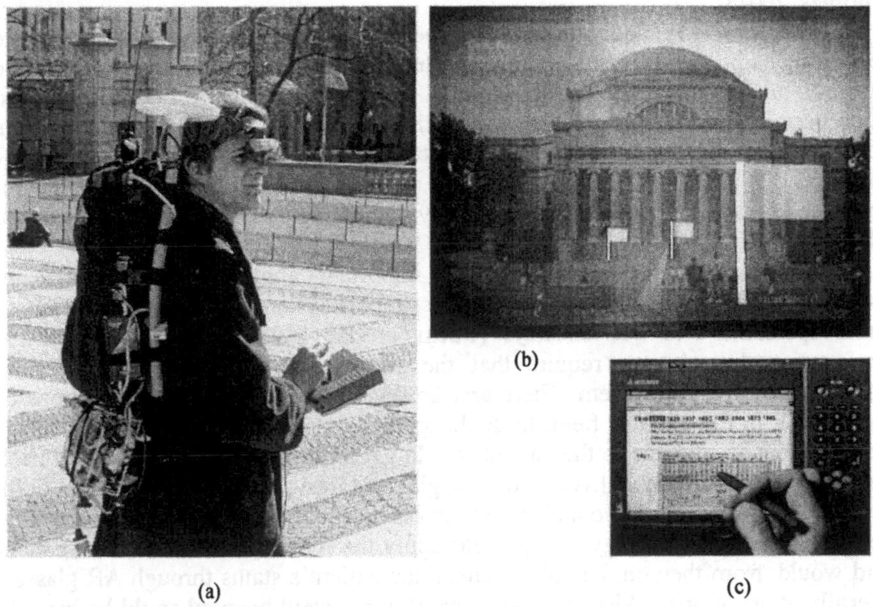

(a) (b) (c)

Figure 8.4 Situated Documentaries. (a) User with backpack MARS. (b) View through head-worn display. Virtual flags representing points of interest. (c) Additional handheld interface for interacting with virtual material

could be used to leave notes in the scene for other collaborating journalists and photographers to view and act upon. The *Situated Documentaries* project at Columbia University (Höllerer *et al.*, 1999a), shown in Figures 8.4 and 8.5, is a collaboration between computer science and journalism, and uses MARS for storytelling and presentation of historical information.

Architecture and archaeology. AR is also especially useful to visualize the invisible: architects' designs of bridges or buildings that are about to be constructed on a particular site; historic buildings, long torn down, in their original location; or reconstructions of archaeological sites. Figure 8.5, which was shot through a see-through head-worn display presenting a situated documentary (Höllerer *et al.*, 1999a) of the history of Columbia, shows a model of part of the Bloomingdale Insane Asylum, which once occupied the main campus. The European sponsored project ARCHEOGUIDE (Vlahakis *et al.,* 2002) aims to reconstruct a cultural heritage site in AR and let visitors view and learn about the ancient architecture and customs. The first place selected as a trial site is the ancient town of Olympia in Greece.

Urban modeling. AR will not be used solely for passive viewing or information retrieval via the occasional mouse-button (or should we say shirt-button) click. Many researchers are exploring how AR technologies could be used to enter information to the computer (Rekimoto *et al.*, 1998). One practical example is 3D modeling of outdoor scenes: using the mobile platform to create 3D renderings of buildings and other objects that model the very environment to be used later as a backdrop for AR presentations (Baillot *et al.*, 2001; Piekarski and Thomas, 2001).

Entertainment. The situated documentaries application also suggests the technology's potential for entertainment purposes. Instead of delivering 3D movie "rides," such as the popular *Terminator 2* presentation at Universal Studios, to audiences in special purpose theme park theatres, virtual actors in special effects scenes could one day populate the very streets of the theme parks, engaging AR outfitted guests in spectacular action. As an early start in this direction, several researchers have experimented with applying mobile AR technology to gaming (Thomas *et al.*, 2000; Starner *et al.*, 2000).

Medicine. Augmented reality has important application possibilities in medicine. Many of these, such as surgery support systems that assist surgeons in their operations via live overlays (Fuchs *et al.*, 1998), require very precise registration, but do not require that the surgeon be extremely mobile while supported by the AR system. There are, however, several possible applications of mobile AR in the medical field. In the hospital or nursing home, doctors or nurses on their rounds of visits to the patients could get important information about each patient's status directly delivered to their glasses (Hasvold, 2002). Out in the field, emergency medicine personnel could assess a situation quicker with wearable sensing and AR technology. They could apply the wearable sensors to the patient and would, from then on, be able to check the patient's status through AR glasses, literally at one glance. Also, a remote expert at a distant hospital could be brought into the loop and communicate with the field worker via the AR system, seeing through camera feeds what the field worker is seeing, which could be important to prepare an imminent operation at the hospital.

Monitoring the health information of a group of people at the same time could be advantageous for trainers or coaches during athletic training or competition. The military also has potential medical uses for mobile AR technologies. The health status of soldiers on the battlefield could be monitored, and in case of any injuries, the commanding officer could get live overview visualizations of location and status of the wounded.

Military training and combat. Military research led to the development of satellite navigation systems and heads-up displays for combat pilots. Military research laboratories have also been exploring the potential of mobile AR technology for land warriors for some time now (Tappert *et al.*, 2001). In terms of the possible use of AR in military operations, there is considerable overlap with civilian applications on a general level. Navigational support, enhancement of communications, repair and maintenance, and emergency medicine, are important topics in civilian and military life. There are, however, specific benefits that AR technology could bring to the military user. Most missions take place in unfamiliar territories. Map views, projected naturally into a warrior's limited view of the battle scene, can provide additional information about terrain that cannot easily be overseen. Furthermore, reconnaissance data and mission planning information can be integrated into these information displays, clarifying the situation and outlining specific sub-missions for individual troops. Ongoing research at the Naval Research Laboratory is concerned with how such information displays can be delivered to the warriors most effectively (Julier *et al.*, 2000). Apart from its use in combat, mobile AR might also prove a valuable military tool for training and simulation purposes.

Figure 8.5 Situated Documentaries: Historic building overlaid at its original location on Columbia's campus

For example, large-scale combat scenarios could be tested with simulated enemy action in real training environments.

Personal Information Management and Marketing. It is anybody's guess which endeavors in mobile AR might eventually lead to commercial success stories. Some small companies already offer specialized AR solutions (TriSense). Whatever might trigger widespread use, the biggest potential market for this technology could prove to be personal wearable computing. AR could serve as an advanced and immediate UI for wearable computing. In personal, daily use, AR could support and integrate common tasks, such as email and phone communication with location-aware overlays, provide navigational guidance, enable individuals to store personal information coupled with specific locations, and provide a unified control interface for all kinds of appliances in the home (Feiner, 2002). Of course, such a personal platform would be very attractive for direct marketing agencies. Stores could offer virtual discount coupons to passing pedestrians. Virtual billboards could advertise products based on the individual's profile. Virtual 3D product prototypes could pop up in the customer's eyewear (Zhang and Navab, 2000). To protect the individual from unwanted information, an AR platform would need to incorporate appropriate filtering and view management mechanisms (see Section 8.4.3 and Chapter 7).

8.2.2 Challenges

In spite of the great potential of mobile AR in many application areas, progress in the field has so far almost exclusively been demonstrated through research prototypes. The time is obviously not quite ripe yet for commercialization. When asking for the reasons why, one has to take a good look at the dimension of the task. While increasingly better solutions to the technical challenges of wearable computing are being introduced, a few problem areas remain, such as miniaturization of input/output technology, power sources, and thermal dissipation, especially in small high-performance systems. Ruggedness is also required. With some early wearable systems the phrase 'wear and tear' seemed to rather fittingly indicate the dire consequences of usage. In addition to these standard wearable computing requirements, mobile AR adds many more: reliable and ubiquitous wide area position tracking, accurate and self-calibrating (head-) orientation tracking; ultra-light, ultra-bright, ultra-transparent, display eyewear with wide field of view; fast 3D graphics capabilities, to name just the most important requirements. We will look at current technologies addressing these problem areas in the following section. If one were to select the best technologies available today for each of the necessary components, one could build a powerful (though large) MARS. Section 8.3.7 will take a closer look at such a hypothetical device.

AR in the outdoors is a particular challenge since there is a wide range of operating conditions to which the system could be exposed. Moreover, in contrast to controlled environments indoors, one has little influence over outdoor conditions; for example, lighting can range from direct sunlight, possibly exacerbated by a reflective environment (e.g., snow), to absolute darkness without artificial light sources during the night. Outdoor systems should withstand all possible weather conditions, including wind, rain, frost, and heat.

The list of challenges does not end with the technology on the user's side. Depending on the tracking technology, AR systems either need to have access to a model of the environment they are supposed to annotate, or require that such environments be *prepared* (e.g., equipped with visual markers or electronic tags). Vision-based tracking in unprepared environments is currently not a viable general solution, but research in this field is trying to create solutions for future systems. The data to be presented in AR overlays needs to be paired with locations in the environment. A standard access method needs to be in place for retrieving such data from databases responsible for the area the MARS user is currently passing through. This requires mechanisms such as automatic service detection and the definition of standard exchange formats that both the database servers and the MARS software support. It is clear from the history of protocol standards, that without big demand and money-making opportunities on the horizon, progress on these fronts can be expected to be slow. On the other hand, the World Wide Web, HTML, and HTTP evolved from similar starting conditions. Some researchers see location-aware computing on a global scale as a legitimate successor of the World Wide Web as we know it today (Spohrer, 1999).

In the movie industry, special effects seamlessly merge computer-generated worlds with real scenes. Currently these efforts take days and months of rendering time and very carefully handcrafted integration of the virtual material into real-world footage. In AR, not only does the rendering need to be performed in real time, but the decisions about what to display and the generation of this material must be triggered and controlled on the fly. Making the visuals as informative as possible, and, in some cases, also as realistic as possible, rendered with the correct lighting to provide a seamless experience, is an open-ended challenge for visual AR. Section 8.4 will present ongoing research in sorting out some UI challenges.

8.3 COMPONENTS AND REQUIREMENTS

In this section, we review in greater depth the basic components and infrastructure required for MARS, as outlined before in Section 8.1.3. We take a look at mobile computing platforms, displays for MARS, tracking and registration issues, environmental modeling, wearable input and interaction techniques, wireless communication, and distributed data storage and access. We give brief overviews of important historic developments in these areas, and point out technologies that have successfully been employed in MARS prototypes, or that have great potential to be employed in future systems. Finally, we summarize this material by describing a hypothetical top-of-the-line MARS, assembled from the most promising components currently available.

8.3.1 Mobile Computing Platforms

Mobile computing platforms have seen immense progress in miniaturization and performance over recent years, and are sold for increasingly less. Today, high-end notebook computers catch up in computing power with available desktop solutions very shortly after a new processor model hits the market. The trend towards more mobility is clearly visible. The wearable technology market, even though still in its infancy, has a growing customer base in industry, government, and military. Wearable computing solutions for personal use can now be purchased from various sources.

There are several decision factors when choosing a computing platform for mobile AR research, including the *computing power* needed, the *form factor* and *ruggedness* of the overall system, *power consumption*, the *graphics and multimedia capabilities*, availability of *expansion and interface ports*, available *memory and storage space*, *upgradeability* of components, the *operating system* and *software development environments*, availability of *technical support*, and last but not least, *price*. Quite clearly, many of these are interdependent. The smaller the computing device, the less likely it is to have the highest computing power and graphics and multimedia capabilities. Expansion and interface ports normally come at the price of increased size. So does upgradeability of components: if you have miniaturized functionality (e.g., graphics) by using special purpose integrated circuits, you no longer have the luxury of being able to easily replace that component with a newer model. Additionally, it takes a lot of effort and ingenuity to scale down any kind of technology to a size considerably smaller than what the competition is offering, so one can expect such equipment to be sold at a higher price.

The Columbia University MARS project (Höllerer *et al.*, 1999b) provides a concrete example of some of the tradeoffs involved in assembling a mobile AR platform. The hardware platform for this project, illustrated in Figures 1 (a) and 2 (a), was assembled from off-the-shelf components for maximum performance, upgradeability, ease of maintenance, and software development. These choices were made at the cost of the size and weight of the prototype system, whose parts were mounted on a backpack frame. From 1996 to 2002, every single component of the prototype was upgraded multiple times to a more powerful or otherwise more advantageous version, something that would not have been possible if a smaller, more integrated system had initially been chosen. The computing power tracked high-end mobile processing technologies, ranging from a 133MHz Pentium-based system in 1996 to a 2.2 GHz Pentium IV notebook in 2002. During the same time, the 3D graphics capabilities grew from a GLINT 500DTX chip with a claimed fill rate of about 16.5M pixels per second to an NVIDIA Quadro4 500 Go with announced 880M pixels per second.

The smallest wearable computing platforms currently available (Windows CE-based Xybernaut Poma, or the soon-to-be-available higher performance OQO Ultra-Personal Computer, Tiqit eightythree, and Antelope Technologies Mobile Computer Core) provide only modest graphics performance and their main processors do not have enough power for software renderings of complex 3D scenes at interactive speeds. Decision factors in choosing a 3D graphics platform for mobile AR include the *graphics performance* required, *video and texture memory*, *graphics library* support (OpenGL or Direct-X), availability of *stereo drivers*, *power consumption*, and *price*. The most practical solution for a mobile AR system that can support complex 3D interactive graphics comes in the form of small notebook computers with integrated 3D graphics chip. The display could be removed if the computer is exclusively used with a near-eye display. However, in our experience with the Touring Machine, it can be put to good use in prototype systems for debugging purposes and for providing a view for onlookers during technology demonstrations.

Specific application requirements can drastically limit the choices for a MARS computing platform. For example, in late 2002 there are no integrated wearable computing solutions available that support rendering and display of complex graphical scenes in stereo. A system designer targeting such applications either has to assemble their own hardware to create a small form-factor solution, or resort to the smallest available power notebook that has sufficient graphics performance and a graphics chip supporting stereo.

Mobile systems do not necessarily have to follow the pattern of one standalone device, carried or worn, that generates and presents all the information to the user. Instead, there can be varying degrees of "environment participation," making use of resources that are not necessarily located on the user's body. In the most device-centric case, all information is generated and displayed on one single device that the user wears or carries. Examples include portable audio players and hand-held organizers without wireless communication option.

Departing one step from the device-centric approach, functionality can be distributed over multiple devices. Wireless connectivity technologies, such as IEEE 802.11b or Bluetooth come in handy for data exchange between different devices. For example, a personal organizer or wearable computer can send data over a wireless connection to an audio/video headset. With the addition of a Bluetooth-enabled cell phone for global communication purposes, such a combination would constitute a complete wireless mobile computing solution. Not all devices need to be carried by the user at all times. Suppose that we want to minimize the wearable computing equipment's size and power consumption. Lacking the computational power to generate and process the information that is to be displayed, we can turn the wearable device into a so-called *thin client*, relying on outside servers to collect and process information and feed it to the portable device as a data stream that can be comfortably presented with the limited resources available.

In the extreme case, a mobile user would not need to wear or carry *any* equipment and still be able to experience mobile AR. All the computation and sensing could occur in the environment. A grid of cameras could be set up so that multiple cameras would cover any possible location the person could occupy. Information could be stored, collected, processed, and generated on a network of computing servers that would not need to be in view, or even nearby. Displays, such as loudspeakers, video-walls, and projected video could bring personalized information to the people that are standing or moving nearby. However, the infrastructure needed for such a scenario is quite high. The task of making mobile AR work in unprepared environments requires solutions closer to the "one-device" end of the spectrum.

8.3.2 Displays for Mobile AR

There are various approaches to display information to a mobile person and a variety of different types of displays can be employed for this purpose: personal hand-held, wrist-worn, or head-worn displays; screens and directed loudspeakers embedded in the environment; image projection on arbitrary surfaces, to name but a few. Several of these display possibilities may also be used in combination.

In general, one can distinguish between displays that the person carries on the body and displays that make use of resources in the environment. Wearable audio players and personal digital organizers use displays that fall in the first category, as do wearable computers with head-worn displays (see Figure 8.6). An example of the second category would be personalized advertisements that are displayed on video screens that a person passes. For such a scenario, one would need a fairly sophisticated environmental infrastructure. Displays would need to be embedded in walls and other physical objects, and they would either have to be equipped with sensors that can detect a particular individual's presence, or they could receive the tracking information of passersby via a computer network. Such environments do not yet exist outside of research laboratories, but several research groups have begun exploring ubiquitous display environments as part of *Smart Home* or *Collaborative Tele-Immersion* setups, such as the Microsoft Research EasyLiving project (Brumitt *et al.*, 2000), or the University of North Carolina Office of the Future (Raskar *et al.*, 1998).

Another display type that is being explored in AR research is the *head-worn projective display* (Hua *et al.*, 2001). This type of head-worn display consists of a pair of micro displays, beam splitters, and miniature projection lenses. It requires that retroreflective sheeting material be placed strategically in the environment. The head-worn display projects images out into the world, and only when users look at a patch of retroreflective material, they see the image that was sent out from their display. This approach aims to combine the concept of physical display surfaces (in this case: patches of retroreflective material) with the flexibility of personalized overlays with AR eyewear. A unique personalized image can be generated for each person in a set of people looking at the same object with retroreflective coating, as long as their viewing angles are not too close to each other.

One promising approach for mobile AR might be to combine different display technologies. Head-worn displays provide one of the most immediate means of accessing graphical information. The viewer does not need to divert his or her eyes away from their object of focus in the real world. The immediateness and privacy of a personal head-worn display is complemented well by the high text readability of hand-held plasma or LCD displays, and by the collaboration possibilities of wall-sized displays. For example, mobile AR research at Columbia University experimented early on with head-worn and hand-held displays (cf. Figure 8.2) used in synergistic combination (Feiner *et al.*, 1997; Höllerer *et al.*, 1999a).

As mentioned in the historical overview in Section 8.1.2, the concept of *see-through head-worn computer graphics displays* dates back to Ivan Sutherland's work on a head-worn 3D display (Sutherland, 1968). Some time before that, in 1957, Morton Heilig had filed a patent for a head-worn display fitted with two color TV units. In later years, several head-worn displays were developed for research in computer simulations and the military, including Tom Furness's work on heads-up display systems for fighter pilots. VPL Research and Autodesk introduced a commercial head-worn display for VR in 1989. In the same year, Reflection Technology introduced a small personal near-eye display, the P4 Private Eye. This display is noteworthy, because it gave rise to a number of wearable computing and AR and VR efforts in the early 1990s (Pausch, 1991, Feiner *et al.*, 1991). It sported a resolution of 720x280 pixels, using a dense column of 280 red LEDs and a vibrating mirror. The display was well suited for showing text and simple line drawings.

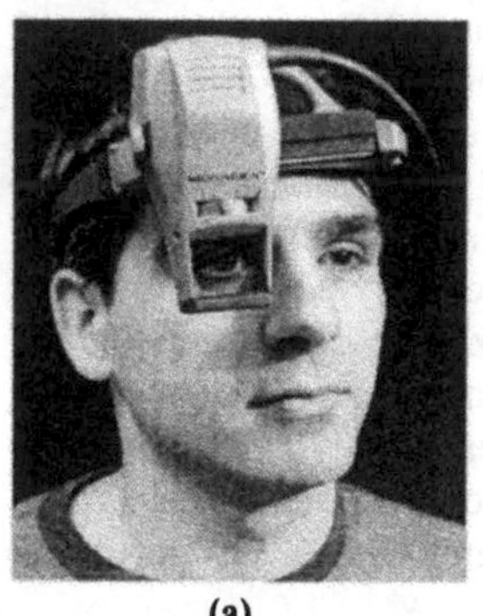
(a)

(b)

(c)

Figure 8.6 Monocular and Binocular Optical See-through Head-Worn Displays. (a) Microvision Nomad. (b) Microoptical Clip-on. (c) Sony LDI-D100B (retrofit onto a customer-provided mount)

RCA made the first experimental liquid crystal display (LCD) in 1968, a non-emissive technology (requiring a separate light source) that steadily developed and later enabled a whole generation of small computing devices to display information. Today, many different technologies, especially emissive ones, are being explored for displays of a wide variety of sizes and shapes. *Plasma displays* provide bright images and wide viewing angles for medium-sized to large flat panels. *Organic light emitting diodes* (OLED) can be used to produce ultra-thin displays. Certain types of OLED technology, such as *light emitting polymers*, might one day lead to display products that can be bent and shaped as required. Of high interest for the development of personal displays are display technologies that are so small, that optical magnification is needed to view the images. These are collectively referred to as *microdisplays.* OLED on silicon is one of the most promising approaches to produce such miniature displays. Non-emissive technologies for microdisplays include transmissive *poly-silicon LCDs*, and several reflective technologies, such as *liquid crystal on silicon* (LCoS) and *digital micromirror devices* (DMD).

One technology that is particularly interesting for mobile AR purposes is the one employed in Microvision's monochromatic, single-eye, Nomad *retinal scanning display,* shown in Figure 8.6 (a). It is one of the few displays that can produce good results in direct sunlight outdoors. It works by pointing a red laser diode towards an electromagnetically controlled pivoting micromirror and diverting the beam via an optical combiner through the viewer's pupil into the eye, where it sweeps across the retina to recreate the digital image. This technology produces a very crisp and bright image, and exhibits the highest transparency any optical see-through display offers today. Microvision has also prototyped a much larger, full-color and optionally stereoscopic display.

(a)

(b)

Figure 8.7 (a) Optical See-through and (b) Video See through Indoor AR

When choosing a head-worn display for mobile AR, several decision factors have to be considered. One of the more controversial issues within the AR research community is the choice between optical see-through and video see-through displays. Optical see-through displays are transparent, the way prescription glasses or sunglasses are. They use optical combiners, such as mirror beam-splitters, to layer the computer generated image on top of the user's view of the environment. Figure 8.7 (a) shows an image shot through such glasses. In contrast, video see-through displays present a more indirect, mediated view of the environment. One or two small video cameras, mounted on the head-worn display, capture video streams of the environment in front of the user, which are displayed on non-transparent screens with suitable optics, right in front of the user's eyes. The computer can modify the video image before it is sent to the glasses to create AR overlays. An example is shown in Figure 8.7 (b). More details and a discussion of the advantages and disadvantages of both approaches are given in Azuma (1997) and Feiner (2002). Non-AR wearable computing applications often use monocular displays. Even if the display is non-transparent, the user is able to see the real world with the non-occluded eye. However, perceiving a true mixture of computer overlay and real world can be somewhat of a challenge in that case.

For mobile AR work, the authors of this chapter prefer optical see-through displays. We believe that a person walking around in the environment should be able to rely on their full natural sense of vision. While AR can enhance their vision, it should not unduly lessen it. In our opinion, several drawbacks of current video see-through technology stand in the way of their adoption in truly mobile applications: Seeing the real world at video resolution and at the same small field of view angle used for the graphical overlays, having to compensate for image distortions introduced by the cameras, the risk of latency in the video feed to the display, and safety concerns about seeing the world solely through cameras.

In our experience, monocular displays can yield acceptable results for AR if the display is see-through to make it easier for the user to fuse the augmented view with the other eye's view of the real world, as is the case with the MicroVision Nomad. A larger field of view is also helpful. It is hard to discuss such display properties in isolation, however, since quite a few display factors influence the quality of mobile AR presentations, among them monocular vs. biocular (two-eye) vs. binocular (stereo), resolution, color depth, luminance, contrast, field of view, focus depth, degree of transparency, weight, ergonomics, and appearance. Power consumption is an additional factor with extended mobile use.

Stereo displays can greatly enhance the AR experience, since virtual objects can then be better perceived at the same distance as the real world objects they annotate. Note though, that even though stereo allows objects to be displayed with the correct left/right eye disparity, all currently available displays display graphics at the same apparent depth, and hence require the viewer's eyes to accommodate at that particular distance, which leads to an accommodation-vergence conflict. For example, when overlaying a virtual object on a real one in an optical see-through display, unless the real object is located at that particular fixed distance, the viewer needs to adjust accommodation in order to see either the real object or the virtual one in focus.

Currently, the options for optical see-through head-worn displays are quite limited. If stereo is a necessity, the options are even more restricted. The Columbia University MARS prototypes employed several stereo capable optical see-through displays over the years, none of which are on the market anymore. Figure 8.6 (c) shows the Sony LDI-D100B, a display that was discontinued in June 2000.

Displays are for the most part still bulky and awkward in appearance today. Smaller monocular displays, such as the MicroOptical CO-1, pictured in Figure 8.6 (b), or the Minolta 'Forgettable Display' prototype (Kasai *et al.*, 2000), are much more inconspicuous, but do not afford the high field-of-view angles necessary for true immersion nor the brightness of, for example, the Microvision Nomad.

Meanwhile, manufacturers are working hard on improving and further miniaturizing display optics. Microdisplays can today be found in a diverse set of products including viewfinders for cameras, displays for cell phones and other mobile devices, and portable video projectors. Near-eye displays constitute a growing application segment in the microdisplay market. The attractiveness of mobile AR relies on further progress in this area.

8.3.3 Tracking and Registration

Apart from the display technology, the single most important technological challenge to general mobile AR is tracking and registration. AR requires extremely accurate position and orientation tracking to align, or register, virtual information with the physical objects that are to be annotated. It is difficult to convince people that computer-generated virtual objects actually live in the same physical space as the real world objects around us. In controlled environments of constrained size in indoor computing laboratories, researchers have succeeded in creating environments in which a person's head and hands can be motion-tracked with sufficiently high spatial accuracy and resolution, low latency, and high update rates, to create fairly realistic interactive computer graphics environments that seemingly coexist with the physical environment. Doing the same in a general mobile setting is much more challenging. In the general mobile case, one cannot expect to rely on any kind of tracking infrastructure in the environment. Tracking equipment needs to be light enough to wear, fairly resistant to shock and abuse, and functional across a wide spectrum of environmental conditions, including lighting, temperature, and weather. Under these circumstances, there does not currently exist a perfect tracking solution, nor can we expect to find one in the near future. Compromises in tracking performance have to be made, and applications will have to adjust.

Tracking technology has improved steadily since the early days of head-tracked computer graphics. Sutherland's original head-worn display was tracked mechanically through ceiling-mounted hardware, and, because of all the equipment suspended from the ceiling, was humorously referred to as the "Sword of Damocles." Sutherland also explored the use of an ultrasonic head-tracker (Sutherland, 1968). The introduction of the Polhemus magnetic tracker in the late 1970s (Raab *et al.*, 1979) had a big impact on VR and AR research, and the same technology, in improved form, is still in use today. During the 1990s, commercial hybrid tracking systems became available, based on different technologies, all explored separately in experimental tracking systems over the previous decades, such as ultrasonic, magnetic, and optical position tracking, and inertial and magnetometer-based orientation tracking. With respect to global positioning systems, the idea for today's *NAVSTAR* GPS (Getting, 1993) was born in 1973, the first operational GPS satellite was launched in 1978, and the 24-satellite constellation was completed in 1993. Satellites for the Russian counterpart constellation, *Glonass*, were launched from 1982 onwards. The European Union has plans underway to launch a separate 30-satellite GPS, called *Galileo*. Chapter 3 gives a detailed introduction to GPS technology.

In the remainder of this section, we will review the tracking technologies most suited for mobile AR. For a more comprehensive overview of tracking technologies for AR and VR, we refer the reader to existing surveys of motion tracking technologies and techniques, such as Rolland *et al.* (2001), or a recent journal special issue on tracking (Julier and Bishop, 2002).

Visual registration of virtual and physical objects can be achieved in several ways. One common approach is to determine the person's head pose in some global coordinate system, and relate it to a computer model of the current environment. Note that in this case a computer model of the environment has to be created in a step called *environmental modeling* (see Section 8.3.4). This model should use the same global coordinate system as the tracking system, or the necessary conversion transformation has to be known. Determining position and orientation of an object is often referred to as six-degree-of-freedom (6DOF) tracking, for the six parameters sensed: position in x, y, and z, and orientation in yaw, pitch, and roll angles.

Absolute position and orientation of the user's head and the physical objects to be annotated do not necessarily need to be known. In one of the most direct approaches to visual registration, cameras observe specific unique landmarks (e.g., artificial markers) in the environment. If the camera's viewing parameters (position, orientation, field of view) coincide with the display's viewing parameters (e.g., because the display is showing the camera image, as in the case of video see-through displays), and stereo graphics are not employed, the virtual annotations can be inserted directly in pixel coordinates without having to establish the exact geometric relationship between the marker and the camera (Rekimoto and Nagao, 1995). On the other hand, if the precise locations of the landmarks in the environment are known, computer vision techniques can be used to estimate the camera pose. The use of cameras mounted on the display together with landmark recognition is sometimes referred to as *closed-loop tracking,* in which tracking accuracy can be corrected to the nearest pixel, if camera image and graphics display coincide. This is in contrast to *open-loop tracking*, which tries to align the virtual annotations with the physical objects in the real world by relying solely on the sensed 6DOF pose of the person and the computer model of the environment. Any inaccuracies in the tracking devices or the geometrical model will cause the annotation to be slightly off from its intended position in relation to the physical world.

An important criterion for mobile AR tracking is how much tracking equipment is needed on the user's body and in the environment. The obvious goal is to wear as little equipment as possible, and to not be required to prepare the environment in any way. Note that a system such as GPS meets this requirement for all intended purposes, even though a "prepared environment" on a global scale is needed in the form of a satellite constellation. Several tracking approaches require some knowledge about the environment. To create any useful AR annotations, either the objects to be annotated have to be modeled or *geo-referenced* in absolute terms, or their location must be able to be inferred by a known relationship to pre-selected and identifiable landmarks.

The tracking accuracies required for mobile AR depend very much on the application and the distance to the objects to be annotated. If we are annotating the rough outlines of buildings, we can afford some registration error. When trying to pinpoint down the exact location of a particular window, we have to be more accurate. When registration errors are measured as the screen distance between the projected physical target point and the point where the annotation gets drawn, the following observation holds: The further away the object that is to be annotated, the less errors in position tracking impact registration accuracy. The opposite is true for errors in orientation tracking. Since most targets in outdoor mobile AR tend to be some distance away from the viewer, one can assume that errors in orientation tracking contribute much more to overall misregistration than do errors in position tracking (Azuma, 1999). Since there are no standalone sensors that afford general reliable 6DOF tracking in unprepared outdoor environments, mobile AR systems normally resort to hybrid approaches, often employing separate mechanisms for position and orientation tracking.

Position tracking via GPS is a natural candidate for outdoor environments, since it is functional on a global scale, as long as signals from at least four satellites can be received. While the use of GPS navigation has long been restricted to areas that afford direct visibility to the satellites, so-called *assisted GPS* (A-GPS) manages to sidestep that restriction in many cases. A-GPS makes use of a world-wide reference network of servers and base stations for terrestrial signal broadcast. In combination with a large number of parallel correlation reception circuits in the mobile GPS receiver, the area of tracking can be extended to many previously uncovered areas, such as urban canyons and indoor environments in which the signal is sufficiently strong (GlobalLocate, 2002).

Plain GPS without selective availability is accurate to about 10–15 meters. GPS using the wide area augmentation system (WAAS) is typically accurate to 3–4 meters in the US and other countries that adopt this technology. Differential GPS typically yields a position estimate that is accurate to about 1–3 meters with a local base station. Real-time–kinematic GPS (RTK GPS) with carrier-phase ambiguity resolution can produce centimeter-accurate position estimates. The latter two options require the existence of a nearby base station from which a differential error-correction signal can be sent to the roaming unit. Therefore, one cannot really speak of an unprepared environment anymore in that case. Commercial differential services are available, however, with base stations covering most of North America. For a long time, commercial differential GPS receivers provided update rates of up to 5Hz, which is suboptimal for tracking fast motion of people or objects. Newer products provide update rates of up to 20 Hz. (Trimble, 2002). More details on GPS and other global tracking systems can be found in Chapter 3.

Another position tracking system applicable for wide-area mobile AR involves calculating a person's location from signal quality measures of IEEE 802.11b (WiFi) wireless networking. This obviously also requires the deployment of equipment in the environment, in this case the WiFi access points, but if such a wireless network is the communication technology of choice for the mobile AR system, the positioning system can serve as an added benefit. Several research projects, and at least one commercial product, are exploring this concept. The RADAR system uses multilateration and precomputed signal strength maps for this purpose (Bahl and Padmanabhan, 2000), while Castro and colleagues (2001) employ a Bayesian networks approach. The achievable resolution depends on the density of access points deployed to form the wireless network. Ekahau (Ekahau, 2002) offer a software product that allows position tracking of WiFi enabled devices after a manual data collection/calibration step.

Two additional means of determining position are often employed in MARS, mostly as part of hybrid tracking systems: Inertial sensors and vision-based approaches. Accelerometers and gyroscopes are self-contained or *sourceless* inertial sensors. Their main problem is drift. The output of accelerometers needs to be integrated once with respect to time, in order to recover velocity, and twice to recover position. Hence, any performance degradations in the raw data lead to rapidly increasing errors in the resulting position estimate. In practice, this approach to position estimation can only be employed for very small time periods between updates gathered from a more reliable source. Inertial sensors can also be used to detect the act of a pedestrian taking a step. This is the functional principle of *pedometers,* which, when combined with accurate heading information, can provide a practical *dead-reckoning* method (Point Research, 2002; Höllerer *et al.*, 2001b).

Vision-based approaches are a promising option for 6DOF poses estimation in a general mobile setting. One or two tiny cameras are mounted on the glasses, so that the computer can approximately see what the user sees. Model-based vision techniques require an accurate model of the environment with known landmarks that can be recognized in the image feeds. In contrast, move-matching algorithms track dynamically chosen key points along the image sequence, leading to relative, rather than absolute, tracking solutions, which means that further registration of the image sequence coordinate system with the physical world needs to be established to enable 3D graphical overlays. Simultaneous reconstruction of the camera motion and scene geometry is possible, but such computations are highly computationally expensive, and existing algorithms require a "batch bundle adjustment," a global offline computation over the entire image sequence. Finally, 2D image-based feature tracking techniques measure so-called "optical flow" between subsequent video images. Such techniques are comparatively fast, but by themselves cannot estimate 3D camera motion. Combinations of all these approaches are possible. Recent research reports promising results for some test scenarios (Julier and Bishop, 2002). However, in general, computer vision algorithms still lack robustness and require such high amounts of computation that pure vision solutions for general-case real-time tracking are still out of reach. For the time being, hybrids of vision-based tracking and other sensing technologies show the biggest promise.

Orientation tracking also benefits greatly from hybrid approaches. The basic technologies available for orientation sensing are electromagnetic compasses (magnetometers), gravitational tilt sensors (inclinometers), and gyroscopes (mechanical and optical). Hybrid solutions have been developed, both as commercial products and research prototypes. The IS300 and InertiaCube2 orientation sensors by InterSense (2002) combine three micro-electromechanical gyroscopes, three accelerometers (for motion prediction), and an electromagnetic compass in one small integrated sensor. Azuma and colleagues (1999) presented a hybrid tracker that combines a carefully calibrated compass and tilt sensor with three rate gyroscopes. You and colleagues (1999) extended that system by a move-matching vision algorithm, which did not, however, run in real time. Behringer (1999) presented a vision-based correction method based on comparing the silhouette of the horizon line with a model of local geography. Satoh and colleagues (2001) employed a template-matching technique on manually selected landmarks in a real-time algorithm that corrects for the orientation drift of a highly accurate fiber optic gyroscope (Sawada *et al.*, 2001).

In summary, the problem of tracking a person's pose for general mobile AR purposes is a hard problem with no single best solution. Hybrid tracking approaches are currently the most promising way to deal with the difficulties posed by general indoor and outdoor mobile AR environments.

8.3.4 Environmental Modeling

For AR purposes, it is often useful to have access to geometrical models of objects in the physical environment. As mentioned above, one use of such models is in registration. If you want to annotate a window in a building, the computer has to know where that window is located with regard to the user's current position and field of view. Having a detailed hierarchical 3D model of the building, including elements such as floors, rooms, doors, and windows, gives the computer flexibility in answering such questions. Some tracking techniques, such as the model-based computer vision approaches mentioned in Section 8.3.3, rely explicitly on features represented in more or less detailed models of the tracking environment. Geometrical computer models are also used for figuring out *occlusion* with respect to the observer's current view. For example, if portions of a building in front of the observer are occluded by other objects, only the non-occluded building parts should be annotated with the building's name to avoid confusing the observer as to which object is annotated (Bell *et al.,* 2001).

For the purposes mentioned so far, an environment model does not need to be photorealistic. One can disregard materials, textures, and possibly even geometric detail. In fact, in model-based tracking, often only a "cloud" of unconnected 3D sample points is used. More realistic geometric models of real-world structures, such as the ones depicted in Figure 8.8, are often used for annotation purposes, or for giving an overview of the real environment. For example, a building that is occluded from the user's view can be displayed in its exact hidden location via AR, enabling the user, in a sense, to see through walls. Somebody looking for a specific building can be shown a virtual version of it on the AR screen. Having gotten an idea of the building's shape and texture, the person might watch the model move off in the correct direction, until it coincides with the real building in physical space. A three-dimensional map of the environment can be presented to the user to give a bird's-eye overview of the surroundings (Stoakley *et al.*, 1995). Figures 8.8 and 8.9 show examples of such *worlds in miniature* (WIM) used in AR.

Creating 3D models of large environments is a research challenge in its own right. Automatic, semiautomatic, and manual techniques can be employed, among them 3D reconstruction from satellite imagery and aerial photographs, 3D imaging with laser range finders, reconstruction from a set of partly overlapping photographs, surveying with total stations and other telemetry tools, and manual reconstruction using 3D modeling software. Even AR itself can be employed for modeling purposes, as mentioned in Section 8.2.1. Abdelguerfi (2001) provides an overview of 3D synthetic environment reconstruction. The models in Figure 8.8 were reconstructed by extruding 2D map outlines of Columbia University's campus and our research laboratory, refining the resulting models by hand, and texture mapping them selectively with photographs taken from various strategic positions. The WIM of Figure 8.9 shows a 3D model of the currently selected building (Dodge Hall) in the context of an aerial photograph of the current environment.

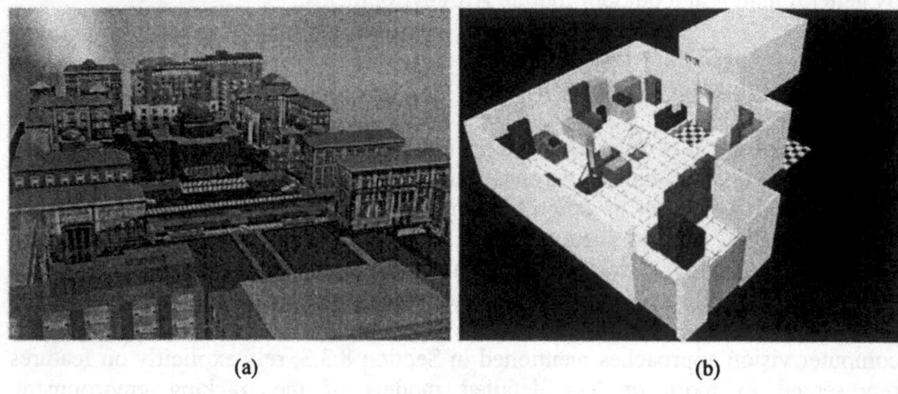

(a) (b)

Figure 8.8 Environmental Modeling: (a) Model of a campus. (b) Model of a laboratory

Figure 8.9 Context-Overview: World-in-miniature map

3D spatial models can be arbitrarily complex. Consider, for example, the task of completely modeling a large urban area, down to the level of water pipes and electric circuits in walls of buildings. There are significant research problems involved in the modeling, as well as the organization and storage of such data in spatial databases and data structures optimized for specific queries. Finally, environmental modeling does not end with a static model of the geometry. Most environments are dynamic: changes in the geometric models (due to moving objects, construction, or destruction) need to be tracked, and reflected in the environmental model. The databases may need to change quite rapidly, depending on the level of detail considered.

8.3.5 Wearable Input and Interaction Technologies

How to interact with wearable computers effectively and efficiently is another open research question. The desktop UI metaphor, often referred to as WIMP (windows, icons, menus, and pointing), is not a good match for mobile and wearable computing, mostly because it places unreasonable motor skill and attention demands on mobile users interacting with the real world.

As a general UI principle, AR can provide a user with immediate access to the physical world. Visual attention does not need to be divided between the task in the physical world and a separate computer screen. However, interacting seamlessly with such a computing paradigm is a challenge. In this section, we review interaction technologies that have been tried for MARS.

Basic interaction tasks that graphical UIs handle, include *selecting, positioning,* and *rotating* virtual objects, *drawing paths* or *trajectories,* assigning quantitative values, referred to as *quantification,* and *text input.* AR UIs deal as much with the physical world as with virtual objects. Therefore, selection, annotation, and, possibly, direct manipulation of physical objects also play an important role in these kinds of UIs.

We already mentioned one class of input devices, namely the sensors that afford tracking and registration. Position tracking determines the user's locale, and head orientation tracking assists in figuring out the user's focus. Establishing the user's context in this fashion can effectively support user interaction. The UI can adapt to such input by limiting the choices for possible courses of action to a context-relevant subset. Both, position and head orientation tracking can also be employed for object selection. Suppose that the task is to select a building in an urban neighborhood. With position tracking only, the closest building, or a list of the *n* closest ones, might be listed on the display for direct selection via a button or scrolling input device. With head orientation, a user can point his or her head in the direction of the object to be selected. Selection can take place by dwelling for a certain time period on the object in view, or by active selection via button-like devices. Höllerer *et al.* (1999a) discuss several tracking-prompted selection mechanisms for a mobile AR system. Additional orientation trackers can provide hand tracking, which can be used to control pointers or manipulate virtual objects on the AR screen. Tracking hand or finger position for full 6DOF hand tracking, as is common in indoor virtual or augmented environments, would be a great plus for MARS, but is hard to achieve with mobile hardware in a general setting. Research prototypes for this purpose have experimented with vision-based approaches, and ultrasonic tracking of finger-worn acoustic emitters using three head-worn microphones (Foxlin and Harrington, 2000).

Quite a few mobile input devices tackle continuous 2D pointing. Pointing tasks, the domain of mice in desktop systems, can be performed in the mobile domain by trackballs, track-pads, and gyroscopic mice, many of which transmit data wirelessly to the host computer. It should be mentioned, however, that these devices are popular in large part because, lacking a better mobile UI standard, many researchers currently run common WIMP UIs on their mobile and wearable platforms. Accurate 2D pointing poses a big challenge for a mobile user's motor skills. However, 2D pointing devices can also be used to control cursor-less AR UIs (Feiner *et al.*, 1997). When user interaction mostly relies on discrete 2D pointing events (e.g., selecting from small lists of menu items), then small numeric keypads with arrow keys, or arrow keys only, might provide a solution that is more easily handled on the run, and more easily worn on the body.

Mobile UIs should obviously try to minimize encumbrance caused by UI devices. The ultimate goal is to have a free-to-walk, eyes-free, and hands-free UI with miniature computing devices worn as part of the clothing. As should be clear from our overview so far, this ideal cannot always be reached with current mobile computing and UI technology. Some devices, however, already nicely meet the size and ergonomic constraints of mobility. Auditory UIs, for example, can already be realized in a relatively inconspicuous manner, with small wireless earphones tucked into the ear, and microphones worn as part of a necklace or shoulder pad. There is a growing body of research on wearable audio UIs, dealing with topics such as speech recognition, speech recording for human-to-human interaction, audio information presentation, and audio dialogue. It is clear, however, that a standalone audio UI cannot offer the best possible solution for every situation. Noisy surroundings and environments that demand complete silence pose insurmountable problems to such an approach. On the other hand, audio can be a valuable medium for multimodal and multimedia UIs.

Other devices are more impractical for brief casual use, but have successfully been employed in research prototypes. *Glove-based* input devices, for example, using such diverse technologies as electric contact pads, flex sensors, accelerometers, and even force-feedback mechanisms, can recognize hand gestures, but have the drawback of looking awkward and impeding use of the hands in real-world activities. Nevertheless, the reliability and flexibility of glove gestures has made the computer glove an input device of choice for some MARS prototypes (Thomas and Piekarski, 2002). Starner *et al.* (1997b), on the other hand, explore vision-based hand gesture recognition, which leaves the hands unencumbered, but requires that a camera be worn on a hat or glasses, pointing down to the area in front of the user's body, in which hand gestures are normally made.

We already discussed the use of cameras for vision-based tracking purposes (Section 8.3.3). Apart from that purpose, and the potential of finger and hand gesture tracking, cameras can be used to record and document the user's view. This can be useful as a live video feed for teleconferencing, for informing a remote expert about the findings of AR field workers, or simply for documenting and storing everything that is taking place in front of the MARS user. Recorded video can be an important element in human-to-human interfaces, which AR technology nicely supports.

A technology with potential for mobile AR is gaze tracking. Eye trackers observe a person's pupils with tiny cameras to determine where that person's gaze is directed. Drawbacks are the additional equipment that needs to be incorporated into the eyewear, the brittleness of the technology (the tracker needs to be calibrated and the cameras must be fixed with respect to the eye), and the overwhelming amount of involuntary eye movement that needs to be correctly classified as such. With the right filters, however, gaze control could provide a very fast and immediate input device. As a pointing device, it could eliminate the need for an entire step of coordinated muscle activity that other pointing devices require in order to move a pointer to a location that was found through eye movement in the first place. Even without gaze control, gaze tracking provides a dynamic history of where a user's attention is directed. As computers gather more and more such knowledge about the user's interests and intentions, they can adapt their UIs better to suit the needs of the current context (see Section 8.4.1).

Other local sensors that can gather information about the user's state include biometric devices that measure heart rate and bioelectric signals, such as galvanic skin response, electroencephalogram (neural activity), or electromyogram (muscle activity) data. Employing such monitored biological activity for computer UI purposes is an ambitious research endeavor, but the hopes and expectations for future applicability are quite high. *Affective computing* (Picard, 1997) aims to make computers more aware of the emotional state of their users and able to adapt accordingly.

As we can see, UI technology can be integrated with the user more or less tightly. While the previous paragraph hinted at possible future human-machine symbioses (Licklider, 1960), current wearable computing efforts aim to simply make computing available in as unencumbered a form as possible. One item on this agenda is to make clothes more computationally aware; for example, by embroidering electronic circuits (Farringdon *et al.*, 1999). On the other hand, not every UI technology needs to be so tightly integrated with the user. Often, different devices that the user would carry, instead of wear on the body, can support occasionally arising tasks very efficiently. For example, hand-held devices such as palmtop or tablet computers are good choices for reading text, assuming high-contrast and high-resolution displays, and are well suited for pen-based input, using handwriting recognition and marking gestures. *Hybrid user interfaces*, as Feiner and colleagues (1997) explored them for mobile AR purposes, aim to employ different display and input technologies and reap the benefits of each technology for the purposes for which it is best suited. The applicability of a wide variety of input technologies is utilized nicely by *multimodal* UI techniques. Such techniques employ multiple input and output modes in time-synchronized combination (e.g., gestures, speech, vision, sound, and haptics), using different media to present the user with a more natural and robust, yet still predictable, UI.

Finally, fast and reliable text input to a mobile computer is hard to achieve. The standard keyboard, which is the proven solution for desktop computing, requires too much valuable space and a flat typing surface. Small, foldable, or even inflatable keyboards, or virtual ones that are projected by a laser onto a flat surface, are current commercial options or product prototypes. Chording keyboards, which require key combinations to be pressed to encode a single character, such as the one-handed Twiddler2 (Handykey, 2001), are very popular choices for text input in the wearable computing community. Cell phones provide their own alphanumeric input techniques via a numeric keypad. We already mentioned handwriting recognition, pen-based marking, and speech recognition, which experienced major improvements in accuracy and speed over the last decade, but cannot be applied in all situations. Soft keyboards enable text input via various software techniques, but use valuable display screen space for that purpose. Glove-based and vision-based hand gesture tracking do not provide the ease of use and accuracy necessary for serious adoption yet. It seems likely that speech input and some kind of fallback device (e.g., pen-based systems, or special purpose chording or miniature keyboards) will share the duty of providing text input to mobile devices in a wide variety of situations in the near future.

8.3.6 Wireless Communication and Data Storage Technologies

We already discussed the mobility of a computing system in terms of its size, ergonomics, and input/output constraints. Another important question is how connected such a system is in the mobile world. This question concerns the electronic exchange of information with other, mobile or stationary, computer systems. The degree of connectivity can vary from none at all to true global wireless communication. Most likely is a scenario where the mobile client has different options to get connected to the Internet, currently ranging in the area covered and connection speed from a fast direct cable connection (when used in a stationary office environment) to slightly slower wireless local area networks (WLANs), which offer full connectivity in building- or campus-sized networks of wireless access points, to wireless phone data connections with nationwide or international coverage, but much slower transmission speeds.

The first packet-based WLAN was ALOHANET at the University of Hawaii in 1971. Today, WLANs provide bandwidths ranging between 2 and 54 Mbps, and are quite common for providing coverage in campuses and homes. At least one US telecommunications consortium has plans for nationwide support of IEEE 802.11b (WiFi) networks (Cometa, 2002). During the first two years of the current century, US phone service providers began to roll out new nationwide networks based on third generation wireless technology (at bandwidths of 144 Kbps, and higher in some selected test areas), which nicely complement smaller sized community WLANs.

For close-range point-to-point connections between different devices, the Bluetooth consortium (Bluetooth, 1998) has established an industry standard for low-power radio frequency communication. Using this technology, wearable computers connect with input/output devices that a user can carry or wear on the body, or walk up to, as in the case of stationary printers. Bluetooth-enabled cellular phones provide access to nationwide wireless connectivity whenever faster networking alternatives are not available.

From the perspective of MARS, the integration of LBS with communication systems is an important issue. Whereas it might be sufficient for special purpose AR systems to store all related material on the client computer, or retrieve it from one single task-related database server, this is not true anymore in the mobile case. For true mobility, the AR client will need to connect to multiple distributed data servers in order to obtain the information relevant to the current environment and situation. Among the data that need to be provided to the client computer are the geometrical models of the environment (see Section 8.3.4), annotation material (object names, descriptions, and links), as well as conceptual information (object categorization and world knowledge) that allows the computer to make decisions about the best ways how to present the data. Some formalism is needed to express and store such meta-knowledge. Markup languages, such as various XML derivatives, are well suited for this purpose. XML offers the advantages of a common base language that different authors and user groups can extend for their specific purposes.

For interactive applications, as required by AR technology, as much as possible of the data that is to be displayed in world overlays should be stored, or cached on the local (worn or carried) client computer. This raises the question of how to upload and "page in" information about new environments that the mobile user is ready to roam and might want to explore. Such information can be loaded preemptively from distributed databases in batches of relative topical or geographical closure (e.g., all restaurants in a certain neighborhood close to the user's current location). We would like to emphasize that currently no coherent global data repository and infrastructure exists that would afford such structured access to data. Instead, different research groups working on mobile AR applications have established their own test infrastructures for this purpose. For example, in our own AR work, data and meta-data are stored and accessed dynamically in relational databases, and distributed to various clients via a data replication infrastructure. The database servers effectively turn into *AR servers*, responsible for providing the material used by the client for overlays to particular locations.

Research from grid computing, distributed databases, middleware, service discovery, indexing, search mechanisms, wireless networking, and other fields will be necessary to build examples of new communication infrastructures that enable such semantically prompted data access. The Internet offers the backbone to experiment with such data distribution on virtually any level of scale.

8.3.7 Summary: A Top-of-the-line MARS Research Platform

Now that we have reviewed the technological and data requirements of mobile AR, here is the summary of a hypothetical MARS research platform made from the most promising components that are available today to researchers. Some of the components currently easily exceed all but the most generous budgets. However, prices keep falling steadily, and new technologies enter the market at a rapid pace.

The type of base unit we would select for our MARS depends on the focus of our applications. If detailed 3D graphics is not a strict requirement, we would pick a small wearable computer. According to current product announcements, we will be able in early 2003 to buy hand-held computers with the processing power of a 1GHz Transmeta Crusoe chip, 256 MB main memory, about 10–20 GB disk space, and a full arsenal of I/O interfaces (e.g., OQO Ultra-Personal Computer, Tiqit eightythree). If 3D graphics is a must, we would settle for a (larger) notebook computer with integrated 3D graphics chip. Either solution would come equipped or could be extended with WiFi wide area and Bluetooth personal networking. For nationwide connectivity outside of WiFi networks, we would use a Bluetooth-enabled cell phone with a 3G service plan.

For a display we would pick a Microvision retinal scanning display: monocular, monochromatic, and far larger and heavier than desired, but sufficiently transparent indoors and bright outdoors.

For coarse position tracking, we would use an A-GPS receiver, employing the new chip technology that promises to achieve signal reception even in commonly untracked areas, such as beneath thick foliage, or indoors. Position tracking using WiFi signal quality measures is also a possibility. Higher-precision position tracking indoors is dependent on the availability of special purpose tracking infrastructures. If higher-precision outdoor position tracking is a must, we additionally use an RTK GPS receiver in areas equipped with an RTK GPS base station. For orientation tracking, we would choose fiber optic gyroscope (FOG) sensors (Sawada, 2001) in combination with a magnetometer, if the budget allows it. Small FOG-based sensors, customized for use as a head tracker, can currently cost up to $100,000. A much more affordable and smaller, but far less accurate, backup solution would be a small hybrid inertial and magnetometer-based sensor, such as the Inertiacube2 by InterSense (2002). Ideally, these position and orientation tracking technologies should be used to provide first guess for state-of-the-art vision-based tracking, which requires the addition of one or more tiny cameras.

We select a large set of input devices, so that we can make use of different technologies in different situations: Bluetooth-enabled earphones and microphone, wrist-worn keypad, Twiddler2 chord keyboard and mouse, and, if the selected base computer does not have a pen-operated screen already, or is too big for occasional hand-held use, an additional palm-sized tablet computer/display.

We will also need rechargeable batteries for powering all these devices. By relying on built-in batteries and two additional lithium-ion batteries, we can realistically expect such a system to have an up time of about three hours. For extended operation, we would need to add additional batteries.

8.4 MARS UI CONCEPTS

As described in the previous section, one significant impediment to the immediate widespread use of MARS UIs is technological. However, if we look at the progress in the field over the past ten years, and extrapolate into the future, we can be optimistic that many of the current hardware flaws will be resolved. In the remainder of this chapter, we will assume a faultless MARS device, as far as hardware and tracking is concerned, and take a look at the UI concepts of mobile AR, as facilitated by the software. The kind of computing that MARS make possible is quite different from the current static work and play environments of our offices and homes. In contrast, the world is the interface, which means that in dealing with a MARS we will rarely focus exclusively on the computer anymore. In fact, while we go about our day-to-day activities, we would not even be able, let alone want, to pay attention to the computer. At the same time, however, we will expect the system to provide assistance and augmentation for many of our tasks. Broll and colleagues (2001) describe a futuristic scenario of using such a mobile helper UI.

8.4.1 Information Display and Interaction Techniques

AR allows the user to focus on computer-supplied information and the real world at the same time. A UI for visual mobile AR can combine screen-stabilized, body-stabilized, and world-stabilized elements (Feiner *et al.*, 1993; Billinghurst *et al.*, 1998). Figure 8.10 shows a UI from the authors' mobile AR work (Höllerer *et al.*, 1999a), photographed through optical see-through head-worn displays. The virtual flags and labels are *world-stabilized* objects, residing in the world coordinate system, denoting points of interest in the environment. They are displayed in the correct perspective for the user's viewpoint, so the user can walk up to and around these objects just like physical objects. The labels face the user and maintain their size irrespective of distance to ensure readability. The blue and green menu bars on the top are *screen-stabilized*, meaning that they occupy the same position on the screen no matter where the user is looking, as is the cone-shaped pointer at the bottom of the screen, which is always pointing towards the currently selected world object. The two images of Figure 8.10 (a) and (b) show the same *in-place* menu options associated with the red flag in front of the columns, realized in two different ways. In part (a), our initial implementation, the menu was arranged in a circular world-stabilized fashion around the flag. This caused problems when the user turned his or her head during menu selection. In the design shown in part (b), the menu is a *screen-stabilized* element, linked back to its associated flag by a leader line, so that the user can easily turn back to it. We made the menu semi-transparent, so that the view of other virtual elements is not completely obstructed.

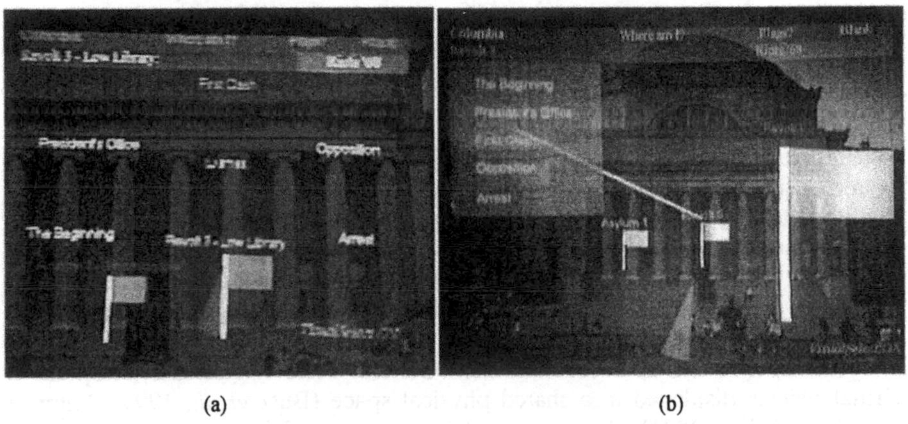

(a) (b)

Figure 8.10 World-Stabilized and Screen-Stabilized UI Elements

Body-stabilized information, unsurprisingly, is stored relative to the user's body, making it accessible at a turn of the head, independent of the user's location. Note, that in order to store virtual objects relative to the body with respect to yaw (e.g., consistently to the user's left), the body's orientation needs to be tracked in addition to head orientation. One can extend the notion of body stabilized objects to using general head-gestures for virtual object control. Figure 8.11 shows a WIM displayed in front of a mobile user who views the scene through a head-worn display. In part (a) the user is looking straight ahead. As the user looks down towards the ground, the WIM is shown in successively more detail and from an increasingly top-down perspective (parts b and c). Note that this is not a body-stabilized element: the user's head orientation alone is used to control the WIM. In this case, the WIM is always kept visible on the screen, aligned in yaw with the surrounding environment it represents, with head pitch used to control the WIM's pitch, size, position, and level of annotation (Bell *et al.*, 2002).

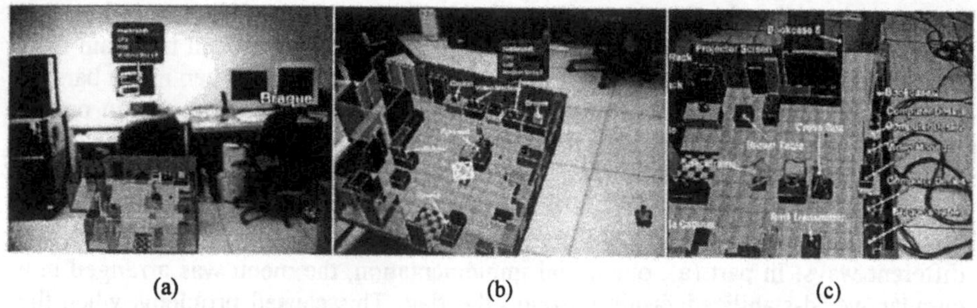

(a) (b) (c)

Figure 8.11 Head-Pitch Control Of WIM. (a) User looking straight ahead. (b) WIM scales up, tilts, und moves up the screen as user tilts head downwards. (c) Near top-down view of WIM as user looks further down

Mobile AR agrees well with the notion of non-command interfaces (Nielsen, 1993), in which the computer is reacting to sensed user context rather than explicit user commands. For a person trying to focus on a real-world task, and not on how to work a particular computer program, it is desirable that computers understand as much as possible about the task at hand without explicitly being told. Often, much of the needed interaction can be reduced to the user's answering several prompted questions (Pascoe *et al.*, 2000). Some tasks, however, such as placing or moving virtual objects in the environment or modeling them in the first place from physical examples, require extended user interaction with the AR UI. UIs that have been tried for such tasks range from a 2D cursor and head motion (Baillot *et al.*, 2001), to a tracked glove (Thomas and Piekarski, 2002), to a tracked graphics tablet on which UI elements can be overlaid (Reitmayr and Schmalstieg, 2001). Simple interaction with virtual material has also been achieved using vision-based hand tracking (Kurata *et al.*, 2001).

Mobile AR UIs invite collaboration. Several users can discuss and point to virtual objects displayed in a shared physical space (Butz *et al.*, 1999; Reitmayr and Schmalstieg, 2001). At the same time, every participant can see their own private version of the shared data, for example to see annotations optimized for their specific viewing angle (Bell *et al.*, 2001). Multiple users can collaborate in the field, and remote experts with a top-down overview of the user's environment can communicate and share information with the field worker (Höllerer *et al.*, 1999b).

8.4.2 Properties of MARS UIs

Mobile AR presents a way for people to interact with computers that is radically different from the static desktop or mobile office. One of the key characteristics of MARS is that both virtual and physical objects are part of the UI, and the dynamic context of the user in the environment can influence what kind of information the computer needs to present next. This raises several issues:

Control: Unlike a stand-alone desktop UI, where the only way the user can interact with the presented environment is through a set of well-defined techniques, the MARS UI needs to take into account the unpredictability of the real world. For example, a UI technique might rely on a certain object being in the user's field of view and not occluded by other information. Neither of the properties can be guaranteed: the user is free to look away, and other information could easily get in the way, triggered by the user's own movement or an unforeseen event (such as another user entering the field of view). Thus, to be effective, the UI technique either has to relax the non-occlusion requirement, or has to somehow guarantee non-occlusion in spite of possible contingencies.

Consistency: People have internalized many of the laws of the physical world. When using a computer, a person can learn the logic of a new UI. As long as these two worlds are decoupled (as they are in the desktop setting), inconsistencies between them are often understandable. In the case of MARS, however, we need to be very careful to design UIs in which the physical and virtual world are consistent with each other.

Need for embedded semantic information: In MARS, virtual material is overlaid on top of the real world. Thus we need to establish concrete semantic relationships between virtual and physical objects to characterize UI behavior. In fact, since many virtual objects are designed to annotate the real world, these virtual objects need to store information about the physical objects to which they refer (or at least have to know how to access that information).

Display space: In terms of the available display space and its best use, MARS UIs have to deal with a much more complicated task compared to traditional 2D UIs. Instead of one area of focus (e.g., one desktop display), we have to deal with a potentially unlimited display space surrounding the user, only a portion of which is visible at any time. The representation of that portion of augmented space depends on the user's position, head orientation, personal preferences (e.g., filter settings) and ongoing interactions with the augmented world, among other things. Management of virtual information in this space is made even more difficult by constraints that other pieces of information may impose. Certain virtual or physical objects may, for example, need to be visible under all circumstances, and thus place restrictions on the display space that other elements are allowed to obstruct.

The display management problem is further complicated by the possibility of taking into account multiple displays. MARS, as a nonexclusive UI to the augmented world, may seamlessly make use of other kinds of displays, ranging from wall-sized, to desk-top, to hand-held. If such display devices are available and accessible to the MARS, questions arise as to which display to use for what kind of information and how to let the user know about that decision.

Scene dynamics: In a head-tracked UI, the scene will be much more dynamic than in a stationary UI. In MARS, this is especially true, since in addition to all the dynamics due to head motion, the system has to consider moving objects in the real world that might interact visually or audibly with the UI presented on the head-worn display. Also, we have to contend with a potentially large variability in tracking accuracy over time. Because of these unpredictable dynamics, the spatial composition of the UI needs to be flexible and the arrangement of UI elements may need to be changed. On the other hand, traditional UI design wisdom suggests minimizing dynamic changes in the UI composition (Shneiderman, 1998).

One possible solution to this dilemma lies in the careful application of automated UI management techniques.

8.4.3 UI Management

In our own work on MARS, we adapt and simplify the UI through a set of management techniques, including the following steps: information filtering, UI component design, and view management.

The large amount of virtual information that can be displayed, coupled with the presence of a richly complex physical world, creates the potential for clutter. Cluttered displays can overwhelm the user with unneeded information, impacting her ability to perform her tasks effectively. Just as in desktop information visualization (Shneiderman, 1998), we address clutter through information filtering. For our MARS work, *information filtering* (Julier *et al.*, 2000) means the act of culling the information that can potentially be displayed by identifying and prioritizing what is relevant to a user at a given point in time. The priorities can be based on the user's tasks, goals, interests, location, or other user context or environmental factors.

While information filtering determines the subset of the available information that will be displayed, it is still necessary to determine the format in which this information is to be communicated, and how to realize that format in detail. Registration accuracy, or how accurately the projected image of a virtual object can be positioned, scaled, and oriented relative the real world, is an important factor in choosing the right UI format. Registration accuracy is determined by tracking system accuracy, which, as the mobile user moves about, may vary for a variety of reasons that depend on the tracking technologies used. Therefore, if information is always formatted in a way that assumes highly accurate registration, that information will not be presented effectively when registration accuracy decreases. To address this issue, *UI component design* (Höllerer *et al.*, 2001b) determines the format in which information should be conveyed, based on contextual information, such as the available display resources and tracking accuracy. This technique determines the concrete elements that comprise the UI and information display.

Filtering and formatting information is not enough — the information must be integrated with the user's view of the physical world. For example, suppose that annotations are simply projected onto the user's view of the world such that each is collocated with a physical object with which it is associated. Depending on the user's location in the world (and, thus, the projection that they see), annotations might occlude or be occluded by other annotations or physical objects, or appear ambiguous because of their proximity to multiple potential referents. *View management* (Bell *et al.*, 2001) attempts to ensure that the displayed information is arranged appropriately with regard to the projections on the view plane of it and other objects; for example, virtual or physical objects should not occlude others that are more important, and relationships among objects should be as unambiguous as possible.

More detail about this suggested UI management pipeline can be found in Höllerer *et al.* (2001a).

8.5 CONCLUSIONS

In this chapter, we presented an overview of the field of mobile AR, including historical developments, future potential, application areas, challenges, components and requirements, state-of-the-art systems, and UI concepts. AR and wearable computing are rapidly growing fields, as exemplified by the soaring number of research contributions and commercial developments since the mid 1990s. We have reached an important point in the progress toward mobile AR, in that the available technology is powerful enough for an increasing number of impressive research prototypes, but not yet sufficiently reliable, general, and comfortable for mass adoption. Compared to other applications described in this book, which are immediately realizable using today's technology, it will take more time for mobile AR to reach the computing mainstream. However, mobile AR will have an enormous impact when it becomes commonplace.

We are looking forward to further progress in the areas of computing hardware miniaturization, battery design, display technology, sensor technology, tracking accuracy and reliability, general vision-based tracking and scene understanding, and overall comfort. We anticipate the emergence of distributed data infrastructures for context-based computing in general, and AR in particular, leading to much improved data access capabilities for mobile users.

Finally, we hope that the benefits of mobile AR will be achieved without compromising privacy and comfort (Feiner, 1999). Research and development in a field that could have a significant impact on social structures and conventions should be accompanied by careful consideration of how the commendable aspects of our social equilibrium can be protected and strengthened.

8.6 ACKNOWLEDGMENTS

The research described here is funded in part by ONR Contracts N00014-99-1-0249, N00014-99-1-0394, and N00014-99-1-0683, NSF Grant IIS-00-82961, and gifts from Intel, Microsoft Research, Mitsubishi Electric Research Laboratories, and Pacific Crest. We would also like to thank Blaine Bell, Drexel Hallaway, Ryuji Yamamoto, Hrvoje Benko, Sinem Güven, Tiantian Zhou, Alex Olwal, Gabor Bláskó, Simon Lok, Elias Gagas, Tachio Terauchi, and Gus Rashid for their invaluable contributions to Columbia University's MARS infrastructure over the years.

REFERENCES

Abdelguerfi, M., 2001, *3D Synthetic Environment Reconstruction*. Kluwer Academic Publishers, New York, NY.

Azuma, R., 1997, A survey of augmented reality. *Presence: Teleoperators and Virtual Environments*, 6(4): pp. 355–385.

Azuma, R., 1999, The challenge of making augmented reality work outdoors. In Ohta, Y. and Tamura, H., editors, *Mixed Reality, Merging Real and Virtual Worlds*, pp. 379–390. Ohmsha/Springer, Tokyo/New York.

Azuma, R., Baillot, Y., Behringer, R., Feiner, S., Julier, S., and MacIntyre, B., 2001, Recent advances in augmented reality. *IEEE Computer Graphics and Applications*, 21(6): pp. 34–47.

Azuma, R., Hoff, B., Neely III, H., and Sarfaty, R., 1999, A motion-stabilized outdoor augmented reality system. In *Proc. IEEE Virtual Reality '99*, pp. 242–259.

Bahl, P. and Padmanabhan, V., 2000, RADAR: an in-building RF-based user location and tracking system. In *Proc. IEEE Infocom 2000*, pages 775–784, Tel Aviv, Israel.

Baillot, Y., Brown, D., and Julier, S., 2001, Authoring of physical models using mobile computers. In *Proc. ISWC '01 (Fifth IEEE Int. Symp. on Wearable Computers)*, pp. 39–46, Zürich, Switzerland.

Behringer, R., 1999, Registration for outdoor augmented reality applications using computer vision techniques and hybrid sensors. In *Proc. IEEE Virtual Reality '99*, pp. 244–251.

Bell, B., Feiner, S., and Höllerer, T., 2001, View management for virtual and augmented reality. In *Proc. ACM UIST 2001 (Symp. on User Interface Software and Technology)*, pp. 101–110, Orlando, FL. (CHI Letters, vol. 3, no. 2).

Bell, B., Höllerer, T., and Feiner, S., 2002, An annotated situation-awareness aid for augmented reality. In *Proc. ACM UIST 2002 (Symp. on User Interface Software and Technology)*, pp. 213–216, Paris, France.

Billinghurst, M., Bowskill, J., Dyer, N., and Morphett, J., 1998, Spatial information displays on a wearable computer. *IEEE Computer Graphics and Applications*, 18(6): pp. 24–31.

Broll, W., Schäfer, L., Höllerer, T., and Bowman, D., 2001, Interface with angels: The future of VR and AR interfaces. *IEEE Computer Graphics and Applications*, 21(6): pp. 14–17.

Brumitt, B., Meyers, B., Krumm, J., Kern, A., and Shafer, S., 2000, Easyliving: Technologies for intelligent environments. In *Proc. Second Int. Symp. on Handheld and Ubiquitous Computing (HUC 2000)*, pp. 12–29, Bristol, UK. Springer Verlag.

Butz, A., Höllerer, T., Feiner, S., MacIntyre, B., and Beshers, C., 1999, Enveloping users and computers in a collaborative 3D augmented reality. In *Proc. IWAR '99 (IEEE and ACM Int. Workshop on Augmented Reality)*, pp. 35–44, San Francisco, CA.

Castro, P., Chiu, P., Kremenek, T., and Muntz, R., 2001, A probabilistic room location service for wireless networked environments. In *Proc. ACM UbiComp 2001: Ubiquitous Computing*, volume 2201 of *Lecture Notes in Computer Science*, pp. 18–35.

Caudell, T. P. and Mizell, D. W., 1992, Augmented Reality: An Application of Heads-Up Display Technology to Manual Manufacturing Processes. In *Proceedings of 1992 IEEE Hawaii International Conference on Systems Sciences*. IEEE Press.

Cheverst, K., Davies, N., Mitchell, K., and Blair, G. S., 2000, Developing a Context-aware Electronic Tourist Guide: Some Issues and Experiences. In *Proceedings of ACM CHI' 00*, Netherlands.

Cometa Networks, 2002, Wholesale nationwide broadband wireless internet access. <http://www.cometanetworks.com>.

Drascic, D. and Milgram, P., 1996, Perceptual issues in augmented reality. *Proc. SPIE: Stereoscopic Displays and Virtual Reality Systems III*, 2653: pp. 123–134.

Ekahau 2002. Location in wireless networks. <http://www.ekahau.com>.

Ericsson, IBM, Intel, Nokia, and Toshiba, 1998. Bluetooth mobile wireless initiative. <http://www.bluetooth.com>.

Farringdon, J., Moore, A. J., Tilbury, N., Church, J., and Biemond, P. D., 1999, Wearable sensor badge and sensor jacket for context awareness. In *Proc. ISWC '99 (Third Int. Symp. on Wearable Computers)*, pp. 107–113, San Francisco, CA.

Feiner, S., 1999, The importance of being mobile: Some social consequences of wearable augmented reality systems. In *Proc. IWAR '99 (IEEE and ACM Int. Workshop on Augmented Reality)*, pp. 145–148, San Francisco, CA.

Feiner, S., 2002, Augmented Reality: A new way of seeing. *Scientific American*, 286(4):48–55.

Feiner, S., MacIntyre, B., Haupt, M., and Solomon, E., 1993, Windows on the world: 2D windows for 3D augmented reality. In *Proc. UIST '93 (ACM Symp. on User Interface Software and Technology)*, pp. 145–155, Atlanta, GA.

Feiner, S., MacIntyre, B., and Höllerer, T., 1999, Wearing it out: First steps toward mobile augmented reality systems. In Ohta, Y. and Tamura, H., editors, *Mixed Reality: Merging Real and Virtual Worlds*, pp. 363–377. Ohmsha (Tokyo)–Springer Verlag, Berlin.

Feiner, S., MacIntyre, B., Höllerer, T., and Webster, A., 1997, A touring machine: Prototyping 3D mobile augmented reality systems for exploring the urban environment. In *Proc. ISWC '97 (First IEEE Int. Symp. on Wearable Computers)*, pp. 74–81, Cambridge, MA.

Feiner, S. and Shamash, A., 1991, Hybrid user interfaces: virtually bigger interfaces for physically smaller computers. In *Proc. ACM UIST '91*, pp. 9–17. ACM press.

Foxlin, E. and Harrington, M., 2000, Weartrack: A self-referenced head and hand tracker for wearable computers and portable vr. In *Proc. ISWC '00 (Fourth IEEE Int. Symp. on Wearable Computers)*, pp. 155–162, Atlanta, GA.

Fuchs, H., Livingston, M., Raskar, R., Colucci, D., Keller, K., State, A., Crawford, J., Rademacher, P., Drake, S., and Meyer, A., 1998, Augmented Reality Visualization for Laparoscopic Surgery. In *Proceedings of the First International Conference on Medical Image Computing and Computer-Assisted Intervention*.

Furmanski, C., Azuma, R., and Daily, M., 2002, Augmented-reality visualizations guided by cognition: Perceptual heuristics for combining visible and obscured information. In *Proc. ISMAR '02 (Int. Symposium on Augmented Reality)*, pp. 215–224, Darmstadt, Germany.

Furness, T. 1986. The super cockpit and its human factors challenges. In *Proc. Human Factors Society 30th Annual Meeting*, pp. 48–52, Santa Monica, CA.

Getting, I. 1993. The global positioning system. *IEEE Spectrum*, 30(12): pp. 36–47.

Global Locate 2002. Global Locate/Fujitsu GL-16000 Indoor GPS chip. <http://www.globallocate.com>.

Handykey, 2001, Handykey Twiddler2 chord keyboard. <http://www.handykey.com>.

Hasvold, P., 2002, In-the-field health informatics. In *The Open Group conference*, Paris, France. <http://www.opengroup.org/public/member/q202/documentation/plenary/hasvold.pdf>.

Höllerer, T., Feiner, S., Hallaway, D., Bell, B., Lanzagorta, M., Brown, D., Julier, S., Baillot, Y., and Rosenblum, L., 2001a, User interface management techniques for collaborative mobile augmented reality. *Computers and Graphics*, 25(5): pp. 799–810.

Höllerer, T., Feiner, S., and Pavlik, J., 1999a, Situated documentaries: Embedding multimedia presentations in the real world. In *Proc. ISWC '99 (Third IEEE Int. Symp. on Wearable Computers)*, pp. 79–86, San Francisco, CA.

Höllerer, T., Feiner, S., Terauchi, T., Rashid, G., and Hallaway, D., 1999b, Exploring MARS: Developing indoor and outdoor user interfaces to a mobile augmented reality system. *Computers and Graphics*, 23(6): pp. 779–785.

Höllerer, T., Hallaway, D., Tinna, N., and Feiner, S., 2001b, Steps toward accommodating variable position tracking accuracy in a mobile augmented reality system. In *2nd Int. Workshop on Artificial Intelligence in Mobile Systems (AIMS '01)*, pp. 31–37.

Hua, H., Gao, C., Brown, L.D., Ahuja, N., and Rolland, J.P., 2001, Using a head-mounted projective display in interactive augmented environments. In *Proc. ISAR 2001 (IEEE and ACM Int. Symp. on Augmented Reality 2001)*, pp. 217–223, New York, NY.

InterSense, 2002, InterSense Inc., IS-900 Wide Area Precision Motion Tracker, IS300 and InertiaCube2 orientation sensors. <http://www.isense.com>.

Ishii, H. and Ullmer, B., 1997, Tangible bits: Towards seamless interfaces between people, bits and atoms. In *Proc. Conference on Human Factors in Computing Systems (CHI '97), (Atlanta, March 1997)*, pp. 234–241. ACM Press.

Julier, S. and Bishop, G., editors 2002. *IEEE Computer Graphics and Applications, Special Issue on Tracking*, volume 6(22), pp. 22–80. IEEE Computer Society.

Julier, S., Lanzagorta, M., Baillot, Y., Rosenblum, L., Feiner, S., Höllerer, T., and Sestito, S., 2000, Information filtering for mobile augmented reality. In *Proc. ISAR '00 (IEEE and ACM Int. Symposium on Augmented Reality)*, pp. 3–11, Munich, Germany.

Kasai, I., Tanijiri, Y., Endo, T., and Ueda, H., 2000. A forgettable near eye display. In *Proc. ISWC '00 (Fourth Int. Symp. on Wearable Computers)*, pp. 115–118, Atlanta, GA.

Klinker, G., Creighton, O., Dutoit, A. H., Kobylinski, R., Vilsmeier, C., and Brügge, B., 2001, Augmented maintenance of powerplants: a prototyping case study of a mobile AR system. In *Proc. ISAR '01 (IEEE and ACM Int. Symposium on Augmented Reality)*, pp. 124–133, New York, NY.

Kurata, T., Okuma, T., Kourogi, M., Kato, T., and Sakaue, K., 2001, VizWear: Toward human-centered interaction through wearable vision and visualization. *Lecture Notes in Computer Science*, 2195: pp. 40–47.

Licklider, J., 1960, Man-Computer Symbiosis. *IRE Transactions on Human Factors in Electronics,* HFE-1, pp. 4–11.

Loomis, J., Golledge, R., and Klatzky, R., 1993, Personal guidance system for the visually impaired using GPS, GIS, and VR technologies. In *Proc. Conf. on Virtual Reality and Persons with Disabilities*, Millbrae, CA.

Mann, S., 1997, Wearable computing: A first step toward personal imaging. *IEEE Computer*, 30(2): pp. 25–32.

Mizell, D., 2001, Boeing's wire bundle assembly project. In Barfield, W. and Caudell, T., editors, *Fundamentals of Wearable Computers and Augumented Reality*, pp. 447–467. Lawrence Erlbaum Assoc, Mahwah, NJ.

Nielsen, J., 1993, Noncommand user interfaces. *Communications of the ACM*, 36(4):82–99.

Nusser, S. M., Miller, L. L., Clarke, K., and Goodchild, M. F., 2003, Geospatial IT for mobile field data collection. *Communications of the ACM*, 46(1): pp. 63–64.

Pascoe, J., Ryan, N., and Morse, D, 2000, Using while moving: HCI issues in fieldwork environments. *ACM Transactions on Computer-Human Interaction*, 7(3): pp. 417–437.

Pausch, R., 1991, Virtual reality on five dollars a day. *Proc. ACM CHI '91 Conference on Human Factors in Computing Systems*, pp. 265–270.

Pavlik, J. V., 2001, *Journalism and New Media*. Columbia University Press, New York, NY.

Petrie, H., Johnson, V., Strothotte, T., Raab, A., Fritz, S., and Michel, R., 1996, MoBIC: Designing a travel aid for blind and elderly people. *Jnl. of Navigation*, 49(1): pp. 45–52.

Picard, R. W., 1997, *Affective Computing*. MIT Press, Cambridge, MA.

Piekarski, W. and Thomas, B., 2001, Tinmith-Metro: New outdoor techniques for creating city models with an augmented reality wearable computer. In *Proc. ISWC '01 (Fifth IEEE Int. Symp. on Wearable Computers)*, pp. 31–38, Zürich, Switzerland.

Point Research Corporation, 2002, Dead reckoning module DRM-III. Accurate navigation for personnel on foot. <http://www.pointresearch.com>.

Raab, F. H., Blood, E. B., Steiner, T. O., and Jones, H. R., 1979, Magnetic position and orientation tracking system. *IEEE Transaction on Aerospace and Electronic Systems*, AES-15(5): pp. 709–717.

Raskar, R., Welch, G., Cutts, M., Lake, A., Stesin, L., and Fuchs, H., 1998, The office of the future: A unified approach to image-based modeling and spatially immersive displays. In *Proc. SIGGRAPH '98*, pp. 179–188.

Reitmayr, G. and Schmalstieg, D., 2001, Mobile collaborative augmented reality. In *Proc. ISAR '01 (Int. Symposium on Augmented Reality)*, pp. 114–123, New York, NY.

Rekimoto, J., Ayatsuka, Y., and Hayashi, K., 1998, Augment-able reality: Situated communication through physical and digital spaces. In *Proc. ISWC '98 (Second Int. Symp. on Wearable Computers)*, pp. 68–75, Cambridge, MA.

Rekimoto, J. and Nagao, K., 1995, The world through the computer: Computer augmented interaction with real world environments. In *Proc. ACM Symposium on User Interface Software and Technology*, Virtual and Augmented Realities, pp. 29–36.

Rolland, J. P., Davis, L. D., and Baillot, Y., 2001, A survey of tracking technologies for virtual environments. In Barfield, W. and Caudell, T., editors, *Fundamentals of Wearable Computers and Augumented Reality*. Lawrence Erlbaum Assoc, Mahwah, NJ.

Sato, K., Ban, Y., and Chihara, K., 1999, MR aided engineering: Inspection support systems integrating virtual instruments and process control. In Ohta, Y. and Tamura, H., editors, *Mixed Reality, Merging Real and Virtual Worlds*, pp. 347–361. Ohmsha/Springer, Tokyo/New York.

Satoh, K., Anabuki, M., Yamamoto, H., and Tamura, H., 2001, A hybrid registration method for outdoor augmented reality. In *Proc. ISAR '01 (Int. Symposium on Augmented Reality)*, pp. 57–76, New York, NY.

Sawada, K., Okihara, M., and Nakamura, S., 2001, A wearable attitude measurement system using a fiber optic gyroscope. In *Proc. ISMR '01 (Second Int. Symp. on Mixed Reality)*, pp. 35–39, Yokohama, Japan.

Sawhney, N. and Schmandt, C., 1998, Speaking and listening on the run: Design for wearable audio computing. In *Proc. ISWC '98 (Second Int. Symp. on Wearable Computers)*, pp. 108–115, Cambridge, MA.

Shneiderman, B., 1998, *Designing the User Interface*. Addison-Wesley, third edition.

Spohrer, J., 1999, Information in places. *IBM Systems Journal*, 38(4): pp. 602–628.

Starner, T., Leibe, B., Singletary, B., and Pair, J., 2000, MIND-WARPING: towards creating a compelling collaborative augmented reality game. In *Proc. Int. Conf. on Intelligent User Interfaces (IUI '00)*, pp. 256–259.

Starner, T., Mann, S., Rhodes, B., Levine, J., Healey, J., Kirsch, D., Picard, R., and Pentland, A., 1997a, Augmented reality through wearable computing. *Presence*, 6(4): pp. 386–398.

Starner, T., Weaver, J., and Pentland, A., 1997b, A wearable computing based american sign language recognizer. In *Proc. ISWC '97 (First Int. Symp. on Wearable Computers)*, pp. 130–137, Cambridge, MA.

Stoakley, R., Conway, M., and Pausch, R., 1995, Virtual reality on a WIM: Interactive worlds in miniature. In *Proceedings of Human Factors in Computing Systems (CHI '95)*, pp. 265–272.

Sutherland, I., 1968, A head-mounted three dimensional display. In *Proc. FJCC 1968*, pp. 757–764, Washington, DC. Thompson Books.

Tappert, C. C., Ruocco, A. S., Langdorf, K. A., Mabry, F. J., Heineman, K. J., Brick, T. A., Cross, D. M., and Pellissier, S. V., 2001, Military applications of wearable computers and augmented reality. In Barfield, W. and Caudell, T., editors, *Fundamentals of Wearable Computers and Augumented Reality*, pp. 625–647. Lawrence Erlbaum Assoc, Mahwah, NJ.

Thomas, B., Close, B., Donoghue, J., Squires, J., De Bondi, P., Morris, M., and Piekarski, W., 2000, ARQuake: An outdoor/indoor augmented reality first person application. In *Proc. ISWC '00 (Fourth Int. Symp. on Wearable Computers)*, pp. 139–146, Atlanta, GA.

Thomas, B. H. and Piekarski, W., 2002. Glove based user interaction techniques for augmented reality in an outdoor environment. *Virtual Reality: Research, Development, and Applications*, 6(3): pp. 167–180. Springer-Verlag London Ltd.

Thorp, E. O., 1998, The invention of the first wearable computer. In *Proc. ISWC '98 (Second Int. Symposium on Wearable Computers)*, pp. 4–8.

Trimble Navigation Ltd, 2002, Trimble MS750 RTK receiver for precise dynamic positioning. <http://www.trimble.com/ms750.html>.

Vlahakis, V., Ioannidis, N., Karigiannis, J., Tsotros, M., Gounaris, M., Stricker, D., Gleue, T., Daehne, P., and Almaida, L., 2002, Archeoguide: An augmented reality guide for archaeological sites. *IEEE Computer Graphics and Applications*, 22(5): pp. 52–59.

Weiser, M., 1991, The computer for the 21st century. *Scientific American*, 3(265): pp. 94–104.

Zhang, X., Genc, Y., and Navab, N., 2001, Taking AR into large scale industrial environments: Navigation and information access with mobile computers. In *Proc. ISAR '01 (Int. Symposium on Augmented Reality)*, pp. 179–180, New York, NY.

Zhang, X., Navab, N., and Liou, S.-P., 2000, E-Commerce direct marketing using augmented reality. In *Proc. ICME2000 (IEEE Int. Conf. on Multimedia and Exposition)*, New York, NY.

Part Three: Applications

CHAPTER NINE

Emergency Response Systems

Jun Tobita and Nobuo Fukuwa

9.1 OVERVIEW OF EMERGENCY RESPONSE SYSTEMS

9.1.1 General Aspects

In a state of emergency, such as that of a large-scale natural disaster, authorities need to quickly and effectively assess the situation and immediately decide on countermeasures. They need to answer the following questions: What has happened? Where has the greatest damage occurred? What kind of support systems are needed and to what degree are they needed? Do the authorities need to work with other organizations? As the police and fire departments rush to the scene of damage, the situation gradually becomes clearer and the authorities can start to tally up the damage. However, handling a state of emergency in this way generally takes time, and in the case of large-scale disasters, can require up to a day or more. At the same time, there is a continuing need for actions that must be performed promptly, such as search and rescue (SAR) missions, medical treatment, fire extinguishing, providing care and shelter for evacuees, and conducting assessments of building safety. With only limited information available, these activities cannot be performed with suitable efficiency.

The Northridge Earthquake of 1994 in California and the Hyogo-ken Nanbu Earthquake that hit Japan in 1995 both caused severe and extensive damage to highly built-up urban areas of technologically advanced countries. The Emergency department of Kobe city office had no effective way to grasp the full magnitude of the damage. Just after the earthquake, confusion ensued between fire fighting and first aid services, since these services had to rely on judgments made in the field. Government authorities were also very late in taking effective response measures. Not until aerial video images appeared on television in the afternoon, more than half a day after the earthquake struck, did the Japanese government realize the full severity of the situation. Only then could a response system be set up to deal with the crisis. In part, this was because the Kansai region was inadequately prepared for an earthquake, due to the fact that this area had not been struck by a major earthquake for a long time. In contrast, the state of California has been hit by earthquakes quite frequently, including the San Fernando Earthquake in 1971 and the Loma Prieta Earthquake in 1989. Furthermore, California had a wealth of experience in dealing with large-scale flooding and wildfire outbreaks. As a result, California was better prepared in terms of organization and emergency response procedures, and also in terms of public awareness of emergencies and preparation of disaster support systems.

The Emergency Response Systems (ERSs) discussed in this chapter are defined as systems for collecting and analyzing information in the event of such disaster situations, and for implementing appropriate and rapid response measures based on such information, in order to prevent or minimize damage. Emergency response systems are needed not only for implementing real-time response measures immediately following the occurrence of a disaster, but also for the recovery and reconstruction process. In addition, the use of ERSs for preventive purposes, that is, before a disaster occurs, is important. Thus, ERSs should serve to support activities in both non-emergency and emergency situations, and also during recovery times, when efforts are being made to restore an area to normalcy. Emergency response systems should make maximum use of information technology, such as sensors, to monitor earthquakes and other phenomena; communication technology to quickly and reliably transfer information; Geographical Information Systems (GISs) technology to deal with spatial distribution of disasters; mobile computers to collect information; mobile communications; tracking; and remote sensing and database technology to collect detailed information relating to cities and ground conditions. This is a good example of the application of Telegeoinformatics under critical conditions, both in terms of time and in terms of saving human lives.

9.1.2 Structure of ERSs

The structure and technologies used for ERSs vary depending on the situation. As an example, this chapter examines earthquake disasters. The structure of an ERS for earthquake emergencies is illustrated in Figure 9.1.

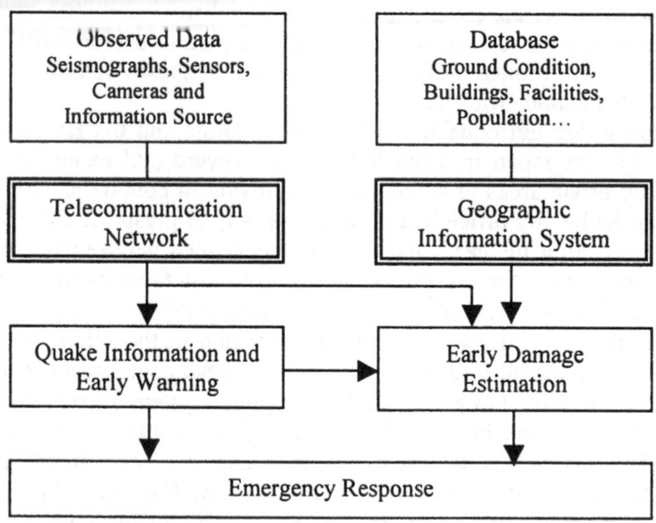

Figure 9.1 Structure of ERSs for Earthquakes and Other Natural Disasters

First, the most basic earthquake information system quickly detects the occurrence of a disaster from data collected using sensors such as seismographs. The system then distributes information such as epicenter position, and the earthquake's magnitude and spatial distribution. Some actions can be triggered automatically by this information. For example, when an earthquake is detected, emergency stop signals can be activated for elevators and trains; tidal wave warnings can be issued to coastal areas; and rescue activities can be preferentially initiated in the areas of greatest seismic intensity, according to a seismic intensity map. In addition, by increasing the swiftness of such a system, warnings can be provided to areas even before seismic waves arrive (early warning), for example, if monitoring data is transmitted from a location near the earthquake's epicenter to a distant location.

Second, an early stage damage estimation system can be set up to provide detailed estimates of seismic intensity distribution and rapid assessments of damage. This system would use data from an earthquake information system, incorporating databases that record data on ground conditions, buildings, city facilities and population, and make use of GISs. Based on these estimates, such a system would be used to launch fast and effective initial response measures immediately after the occurrence of a disaster. The system would also take in further information that is continually updated, according to new assessment information sent from damage scenes, and would evolve into an ERS to support decision-making in response to changing information.

A seismograph network having a density that is as high as possible is desirable in order to measure the distribution of ground vibrations over a given area. Seismic motion can vary greatly even within limited areas depending on ground conditions. Therefore for more crowded urban areas, higher density networks are required. If the area within which a major earthquake may occur is known, seismographs should be positioned primarily in that area (e.g., in the cases of the Tokai Earthquake in Japan and Mexico City, mentioned below). It is also necessary to secure a means of communication for transferring information rapidly and reliably. Although conventional telephone lines have been used, recent years have seen the increasing use of Wide Area Networks (WANs) and the Internet for this purpose. In some cases, authorities have duplicated means of communication in order to secure redundancy in the event of a disaster, as well as examples of securing wireless and satellite services for emergency communication.

The primary technological requirement of earthquake information systems is to provide information regarding earthquake and seismic motion distribution. However, the means for transmitting this information is also important. Dedicated lines and pagers have been used in cases where a limited number of target recipients exist. In addition, the World Wide Web has been used increasingly in recent years due to widespread, ready availability.

Early stage damage estimation systems require the use of GIS technology as a fundamental element in grasping disaster information spatially (Fukuwa *et al.*, 2000). The system prepares a spatial database that connects seismic motion information acquired from the earthquake information system. The database includes data on ground structure in order to enable detailed estimates of vibration distribution; data on the distribution of buildings and their structures and age; and demographic data such as population distribution. Furthermore, documentation and information from past earthquake disasters can be used to create a damage rate curve that links vibration strength with building damage and establishes a relationship between building damage and human casualties. Based on this information, detailed estimates can be made of earthquake vibration distribution, building damage distribution, and human casualty distribution. Note that information other than seismic motion data has been prepared in advance. Thus, this information can also be used as a disaster-prevention information system, for example, in order to examine the factors behind the occurrence of disasters and to develop and carry out preventive measures by forecasting damage under the assumption of specific types of seismic motion.

Damage information collection systems are designed to make the maximum use of mobile systems to support on-site investigation, of remote sensing technology that targets wide areas simultaneously, and of network technology that integrates information on many organizations and individuals. A wide variety of features and technologies are frequently used as part of on-site investigation systems, such as mobile computers, GPS tracking and navigation systems, GISs to sort locally added information such as photos and survey results, and transfer of data to servers.

9.2 STATE-OF-THE-ART ERSS

9.2.1 Strong Motion Instrumentation and ERSs for Earthquake Disasters in California

Due to its location over the San Andreas Fault among other factors, southern California has experienced frequent earthquakes. As a result, public awareness concerning measures to deal with earthquake disasters is high. There is also a long history of seismic motion monitoring in the region. The California Strong Motion Instrumentation Program (CSMIP), which was established within the California Division of Mines and Geology (CDMG) following the San Fernando Earthquake of 1971, performs strong motion measurements on more than 250 objects, including buildings, civil structures, and the ground. The data acquired by this system has been utilized to facilitate earthquake-resistant design of buildings. In addition, with recent advances in communications technology, this measurement data can be collected within a few minutes of the occurrence of an earthquake and applied immediately in an ERS.

Caltech/USGS Broadcast of Earthquakes (CUBE), a project with the goal of achieving real-time earthquake disaster prevention, was started in the late 1980s by the California Institute of Technology (Caltech) and the U.S. Geological Survey (USGS). The monitoring for this project involves several strong motion seismographs, calculates epicenter position and magnitude, and distributes the information to relevant organizations. The system has been put into practical use in Los Angeles. Mobile communication devices such as pagers are used for reporting.

After the Northridge Earthquake in 1994, Caltech, USGS and CDMG launched a project called TriNet to jointly develop a high-quality, high-density

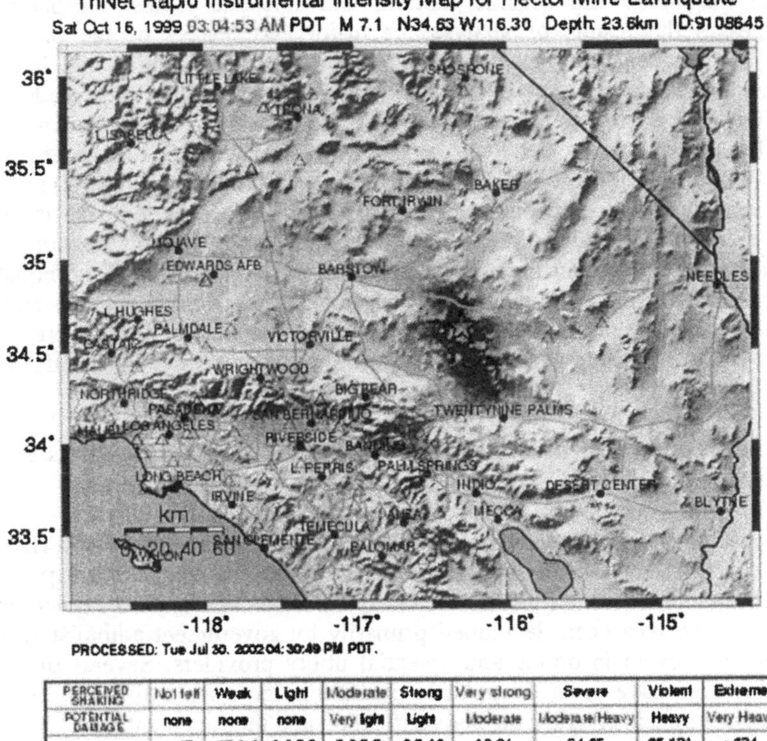

PERCEIVED SHAKING	Not felt	Weak	Light	Moderate	Strong	Very strong	Severe	Violent	Extreme
POTENTIAL DAMAGE	none	none	none	Very light	Light	Moderate	Moderate/Heavy	Heavy	Very Heavy
PEAK ACC.(%g)	<.17	.17-1.4	1.4-3.9	3.9-9.2	9.2-18	18-34	34-65	65-124	>124
PEAK VEL.(cm/s)	<0.1	0.1-1.1	1.1-3.4	3.4-8.1	8.1-16	16-31	31-60	60-116	>116
INSTRUMENTAL INTENSITY	I	II-III	IV	V	VI	VII	VIII	IX	X+

Figure 9.2 An Example of ShakeMap Webpage of TriNet Showing Distribution of Ground Shaking Intensity (TriNet Shake Map Working Group, 2002)

digital strong motion monitoring network in southern California (Heaton *et al.*, 1996). The data obtained through this network is released for use by other organizations. One of the goals of this project is to rapidly gather earthquake data records when an earthquake occurs in order to create ShakeMap, a distribution of vibration strength, and to post this on the World Wide Web. So, within a few minutes of a major earthquake, a map showing peak acceleration, peak velocity and seismic intensity will be published in the form of web pages (Figure 9.2), and e-mails will be sent out. The ShakeMap can be used, together with GISs, for emergency response purposes, such as the indexing of initial response measures and the creation of immediate damage estimates. Various software packages for performing post-earthquake regional loss estimates using ground vibration data from TriNet have been developed, including Hazards U.S. (HAZUS) program (the Federal Emergency Management Agency (FEMA), 2002) and the Early Post-Earthquake Damage Assessment Tool (EPEDAT) (EQE International Inc., 2002). These software packages incorporate information on, for example, ground and buildings, from GISs; make estimates of damage based on epicenter position, earthquake magnitude and seismic motion; and calculate losses. Furthermore, TriNet releases the results of analyses of major earthquakes of the past and "scenario earthquakes" that may occur in the future, all of which can help in formulating disaster-prevention plans. An investigation has also been made into utilizing TriNet for early stage earthquake warning systems, to enable even faster reaction. In the case of large-scale earthquakes in the San Andreas Fault area, it is even possible to send emergency alerts to Los Angeles in advance, because it can take between 10 seconds and one minute for the seismic waves to reach the city from the epicenter. In addition, the highly accurate strong motion measurement records made by TriNet can be used to help in the development of earthquake-resistance standards for buildings, in the structural design of buildings, and in seismology studies. The California Integrated Seismic Network (CISN) is presently working to expand TriNet's monitoring network throughout the state of California.

9.2.2 Strong Motion Instrumentation and ERSs for Earthquake Disasters in Japan

Over the past several years in Japan, the lessons of the Hyogo-ken Nanbu (Kobe) Earthquake in 1995 have been applied to the study, development, rapid expansion and implementation of ERSs (Yamazaki, 1997; Yamazaki *et al.*, 1998). Emergency response systems have been developed primarily by government administrations, governmental research institutes and essential utility providers. Several of these systems, however, have yet to be put into practice in large-scale disasters, so it is likely that these systems require further verification and improvement.

For several years, Japan had maintained a higher number of strong motion seismographs than any other country. However, the devastating Hyogo-ken Nanbu Earthquake in 1995 struck an area containing few observation points, in which the "earthquake disaster belt," experienced a seismic force of intensity 7 on the JMA scale. Within just two or three years of the disaster, however, the density of strong motion seismographs was greatly improved. Previously, the Japan Meteorological Agency used human perception to measure seismic intensity at approximately 150 locations throughout Japan. After the Kobe quake, and also due to the experience of the tidal wave disaster on Okushiri Island, caused by the Hokkaido Nansei-oki Earthquake in 1993 (198 fatalities and missing), the agency introduced a full-scale system of seismic intensity monitoring using seismometers, and expanded its network of monitoring points to approximately 600 points around the country (Doi, 1998). The agency took these steps with the aim of covering the areas damaged by huge seismic motions in detail, and transmitting tidal wave warnings more quickly. For communications, the agency secured multiple routes for their telecommunications lines, and in some places used satellite communications. Similarly, the Fire Defense Agency set up one seismometer in each municipality throughout Japan, for a total of more than 3,000 seismometers nationwide. Information from these measurement points is collected within a few minutes of an earthquake and sent out immediately to relevant organizations in order to put the ERS into action. At the same time, this information is made available to the general public through mass media broadcasts, including TV and Internet broadcasts (Figure 9.3).

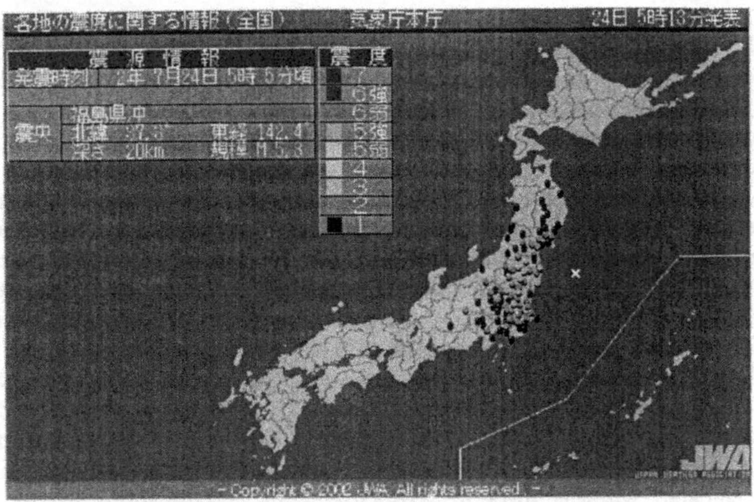

Figure 9.3 An Example of Japan Weather Association (JWA) Webpage Showing Distribution of Ground Shaking Intensity (Japan Weather Association, 2002)

The Early Estimation System (EES) for earthquake damage by the Cabinet Office (previously by the National Land Agency) is an ERS that uses seismic intensity information from over 3,000 points around Japan. The EES uses this seismic intensity data to estimate the seismic intensity distribution of 1-km square meshes, after taking into account ground conditions. The system then estimates building damages and human casualties, as well as damage due to tidal waves in coastal areas. The calculation of damage is performed automatically within approximately 30 minutes of an earthquake. The EES is currently in operation. The system has been improved based on the experience gained using the system for several recent damage-causing earthquakes. Another system that has been implemented is the Emergency Measures Support (EMS) System, used to gather and sort information from ministries and government offices in emergency situations and share this information in GISs. The system has been applied to operations such as wide-area medical treatment and the conveyance of patients.

The National Research Institute for Earth Science and Disaster Prevention (NIED) launched K-net, a network of strong motion seismographs set up at 1,000 locations over Japan (NIED, 1995). K-net covers the entire country with seismographs at a fairly even density of approximately one site per 25 square kilometers. K-net is very effective for monitoring high-level strong seismic intensity data near an epicenter, and for grasping the spatial distribution of seismic motion. NIED has laid out various other monitoring networks such as KiK-net (approximately at 560 locations), a digital strong-motion seismograph network; Hi-net (approximately at 640 locations), a high-sensitivity seismograph network; and F-net (approximately at 70 locations) a full-range seismograph network. Digital data in the form of waveforms is available to the general public on a web site. In their current form, K-net and KiK-net cannot be applied immediately after an earthquake, since the data is collected manually by dial-up connections after an earthquake has hit. However, these systems will be upgraded to provide real-time capability via the Internet.

For smaller areas, there exist examples of ERSs that are based on higher-density earthquake monitoring. Several local governments have their own strong motion monitoring systems. One of these is Yokohama, which has set up strong motion seismographs at 150 locations, which cover the city at approximately 3-km^2 intervals (Midorikawa and Abe, 2000). The city can collect seismic intensity data using an ISDN line within three minutes of an earthquake, to assist in formulating initial response measures. Furthermore, Yokohama is also developing a system that makes use of this data to estimate building damage. Nagoya is developing a system for early-stage damage estimation, by increasing the measurement point density and arranging to have data forwarded from universities and utility providers. As an example of the latter, Tokyo Gas Co. Ltd. has been actively developing its own ERS (Shimizu *et al.*, 2000). In 1994, Tokyo Gas began to operate a conduit network warning system for earthquakes, called Seismic Information Gathering & Network Alert System (SIGNAL). SIGNAL monitors seismic motion distribution using dedicated wireless links to sensors at over 300 locations in the Tokyo Metropolitan area. SIGNAL then combines this data with information on ground conditions, buried conduits, information on households supplied by the gas company, and other factors, sorted out in a GIS. Then, within approximately 10 minutes, SIGNAL estimates the damage to buried conduits and buildings. These results are then used to control the gas supply. Tokyo Gas is currently building a system of Super-dense Real-time Monitoring of Earthquakes (SUPREME), based on high-density monitoring (approximately 3,600 points over a 3,100-km^2 area of the Tokyo metropolis).

A well-known example of an early warning system is the Urgent Earthquake Detection and Alarm System (UrEDAS) of Japan Railways (JR) (Nakamura, 1996). The system detects P-waves (initial tremor) using seismographs set up along the railways, and urgently and automatically stops trains or reduces their speed. This system has been used on the Tokaido Shinkansen (bullet train) and other lines since 1992. In recent years, the Japan Meteorological Agency and National Land Agency have been examining Now Cast, a system for issuing warnings immediately before an earthquake hits, based on data from measurements taken near epicentres (Doi, 1998). The outline of the system is illustrated in Figure 9.4. The epicentre positions of deep trench earthquakes, such as the major earthquake that is widely predicted to occur in the Tokai area (Suruga Bay) in Japan, are quite far from major cities, which means that these kinds of systems could be used quite effectively.

In the case of the predicted Tokai Earthquake, the plan is to announce an earthquake warning immediately before it hits (several hours to several days), based on various observation such as observed changes in the earth's crust. Various restrictions on social activities would then be imposed in order to minimize damage in the event of the earthquake occurring. This is one type of ERS. Of course, judging whether an earthquake will occur based on measurement data is an extremely ambitious scientific task, and such predictions would greatly affect peoples' lives, thus the final decision on issuing warnings and restrictions would have to be made by a committee of specialists.

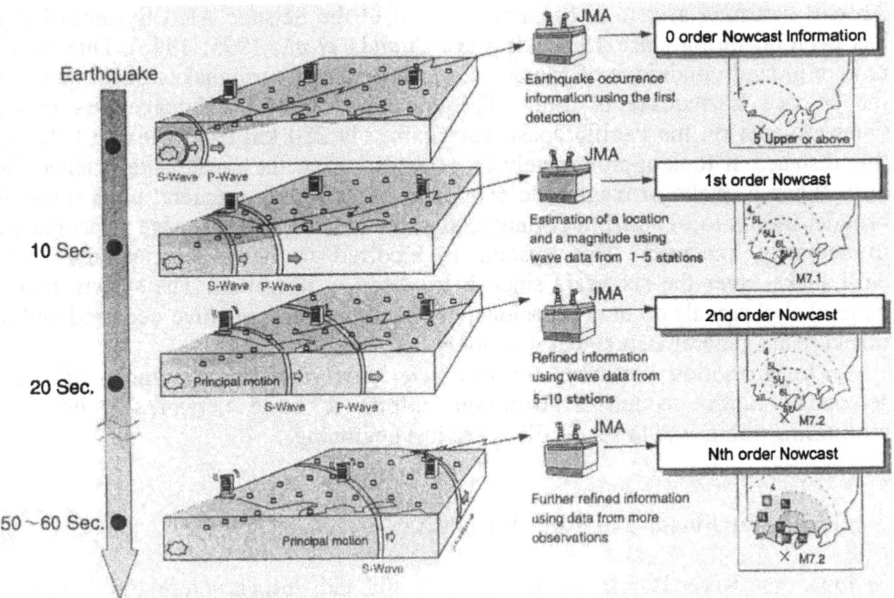

Figure 9.4 Nowcast System by JMA (Ashiya *et al.*, 2002)

9.2.3 Strong Motion Instrumentation and ERSs in Taiwan

Since Taiwan is located directly above a plate boundary, the country experiences frequent earthquakes. As a result, earthquake measurements are actively conducted in Taiwan. Partly because the country is relatively small (one tenth the size of California), Taiwan has a number of mature and quite characteristic measurement systems in place, including multiple strong motion monitoring networks that cover the entire country, and high-density monitoring arrays in certain smaller areas. Examples of national networks include the SMA-1 network (over 100 points) of the Institute of Earth Sciences Academia Sinica (IESAS), which has been in operation since the mid-1970s; the Central Weather Bureau Seismic Network (CWBSN) (75 points) of the Central Weather Bureau (CWB); and the Taiwan Strong Motion Instrumentation Program (TSMIP) (approximately 600 points), which was developed between 1991 and 1997 by the Seismological Center of the CWB (Kuo *et al.*, 1995). A rapid reporting system using CWBSN, connected to a telemeter, has been developed, and the system estimates seismic intensity maps, epicenter position and magnitude within one minute. This time can be reduced to approximately 30 seconds. This means that the system can serve as an early warning system, particularly in the case of large-scale earthquakes (Lee *et al.*, 1996; Shin *et al.*,1996).

9.2.4 Strong Motion Instrumentation and ERSs in Other Countries

An early warning system in Mexico City, called the Seismic Alert System (SAS), has been operating since 1991 (Espinosa-Aranda *et al.*, 1995; 1996). Due to the unique ground conditions of Mexico City, devastating earthquakes occur, such as the Mexico Earthquake of 1985. The epicentre of this earthquake was in the Guerrero area on the Pacific coast, approximately 280 km from Mexico City. At this distance, it took approximately 60 seconds before the earthquake reached the city. When a quake of magnitude 6 or more is detected, a general alert signal is broadcast from local AM/FM commercial radio stations, and response measures are immediately executed. This system is reported to have detected over 600 earthquakes over the six years since it has been in operation. The system issued general alert signals on nine occasions, of which one false positive occurred and in one case the general alert signal was not issued.

Strong motion measurement has been performed in a number of other seismically active countries. However, efforts to build networks using such monitoring points and to set up ERSs are just beginning.

9.2.5 ERSs for Floods and other Disasters

In Japan, the River Bureau of the Ministry of Land, Infrastructure and Transport has set up an ERS to deal with floods and landslides due to heavy rains. The system is designed to provide real-time information on rainfall distribution and the water levels of major rivers throughout Japan. This information is made available on the Internet as well as mobile phone networks. This service is mainly intended to help residents of affected areas to make their own decisions about evacuation. In addition, some local governments also have their own real-time systems to monitor rivers and landslide conditions, and networked live cameras are also used. Meteorological sensors and satellite data are effective for monitoring and mapping heavy rain and the inundated areas.

9.2.6 New Method of Damage Reconnaissance

Collecting information regarding the level of damage in affected areas at the time of severe disasters is very difficult. The situation immediately following the occurrence of a disaster can be assessed by setting up measurement network systems using seismographs, as explained above. However, the next step is to collect damage information, which cannot be automatically and uniformly gathered using sensors. In most cases up to now, this task has been performed by specialists going directly to the damage scene and collecting information by taking photos and making notes, and then sorting this information and sending it into a central office, where all such information would be assessed. At the office, the information to be analysed is entered into computers. However, since information collection at the sites is not computerized, the efficiency is low.

In recent years, various systems have been developed to support on-site investigation work. These systems typically involve GIS software, GPS tracking, digital cameras, mobile telecommunications and computers (Tobita and Fukuwa, 2002; Shibayama and Hisada, 2002; Iwai and Kameda, 2000; Zama, *et al.*, 1996; Kameda *et al.*, 1995). These systems have many significant advantages. For example, investigative reports, damage summation, information releases and utilization of materials can be performed quickly through the direct creation of digital data; locations can be clearly specified and GPS tracking can be used for navigation around the scenes of disasters; and investigation methods are unified, making it easy for work to be shared by multiple people. In addition, these systems can be used by non-specialists. An example of such systems is introduced in the next section. Using non-specialists to collect information at the time of disasters has been found to be very helpful, and as such, Internet support services, such as e-mail and electronic bulletin board systems are being evaluated.

Considerable attention has been given to communications at the time of disasters including satellite phones and data transfer. In addition, the examination of alternatives to wireless methods by which to exchange information (including low-tech methods) is important.

Satellite remote sensing and the use of aerial images taken from aircraft have been under development as effective on-site investigation methods to enable damage conditions to be assessed quickly in highly concentrated urban areas (Ogawa *et al.*, 1999; Matsuoka and Yamazaki, 2000).

9.3 EXAMPLES OF DEVELOPING ERSS FOR EARTHQUAKES AND OTHER DISASTERS

In this section, examples of GIS applications that were developed primarily by the authors are presented along with applications to ERSs.

9.3.1 Facility Management in Nagoya University

Facility management and master planning of the university campus requires synthetic management of facilities based on numerous information such as information on soil, building, facility and equipment. On the Higashiyama campus of Nagoya University, widespread construction is underway, and beneath the ground, subway and urban expressway construction is progressing. Buildings of various ages exist on the campus, which has a complicated ground condition due to land removal and filling. Therefore, utilization of existing buildings and bore data is important in the master planning stage. In addition, ultra-precise equipment on the campus is affected by minute vibration. Thus, maintaining the research environment under construction and traffic vibrations becomes an important problem. In addition, the Hyogo-ken Nanbu Earthquake raises the problem of seismic safety, so that seismic performance evaluation and retrofitting are being carried out. Therefore, the university campus requires appropriate countermeasures based on the multifaceted analysis of various data.

Figure 9.5 shows the GIS called Higashi (after the campus name), which was developed as a means to solve these problems. This system offers various data obtained by the authors to support the planning of new buildings, as well as the maintenance and management of existing buildings. These data include soil data, such as bore and landform data, and building data, such as seismic performance and dynamic characteristics of existing buildings. This system can analyze vibration data due to earthquakes and environment vibration obtained by a vibration monitoring system described later.

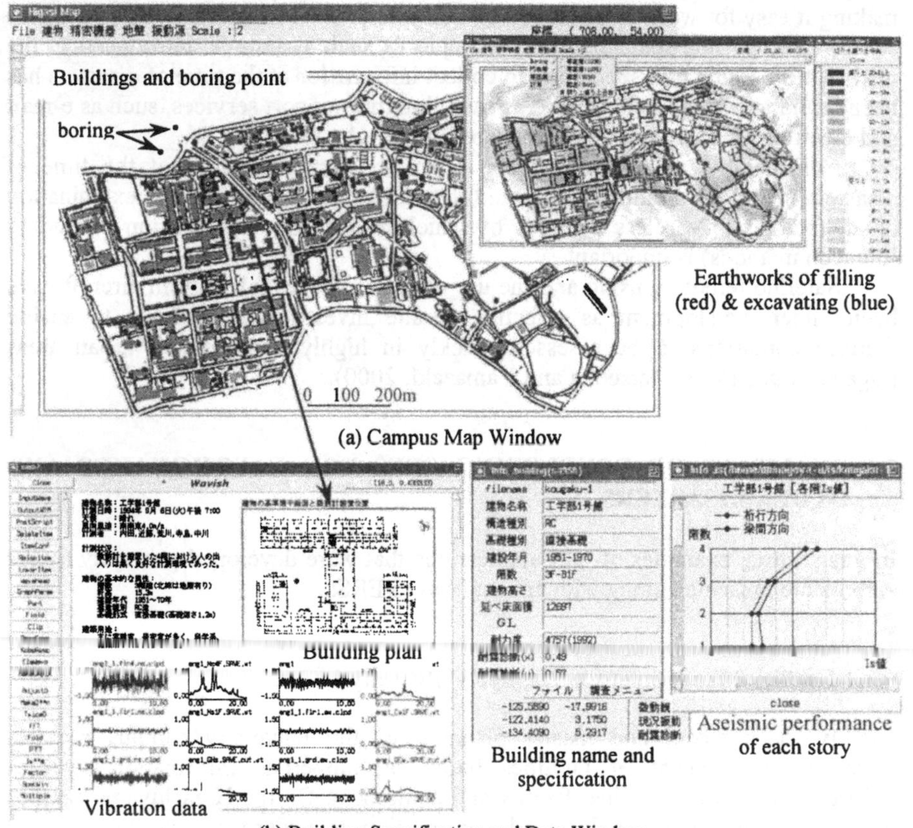

Figure 9.5 GIS for Facility Management of Nagoya University

9.3.2 Seismic Ground Motion Evaluation

In aseismic design and earthquake disaster prevention, the evaluation of seismicity and seismic ground motion is fundamental. The engineer should judge the seismic risk based on surveys of various fault parameters of historical earthquakes, including data such as magnitude and hypocenter data, and activity and hysteresis data of existing active faults. In the estimation of earthquake ground motion, the following items should be considered: (1) specification of earthquakes considered in aseismic design and damage estimation, (2) level evaluation of the earthquake ground motion based on seismic risk analysis, (3) prediction of ground motion at bedrock, (4) grasp of soil amplification based on earthquake response analysis of subsurface subsoil, and (5) evaluation of surface ground motion. Since existing knowledge and data on ground motion evaluation are numerous, appropriate synthesis of these data should contribute to the improvement of accuracy of estimated results.

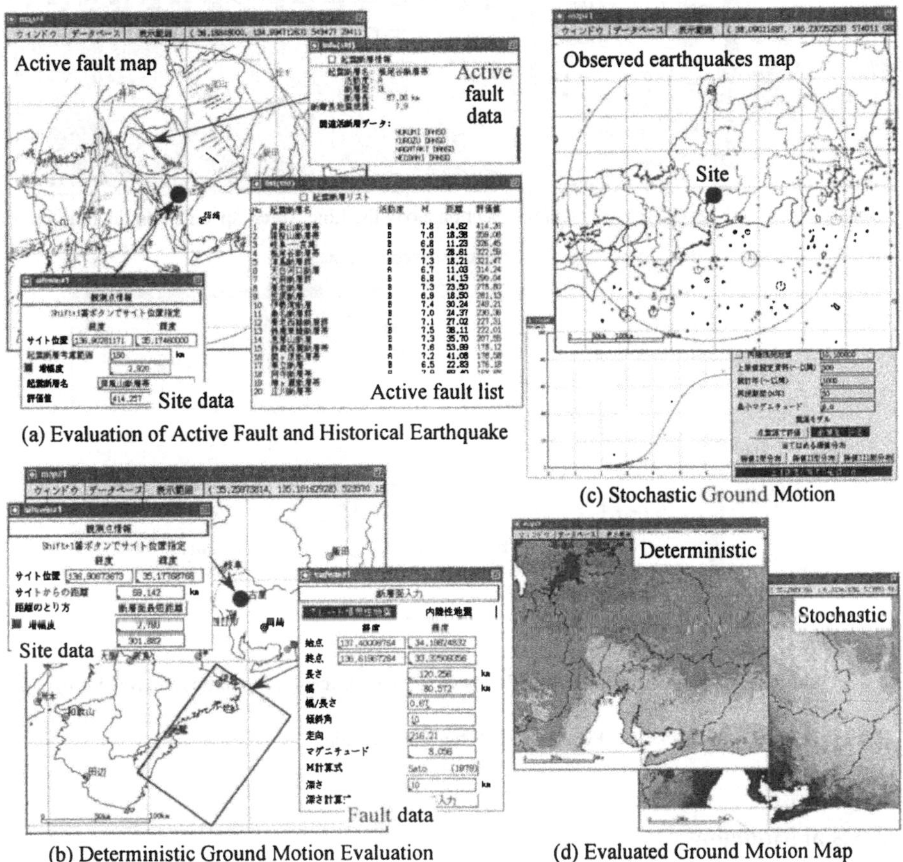

(a) Evaluation of Active Fault and Historical Earthquake

(c) Stochastic Ground Motion

(b) Deterministic Ground Motion Evaluation

(d) Evaluated Ground Motion Map

Figure 9.6 GIS for Evaluation of Seismic Ground Motion

Here, a GIS called Quick Seismic Evaluation (QuSE) System to evaluate earthquake ground motion is constructed, where the user can visualize the fluctuation of the result due to the change of method and data. The user can select various parameters and evaluation techniques through the graphical user interface by consulting a reference window in which various information on seismic source data and knowledge on method applicability is provided. The system is composed of three subsystems as shown in Fig. 9.6, which are (1) listing and evaluation of the active faults and historical earthquakes which affect the site, (2) deterministic evaluation of ground motion for inland active faults and inter-plate earthquakes, and (3) stochastic evaluation of seismic risk. The use of this system also promotes education related to the evaluation of seismic ground motion.

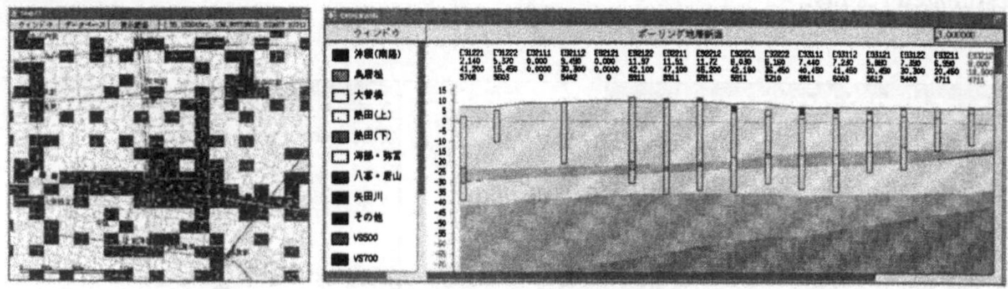

(a) Boring Points on Digital Map (b) Geologic Section along the Line which Specified on the Map in (a)

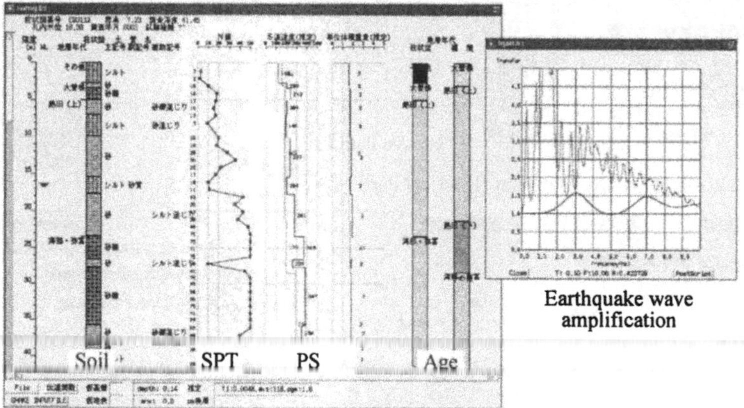

(c) Boring Log Data, Earthquake Wave Velocity and Amplification Properties at a Site in (b)

Figure 9.7 GIS for Soil Modeling

9.3.3 Soil Modeling

A GIS to estimate the dynamic soil model in the Nagoya urban area is constructed, which conjugates maximally all of the existing subsurface exploration data. As shown in Fig. 9.7, the system estimates the soil velocity and density layering structure along an arbitrary section line and evaluates the soil amplification at an arbitrary site. Basic policies to create the dynamic soil model are as follows. (1) Previously obtained PS logging[1] data is utilized. There exist approximately one hundred data of average depth 57m and average bedrock shear wave velocity

[1] A kind of soil exploration for evaluation of earthquake wave velocity in each soil layer.

400m/s. (2) When data exist for a standard penetration test (SPT[2]) near the site, shear wave velocity and density are estimated using SPT data, depth, geological age and soil classification. In the city, there exist 4,200 data of average depth 25 m, which covers 20% of the city area over 125-m mesh data. (3) In the area lacking sufficient bore data, the depth of each stratum at an arbitrary point is estimated by interpolating the depth contour line for each geological layer. The shear wave velocity and density are estimated from the regression formula based on geological age and depth. (4) The deep soil structure is modeled by interpolating shear wave contour line of the same velocity, which is estimated based on the seismic refraction survey, gravity prospecting and deep well survey. (5) Using the above estimated shear wave velocity and density, the soil amplification is calculated by one-dimensional wave propagation analysis. (6) At the locations that lack soil data, the soil amplification is estimated based on the regression formula using digital national land information.

9.3.4 Seismic Damage Estimation

A GIS called Seismic Damage Estimation System (SDES) to estimate building damage for arbitrary earthquake can be constructed by combining the above mentioned GISs with building information from urban areas and the vulnerability function of buildings. The number of buildings according to structural type and building age in each region is calculated based on taxation figures. Then, the number of damaged buildings in each region is estimated using the vulnerability function for each structural type, building age and damage level which was derived from the analysis on damage and ground motion intensity in Kobe. Since the damage for an arbitrary earthquake can be grasped interactively, any type of earthquake can be assumed, and disaster weakened regions can be specified. In addition, the cause of the weakness can be investigated. The system may offer effective information upon which to base disaster prevention measures by adding a human and socioeconomic viewpoint. Figure 9.8 shows an example of wooden building damage estimation in Nagoya city caused by an earthquake.

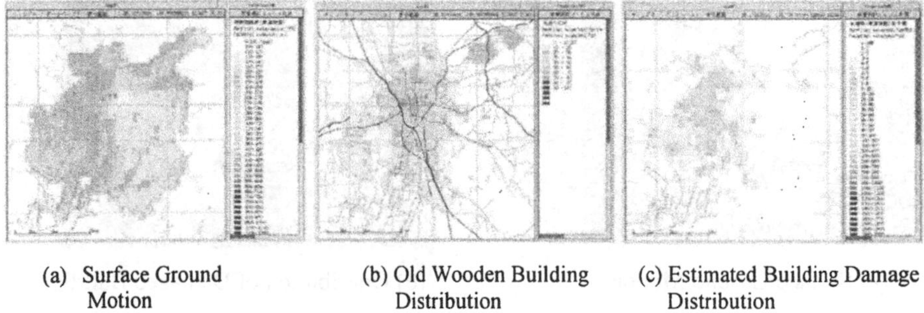

(a) Surface Ground Motion (b) Old Wooden Building Distribution (c) Estimated Building Damage Distribution

Figure 9.8 GIS for Earthquake Damage Estimation

[2] A kind of soil exploration for evaluation of soil stiffness. Less expensive and easier than PS logging.

9.3.5 Early Seismic Damage Estimation

By installing 16 seismometers at three campuses of Nagoya University and by combining this observation system with the seismic damage estimation system (SDES), the early seismic damage estimation system is constructed (Figure 9.9). The basic idea is to connect the seismometers via the Internet, thus enabling the immediate transmission of strong-motion records into an Engineering Workstation (EWS) and to combine these records and the various databases in the SDES.

In the system, just after a quake, first the hypocenter and magnitude are determined automatically, and then the ground motion distribution is estimated using attenuation curves and pre-calculated soil amplification, and finally the building damage is estimated using building database and vulnerability curves. The use of the Internet and EWS allows for high-speed, robust data transfer and analysis. The extension of the seismic network and the immediate acquisition of the strong-motion record and damage prediction result through the web are also technically easy to realize.

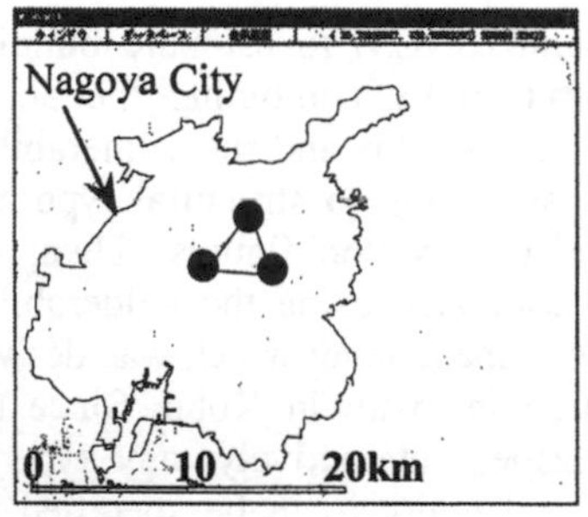

(a) Seismic Observation Sites

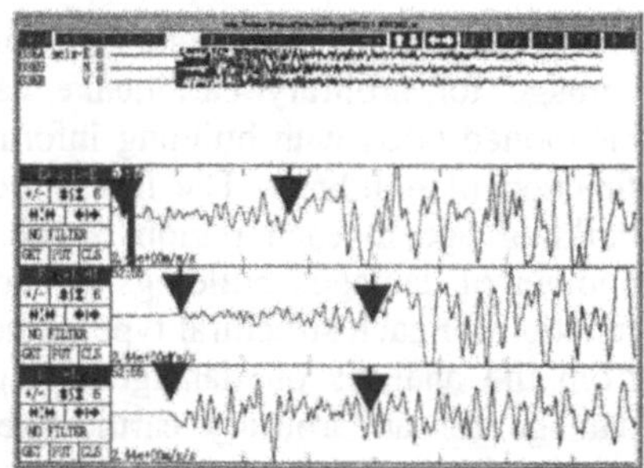

(b) Hypocenter Determination

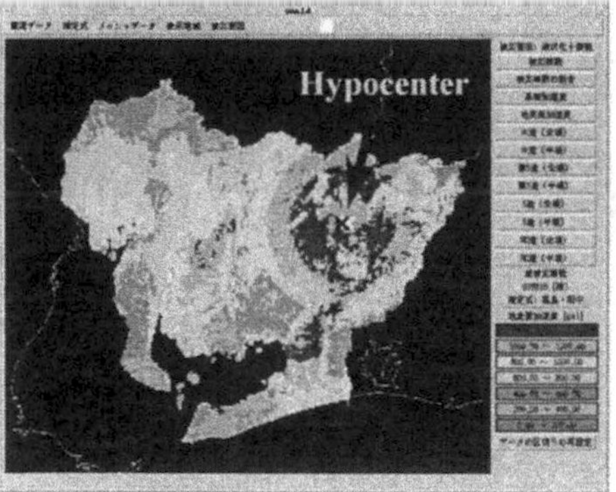

(a) Surface Ground Motion

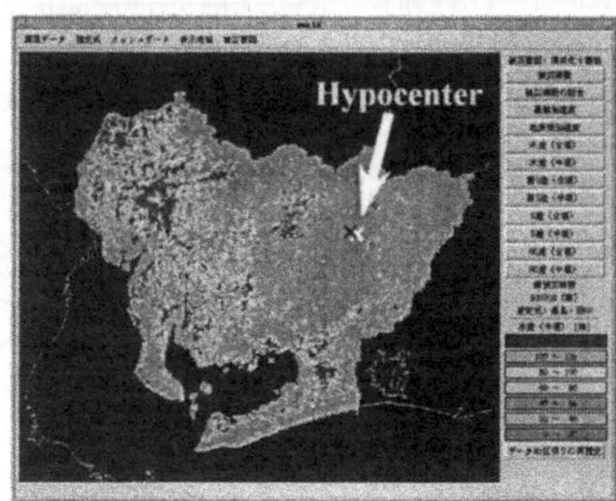

(b) Distribution of Damaged Buildings

Figure 9.9 Early-Damage Estimation System

9.3.6 Environmental Vibration Alarm

In urban areas, environmental vibration, such as traffic- and construction-induced vibration, cannot be avoided. This vibration not only has an adverse impact on the human body, but also causes problems in precision instruments. As mentioned previously, subway and underground expressway construction is now underway on the Nagoya University campus. Therefore, researchers using sensitive equipments such as electron microscopes are worried about construction vibration over the next few years and traffic vibration after these projects are completed. In addition, reconstruction and seismic retrofitting of existing buildings are also planned. Moreover, the amount of ultra-precise equipment for advanced research is increasing. Therefore, the authors constructed the environmental vibration monitoring system called MoVIC (Monitoring system for Vibration Induced by Construction).

This system continually monitors vibration due to construction and traffic using 26 sensors placed throughout the campus and connected to the university LAN system. The basic concept of this system is identical to the early seismic damage estimation system mentioned above, and is combined with the "Higashi" facility management system. The system is intended to deal with the daily environmental vibration problem. This system automatically displays the waveform on an EWS when vibration that exceeds the allowable vibration level of an instrument has been detected. The same information is transmitted automatically to the construction work office through a FAX modem. Researchers who use ultra-precise equipment are also informed of the environmental vibration situation via a web site. This system allows equipment users to understand vibration conditions and helps the general contractor to select low-vibration construction methods. As a result, the various construction projects make progress more smoothly. The system also functions as a high-density strong-motion network system.

9.3.7 "Anshin-System": Intercommunication System for Earthquake Hazard and Disaster Information

After the 1995 Hyogo-ken Nanbu Earthquake, conventional disaster information systems have been found to be insufficient for catastrophic cases. In the conventional systems, top-down one-way information is provided from the administration side to residents; however, collecting on-site information in severely damaged areas is difficult. Disaster information provided by administration offices or researchers is often not sufficiently understood and used by inhabitants or engineers. Consequently, an intercommunications system for administration office, researchers, inhabitants and engineers is required. We have developed various systems for earthquake observation and disaster prevention as shown in the previous sections (Fukuwa et al., 2000). In addition, an on-line data acquisition system for the strong-motion earthquake records of the Tokai district has recently been developed (Tobita et al., 2001). Based on such research and development, a new intercommunication framework for a disaster information system is shown herein.

The Anshin system ("Anshin" means Security in Japanese) is a series of subsystems for collecting, arranging and sharing regional disaster information. The system consists of the following three parts as shown in Figure 9.10.

Anshin Web is a Web GIS system for disaster prevention, earthquake engineering and aseismic design of buildings (Figure 9.11). The server cooperates with systems and data provided by the administration side. Data on earthquake source and faults, recorded ground motions, soil conditions, buildings, urban facilities and damage estimation is stored. Using JAVA applications, users can access the database system via the Internet or LAN, and no specific software is required. The administration can utilize the system for examination of disaster prediction and disaster prevention planning, while inhabitants can obtain regional information on disasters. In addition, engineers can obtain technical information on architectural design and disaster prevention. Information from mobile terminals, such as position, maximum acceleration, and image data, are received via Personal Handy-phone System[3] (PHS) and entered directly into the database. When a disaster occurs, the system collects real-time information. Such information is then fully utilized for disaster relief activities.

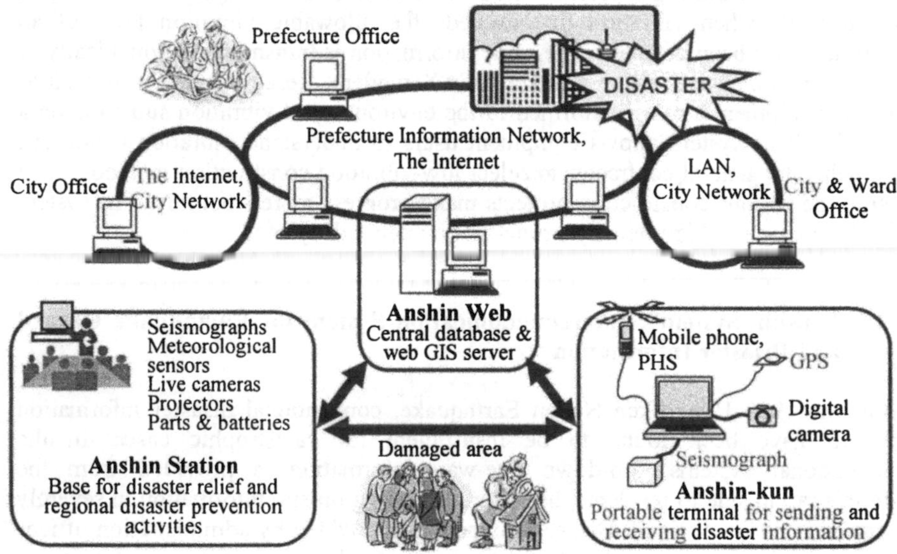

Figure 9.10 Structure of Anshin System

[3] A kind of mobile-phone service in Japan. 128kbps data transfer rate is available by use of 1.9GHz band.

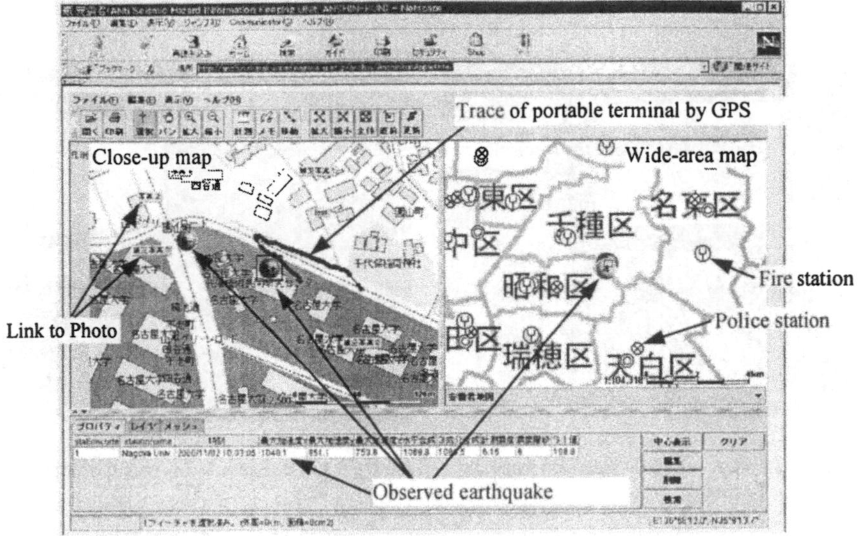

Figure 9.11 Anshin Web: Web GIS System for Disaster Prevention

Anshin-kun is a portable information terminal composed of a PC with a small seismometer, a wireless telecommunication interface, a digital camera, a GPS unit, and GIS with navigation software (Figure 9.12). This unit is used by several users such as delegates of local unions, architects, lifeline engineers, local government staff members, and researchers. Using this terminal, users can access data on the server to be used for disaster prevention activities. When an earthquake occurs, the terminal immediately sends observed earthquake shaking data to the server. A high-density shaking map is obtained from data from several units in the area, and the system functions as real-time observation system. Then, the terminal sends various information, such as answers to simple questionnaires on damage. Field surveys are performed using a GPS for navigation, a GIS software package, a digital camera, and various terminal software. Information collected using the terminal is then sent to the server by a telecommunications device and is added to a database on the server. In the case of damage of the telecommunications facilities, users can access the main server by visiting a base station at shelter facilities.

Anshin Station is a base for area-wide disaster prevention activities located in shelter facilities such as elementary schools and community centers (Figure 9.13). This system consists of sub servers of Anshin-web, which connects city offices and ward offices by dedicated lines or by radio. Spare parts and batteries for these terminals are also maintained at the Anshin base station. Projectors and screens are prepared for the purpose of displaying disaster information. The system supports shelter management during disasters and is usually used for education in science, social studies and disaster prevention.

The Anshin system shown here presents a new framework with bottom-up disaster information collection and distribution, which is opposite to conventional top-down information flow. Thus, the consciousness of regional inhabitants, administration officers and engineers is expected to change.

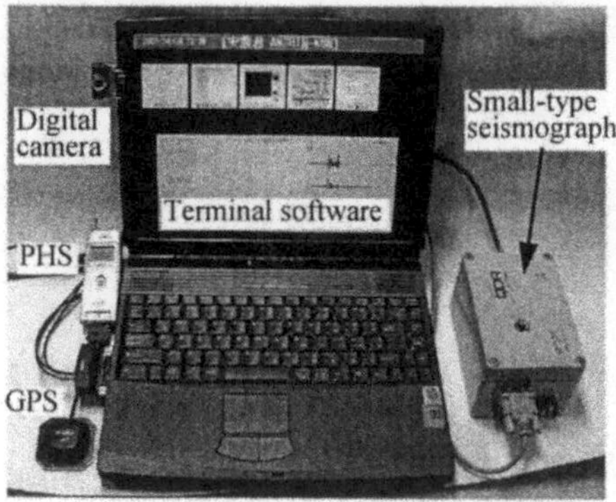

Figure 9.12 Anshin-kun: Disaster Information Terminal

Figure 9.13 An Example of Anshin Station

9.4 FUTURE ASPECTS OF EMERGENCY RESPONSE SYSTEMS

9.4.1 Implementation Issues

This section introduces several challenges to putting ERSs into practical use. ERSs need to be applied consistently, starting with disaster-prevention in non-emergency times, and continuing with the initial response action immediately following a disaster, through the recovery and reconstruction period. It is desirable for an ERS to be developed into a system that can be used in a variety of ways, such as predicting the loss of services and supporting recovery efforts, as well as reflecting newly collected information in the GIS in real time.

In order to achieve this, a GIS used as part of an ERS should be advanced, and of particular importance is the implementation of various effective databases. When we examine early-stage damage estimation, a powerful database is needed to provide detailed data on characteristics such as deep and shallow soil foundations, the structure and age of buildings, the damage function, which indicates the damage ratio of buildings as a function of the level of ground shaking, and population. Furthermore, it is desirable to develop and improve new methods for collecting damage investigation data, as mentioned before, and to develop systems that can quickly collect detailed and highly accurate information, and convert this into data for ERSs. In addition, the use of systems for information sharing from multiple institutions will result in more efficient information collection and reduced overlap.

ERSs should be fast and stable, so that they will operate reliably in the event of an emergency. Thus, an ERS must comply with the requirement of critical systems, in terms of both hardware and software. Here, hardware includes infrastructure such as communications and electric power, and the safety features, such as earthquake-proofing, of facilities at which ERSs are installed. The design of the software should take into account the use under chaotic conditions and so should support user-friendly interfaces and ways to prevent operational errors.

9.4.2 Developing New Technologies for ERSs

This section addresses the future directions in ERSs with respect to both hardware, i.e. sensors, computers and communications devices, and software, i.e. infrastructure, GIS and various types of data.

Different types of sensors will be used for the detection of vibrations from earthquakes. In the past, the cost of such sensors was several tens of thousands of dollars per sensor. However, through advances in semiconductor and other technologies, seismic sensors are now available for prices ranging from several hundreds to several thousand dollars. Cheaper sensors have helped to greatly increase the number of installation points (installation density) of monitoring networks, thereby increasing the speed and accuracy of the ERS. Furthermore, development of sensors that can collect a broad range of disaster data, such as meteorological sensors for heavy rains and strong winds and network cameras to capture visual data, has been progressing.

Advances in computer hardware have also been significant. To meet the requirements of mobility and usability of devices for outdoor use in a situation of emergency, improvement of small, light and wearable computers and devices with long-life batteries will be important.

GIS has been developed as a tool for storage, analysis and utilization of various types of spatial data such as remotely sensed data, distribution of artificial facilities, human activities and policy-oriented information. Various GISs act important roles in ERS as mapping, monitoring, simulating and managing of data on natural disasters and various emergent situations. On the other hand, GIS software is now required to improve its interface targeted to a larger numbers of users, including non-specialists, and to sort out huge amounts of data. The essential point would be to cooperate with communication systems such as Web GIS on the network (*e.g.* Berkeley Digital Library, 1998; U.S.G.S., 2002).

Remote sensing is a very effective observation method to cover a large area at a time, and to obtain data much faster than conventional observation on the ground. Satellite Remote Sensing has been used for assessment of potential natural hazards such as earthquake fault, landslide, liquefaction of soil, floods, drought and fire. It will be effectively used for early warning and monitoring of changing situations during and after the hazards. In the investigation of recent earthquakes, data from satellite sensors and radars are used for building damage evaluation in urban areas. On the other hand, Aerial Remote Sensing covers relatively small area by various types of images and videos with high resolution. Quick and detailed evaluation of building damage distribution will be improved by integration of such remotely sensed data on GISs.

Finally, key to ERS is the improvement and development of communications methods. Improvements have also been made in the speed and stability of circuits, on the assumption that circuits must be tough enough to withstand natural disasters. In some cases, even satellite communications have been used. In addition, mobile phones are used as an effective communications tool for information transfer to relevant people, including the general public. Mobile phones were once used mainly as devices only for receiving voice information. However, recently, e-mails can be sent and received on mobile phones and pagers, and they can be connected to the Internet as well. So access from the user side has become possible. Thus, in addition to an increasing number of web sites that offer disaster information, information can also be posted on special Internet sites for access by mobile phone users, in a similar way they access general pages of Location-Based Services (LBSs) (see Chapter 6 for more details on LBSs). Furthermore, mobile phones equipped with GPS and digital cameras are also available. These devices can be used as simple tracking or navigation devices with map information. As information terminals, these devices have considerable advantages. For example, the devices are inexpensive, can be easily handled by non-specialists, are compact, and can be used in both non-emergency and emergency situations. Therefore, unless infrastructures such as mobile communications base stations are damaged by disasters, these devices can be extremely effective disaster information terminals.

9.5 CONCLUDING REMARKS

This chapter offered an overview of ERSs, with a focus on earthquake disasters. Emergency response systems are aggregations of technologies from various different fields, all relating to the concept of Telegeoinformatics. The difficulty concerning ERSs is that they should be developed with careful consideration for reactions of people in emergency situations as well as the reactions of the society, and even consideration for legal aspects. Research and development are highly advanced particularly on the west coast of the United States and in Japan, which are highly vulnerable to powerful earthquakes. However, measures to deal with massive earthquake disasters in which large areas are devastated in a few moments have yet to be developed.

REFERENCES

Ashiya, K., Okada, T., Tsukada, S., Yokota, T., Kato, T. and Kamigaichi, O., 2002, Nowcast earthquake information and its application to early earthquake alarm system, *Butsuri-Tansa*, 55, No.6 (in press, in Japanese with English summary and captions).

Berkeley Digital Library, 1998, Web GIS and Interactive Mapping Sites, Online. Available HTTP: <http://sunsite.berkeley.edu/GIS/intergis.html> (accessed 17 January 2003).

Doi, K., 1998, Earthquake early warning system in Japan, In *Proceedings of the International Conference on Early Warning Systems for Natural Disaster Reduction*, Potsdam.

EQE International Inc., 2002, Webpage on the Internet, Online. Available HTTP: <http://www.eqe.com/revamp/epedat.htm> (accessed 17 January 2003).

Espinosa-Aranda, J.M., Jimenez, A., Ibarrola, G., Alcantara, F., Aguilar, A., Inostroza, M. and Maldonado, S., 1995, Mexico City seismic alert system, *Seismological Research Letters*, 66, No.6, pp. 42–53.

Espinosa-Aranda, J.M., Jimenez, A., Ibarrola, G., Alcantara, F., Aguilar, A., Inostroza, M. and Maldonado, S., 1996, Results of the Mexico City early warning system, In *Proceedings of the 11th World Conference on Earthquake Engineering*, Acapulco, Paper No. 2132 (CD-ROM).

Federal Emergency Management Agency (FEMA), 2002, Web page on the Internet, Online. Available HTTP: <http://www.fema.gov/hazus/> (accessed 17 January 2003).

Fukushima, S., Takahashi, M. and Yashiro, H., 2000, Development of assembling time map for emergency action, In *Proceedings of the 12th World Conference on Earthquake Engineering*, Auckland, Paper No. 629 (CD-ROM).

Fukuwa, N., Tobita, J., Takai, H. and Ishida, E., 2000, Effective application of Geographic Information System in the field of earthquake engineering and disaster prevention, In *Proceedings of the 12th World Conference on Earthquake Engineering*, Auckland, Paper No. 2229 (CD-ROM).

Heaton, T., Clayton, R., Davis, J., Hauksson, E., Jones, L., Kanamori, H., Mori, J., Porcella, R. and Shakal, T., 1996, The TriNet project, In *Proceedings of the 11th World Conference on Earthquake Engineering*, Acapulco, Paper No. 2136 (CD-ROM).

Iwai, S. and Kameda, H., 2000, Post-event data collection using mobile GIS/GPS and development of seismic evaluation technique for damage, In *Proceedings of the 12th World Conference on Earthquake Engineering*, Auckland, Paper No. 2056 (CD-ROM).

Japan Weather Association, 2002, Webpage on the Internet, Online. Available HTTP: <http://www.tenki.or.jp/qua/> (accessed 17 January 2003).

Kameda, H., Kakumoto, S., Iwai, S., Hayashi, H. and Usui, T., 1995, DiMSIS: A geographic information system for disaster information management of the Hyogoken-nambu earthquake, *Journal of Natural Disaster Science*, 16, No.2, pp. 89–94.

Kuo, K.W., Shin, T.C. and Wen, K.L., 1995, Taiwan strong motion instrumentation program (TSMIP) and preliminary analysis of site effects in Taipei basin from strong motion data, In *Urban Disaster Mitigation: The Role of Engineering and Technology*, edited by Cheng, F.Y. and Sheu, M.S., (Elsevier Science), pp. 47–62.

Lee, W.H.K., Shing, T.C. and Teng, T.L., 1996, Design and implementation of earthquake early warning systems in Taiwan, In *Proceedings of the 11th World Conference on Earthquake Engineering*, Acapulco, Paper No. 2133 (CD-ROM).

Matsuoka, M. and Yamazaki, F., 2000, Interferometric characterization of areas damaged by the 1995 Kobe earthquake using satellite SAR images, In *Proceedings of the 12th World Conference on Earthquake Engineering*, Auckland, Paper No. 2141 (CD-ROM).

Midorikawa, S. and Abe, S., 2000, Real-time assessment of earthquake disaster in Yokohama based on dense strong-motion network, In *Proceedings of the 12th World Conference on Earthquake Engineering*, Auckland, Paper No. 1036 (CD-ROM).

Nakamura, Y., 1996, Real-time information systems for hazards mitigation, In *Proceedings of the 11th World Conference on Earthquake Engineering*, Acapulco, Paper No. 2134 (CD-ROM).

National Research Institute for Earth Science and Disaster Prevention (NIED), 1995, Webpage on the Internet, Online. Available HTTP: <http://www.hinet.bosai.go.jp/jishin_portal/index_e.php> (accessed 17 January 2003).

Ogawa, N., Hasegawa, H., Yamazaki, F., Matsuoka, M. and Aoki, H., 1999, Earthquake damage survey method based on airborne HDTV, photography and SAR, In Proceedings of *5th* U. S. Conference on Lifeline Earthquake Engineering, (ASCE), pp. 322–331.

Shakal, A., Petersen, C., Cramlet, A. and Darragh, R., 1996, Near-real-time CSMIP strong motion monitoring and reporting for guiding event response, In *Proceedings of the 11th World Conference on Earthquake Engineering*, Acapulco, Paper No. 1566 (CD-ROM).

Shibayama, A. and Hisada, Y., 2002, Real-time system for acquiring earthquake damage information, In *Proceedings of the 11th Japan Earthquake Engineering symposium*, Tokyo (in Japanese with English summary and captions).

Shimizu, Y., Watanabe, A., Koganemaru, K., Nakayama, W. and Yamazaki, F., 2000, Super high-density realtime disaster mitigation system, In *Proceedings of the 12th World Conference on Earthquake Engineering*, Auckland, Paper No. 2345 (CD-ROM).

Shin, T.C., Tsai, Y.B. and Wu, Y.M., 1996, Rapid response of large earthquakes in Taiwan using a realtime telemetered network of digital accelerographs, In *Proceedings of the 11th World Conference on Earthquake Engineering*, Acapulco, Paper No. 2137 (CD-ROM).

Tobita, J. and Fukuwa, N., 2002, Anshin-system: intercommunication system for earthquake hazard and disaster information, In *Proceedings of the 2002 Japan–Taiwan Joint Seminar on Earthquake Mechanisms and Hazards*, Nagoya, pp. 66–67.

TriNet ShakeMap Working Group, 2002, Webpage on the Internet, Online. Available HTTP: <http://www.trinet.org/shake/9108645/intensity.html> (accessed 17 January 2003).

U.S.G.S., 2002, Planetary Interactive G.I.S.-on-the-Web Analyzable Database, Online. Available HTTP: <http://webgis.wr.usgs.gov> (accessed 17 January 2003).

Yamazaki, F., 1997, Earthquake monitoring and real-time damage assessment systems in Japan, In *Proceedings of the 5th U.S.–Japan Workshop on Urban Earthquake Hazard Reduction*, (EERI), pp. 397–400.

Yamazaki, F., Noda, S. and Meguro, K., 1998, Developments of early earthquake damage assessment systems in Japan, In *Structural Safety and Reliability*, (A. A. Balkema), pp. 1573–1580.

Zama, S., Hosokawa, M. and Sekizawa, A., 1998, Effective gathering of earthquake damage information, 1998, In *Proceedings of the 10th Japan Earthquake Engineering symposium*, Tokyo, pp. 3479–3484 (in Japanese with English summary and captions).

CHAPTER TEN

Location-Based Computing for Infrastructure Field Tasks

Amin Hammad, James. H. Garrett, Jr. and Hassan A. Karimi

10.1 INTRODUCTION

Infrastructure plays a key role in the economic prosperity and the quality of life of a nation. If not maintained, the quality of service of that infrastructure drops due to the effects of deterioration, and in some cases, results in sudden failure and loss of lives. A recent report from the American Society of Civil Engineers emphasized the poor conditions of the infrastructure in the U.S. (ASCE, 2001). Inspection and maintenance activities are expensive and time-consuming. The type and amount of data to be collected during inspection vary according to the specific approach adopted by an infrastructure management agency. Taking bridges as an example of infrastructure, there are nearly 590,000 bridges in the American National Bridge Inventory that must be inspected at least every two years. The current approach to bridge inspection and assessment is based on a condition rating method, whereby the inspectors go to the bridge and assign a condition rating to the major components of the bridge. The rating of the entire bridge is based on the conditions of all of the elements no matter the relative importance of the elements. This approach, while easier to collect data, does not recognize that the nature and location of damage on an element, and the type of element, will greatly change the reliability of the overall structure. Frangopol *et al.* describe a reliability-based approach for managing highway bridge maintenance and claim that this approach could cost at least 50% less than conventional maintenance strategies if labor and user costs are considered (Frangopol *et al.*, 2001). This reliability-based approach will require that a greater amount of condition information be collected from bridge elements. The location and amount of damage for all elements will have to be determined so that the relationship of the damage to the reliability of the elements and to the bridge can be assessed.

Technological advancements in sensing and wireless communication systems are motivating the development of structural health monitoring systems using embedded sensing technology, such as MicroElectroMechanical Sytems (MEMS) based sensors (Lynch *et al.*, 2002). These automated systems are expected to play a major role in providing the data needed for Frangopol's reliability-based approach. However, these systems will not completely substitute for inspection methods, but will rather complement them. Therefore, inspection methods should be enhanced to support these emerging systems.

Field engineers and technicians working on infrastructure projects need to determine and verify the location of structures and structural elements in order to collect the necessary data. They usually use maps, engineering drawings, databases, and other technical documents for this purpose (Bridge, 1995). In a typical scenario, a bridge inspector who is inspecting a reinforced concrete pier needs to know the number of the pier at which he or she is looking. The inspector may also want to look at the maintenance history of this pier and retrieve the location of cracks that have been found in previous inspections. The inspector also has to take notes about the locations and directions of any new cracks by drawing sketches, taking pictures, and writing a description of the problem.

With the increasing availability of commercial mobile and wearable computers, some of the aforementioned data can be collected and retrieved on-site using specialized data collection systems. One early example of such a system is that developed by the University of Central Florida for the Florida Department of Transportation (FDOT) (Kuo *et al.*, 1994). The system consists of both a field and office set up with a pen-based notebook computer used to collect all field inspection data. The Massachusetts Highway Department is using a system called IBIIS to store and manage all of their bridge documents (Leung, 1996). As part of this system, inspectors are equipped with a video camcorder to take video and still photographs and a notebook computer to enter the rating data for each bridge and commentary. A more recent, Personal Digital Assistant (PDA) -based field data collection system for bridge inspection is *Inspection On Hand* (IOH) (Trilon, 2003). IOH helps inspectors capture all rating information, commentary and sketches using hand-held, pen-based PDAs, and share data with Pontis bridge management system.

Using mobile and wearable computers in the field under severe working and environmental conditions requires new types of interaction that increase the efficiency and safety of field workers. Research on systems aiming to provide information related to infrastructure, at different stages of their life cycle, to mobile workers has been undertaken. Garrett *et al.* (1998, 2000, 2002) discussed the issues in delivering mobile and wearable computer-aided inspection systems for field users. Sunkpho *et al.* (2002) developed the Mobile Inspection Assistant (MIA) that runs on a wearable computer and delivers a voice recognition-based user interface. They also proposed a framework for developing field inspection support systems that explicitly considers the *objects* to be inspected, the *instrument* used to collect data, the *background* inspection knowledge, and the collected inspection *data*. Bürgy and Garrett (2002) developed the Interaction Constraint Model that helps system designers identify relevant user interface concepts for applying mobile and wearable computers to a specific context based on the constraints imposed by the task to be performed (e.g., need for using other tools or for full attention), the type of the data needed for the application (e.g., text-based or drawing-based), the environment where the system will be used (e.g., lighting and noise), the type of the computer, and the abilities and work patterns of the user.

Mobility is a basic characteristic of field tasks. In the above bridge inspection scenario, the inspector has to move most of the time in order to do the job at hand. The inspector walks over, under or around the bridge, or in some cases climbs the bridge. Knowing the exact location of the inspector with respect to the inspected elements can greatly facilitate the task of data collection by automatically identifying the elements, and potentially specifying the locations of defects on these elements. Approximate location information can still help in focusing the scope of data collection by identifying the objects that are of interest. Present methods of capturing location information using paper or digital maps, pictures, drawings and textual description can lead to ambiguity and errors in interpreting the collected data. The following description of the location of cracks in a bridge inspection report is a good example of this ambiguity: "Several cracks are found in the welds and base metal of the floor-beam web-to-top flange welds near the box girder connection of arches."

Location-Based Computing (LBC) is an emerging discipline focused on integrating geoinformatics, telecommunications, and mobile computing technologies (see Chapter 5). LBC utilizes geoinformatics technologies, such as Geographic Information Systems (GISs) and the Global Positioning System (GPS) in a distributed real-time mobile computing environment. LBC is paving the way for a large number of location-based services and is expected to become pervasive technology that people will use in daily activities, such as mobile commerce, as well as in critical systems, such as emergency response systems. In LBC, elements and events involved in a specific task are registered according to their locations in a spatial database, and the activities supported by the mobile and wearable computers are aware of these locations using suitable positioning devices. For example, an inspection system based on LBC would allow the bridge inspector in the above scenario to accurately locate the cracks on a predefined 3D model of the bridge in real time without the need for any post-processing of the data.

One driving force behind LBC is the simultaneous emergence of a variety of applications that need some combination of GIS, GPS, and wireless communications. For example, the Federal Communication Commission mandates that all wireless carriers are supposed to provide the location of the emergency calls to Public Safety Answer Points with the accuracy of 125 m for 67% of calls by the end of 2005 (FCC, 2003). Another example is the advancements in Intelligent Transportation Systems (ITSs), such as advanced vehicle control systems, route navigation, and fleet management (ITSs are discussed in detail in Chapter 11).

Several applications of LBC related to infrastructure data collection have been practically used for a number of years, such as mobile mapping of highway features and the automated survey of pavement surface distress (Wang and Gong, 2002). These applications use high accuracy GPS, Inertial Navigation Systems (INSs), and CCD cameras that are connected to a high-performance computer system for image processing and georeferencing the resulting output data. In these applications, all the instruments are usually installed on a van and no interaction with the system is necessary. The requirements of these automated systems are different from those of data collection systems that need continuous interaction with the field workers.

In this chapter, we present and discuss the concept of a mobile data collection system for engineering field tasks based on LBC that we call ***LBC for Infrastructure field tasks*** or LBC-Infra. In the subsequent sections, we first present the LBC-Infra concept. Then, we review the component technologies of LBC-Infra and discuss the issues involved in the deployment of these technologies. Then, we discuss the requirements of LBC-Infra and identify its most promising system architectures based on available technologies and the modes of interaction. We also explore the specific interaction and communication methods needed in LBC-Infra. Finally, we report on a prototype system that has been developed to test and illustrate the discussed issues.

10.2 LBC-INFRA CONCEPT

The concept of LBC-Infra is to integrate spatial databases, mobile computing, tracking technologies and wireless communications in a computer system that allows infrastructure field workers using mobile and wearable computers to interact with georeferenced spatial models of the infrastructure and to automatically retrieve the necessary information in real time based on their location, orientation, and specific task context. Using LBC-Infra, field workers will be able to access and update information related to their tasks in the field with minimum efforts spent on the interaction with the system, which results in increasing their efficiency and reducing the cost of infrastructure inspection. In addition, the wirelessly distributed nature of LBC will allow field workers to share the collected information and communicate with each other and with personnel at a remote site (office). This feature is of a great value especially in emergency cases.

Figure 10.1 shows the concept of LBC-Infra. In this figure, a bridge inspector, equipped with a wearable computer, is inspecting a large highway bridge searching for damage, such as cracks. The inspector is equipped with a mobile or wearable computer that has a wireless communications card and is connected to tracking devices. Based on the location and orientation of the inspector and the task to be achieved, the system may display information about the parts of interest within his or her focus (e.g., the number of a pier or the date when it was previously inspected) or navigation arrows to the locations where cracks are most likely to be found or the locations found in previous inspections. The inspector compares the changes in conditions by wirelessly accessing and viewing any of the previous inspection reports stored in the office database using spatial queries based on his location and orientation. The spatial database of the bridge and the surrounding environment, and the tracking devices attached to the inspector, make it possible to locate structural elements and detected problems and provide navigation guidance to these objects. In addition, all newly collected information is tagged in space. For example, using a pointing device equipped with a laser range finder, the inspector can point at the location of cracks on the bridge. The exact location can be calculated based on the location and direction of the pointer, the distance to the crack, and the geographically registered 3D model of the bridge.

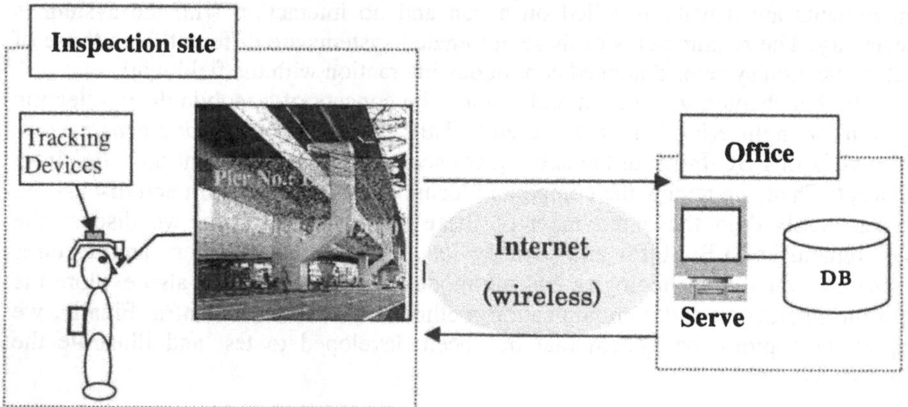

Figure 10.1 Concept of LBC-Infra

The inspector is providing input directly into the database with an inspection report being automatically generated as he or she performs different aspects of the inspection. In addition, the system presents the inspectors with a detailed, context-specific description of the procedure to follow in assessing damage once it has been discovered. The result is a savings in time, as information presentation and collection occur proactively and "just-in-time" without incurring any time penalty due to searching for information or following a complicated data input procedure.

10.3 TECHNOLOGICAL COMPONENTS OF LBC-INFRA

To deliver the LBC-Infra concept, four LBC technologies are required: mobile and wearable computers, spatial databases, positioning technologies, and wireless communications technologies (see Figure 10.2). Integration of these components has been realized at several levels, such as: *Internet GIS*, i.e., GIS services that can be accessed over the Internet by using a web browser (Dodge *et al.*, 1998); *Mobile Internet*, i.e., the ability to wirelessly access the Internet using mobile and wearable computers; and *location-aware mobile computing*, i.e., computing that depends on collecting and utilizing location information of a mobile agent (Beadle *at al.*, 1997). *GPS-enabled mobile computing* is used in some applications that collect and utilize the location information (latitude and longitude) without referring to a spatial database, such as the E-911 service. The technological aspects of these components are discussed in the following subsections.

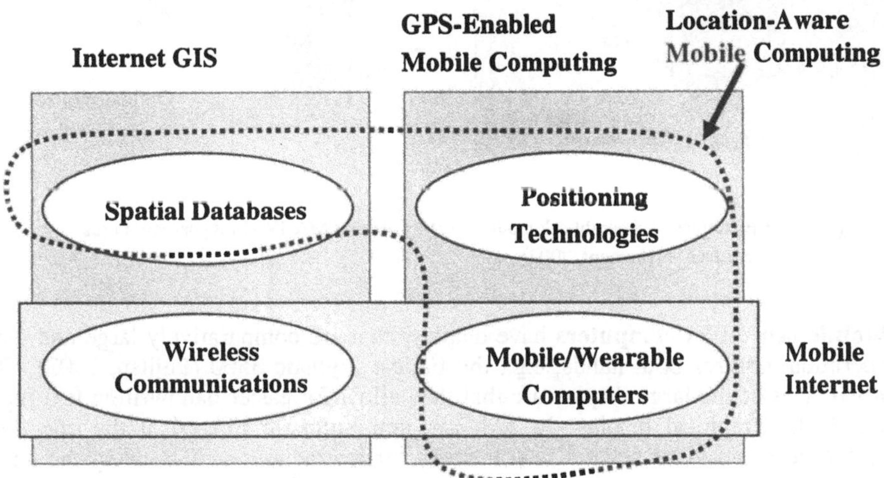

Figure 10.2 Technological Components of LBC

10.3.1 Mobile and Wearable Computers

Our definition of mobile and wearable computers is confined to hand-held and wearable computers that can be used while moving outdoors and are equipped with portable and unobtrusive input/output devices. We can distinguish the following four types of mobile and wearable computers:

Wearable computers have a main unit that includes the CPU, memory, hard disk, etc., and that can be attached to the body of the user. The display unit may be a head-mounted display or a flat panel display that is handheld, arm-worn, or attached to the chest of the user. The flat panel display is usually much smaller than those of laptop computers and should be readable in outdoor lighting conditions. Head-mounted displays can have high resolution comparable to laptops (full color 800x600 pixels) and can be binocular or monocular. Input devices include touch screens, noise-canceling microphones, pen-based systems and wearable keyboards. Examples of wearable computers are Xybernaut (2003) and ViA (2003). There are several groups developing new wearable computer systems, such as the WearableGroup at Carnegie Mellon University (2003) and the MIT Wearable Computing Group (2003). Figure 10.3 shows examples of wearable computers.

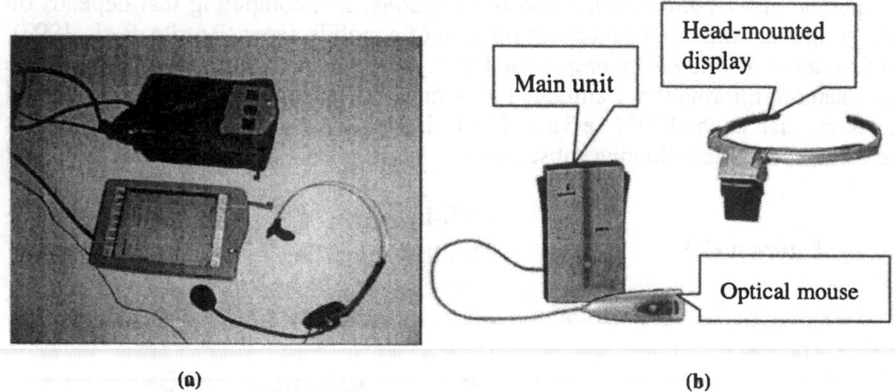

Figure 10.3 Examples of Wearable Computers: (a) Xybernaut MA-IV, (b) Xybernaut Poma (Source: Xybernaut, 2003)

Mobile pen tablet computers have displays that are comparatively large and their operation requires both hands, e.g., the Fujitsu Stylistic 3500 (Fujitsu, 2003). The advantages of the large display are that they allow for easier handwriting text input and better graphical display. In addition, new software offered at the operation system level provides several new features for interaction such as using the tablet computer digital pen to "gesture," e.g., scratching out digital ink on the display with a digital pen to erase the digital ink (Microsoft, 2003).

Personal Digital Assistants (PDAs) are cheaper and lighter than wearable computers and their computing power is limited, e.g., Compaq iPaq (Compaq, 2003). PDAs equipped with a wireless communications card can be used as thin clients for applications running on a remote server. For example, Microsoft Windows Terminal Server allows users to run applications remotely on a thin client

PDA (Microsoft, 2003). However, running desktop applications on a thin client in the field is not always a workable solution.

High-end cellular phones and other devices combining cellular phones and PDAs can also be used as thin clients offering limited services. For example, the Japanese J-Phone wireless telecommunications company provides driving directions, geocoding, and yellow page information on the handset's color graphic display (J-Navi, 2003).

In general, because of mobility requirements, mobile and wearable computers are limited in their computing power, memory, data storage, networking capability, and energy resources. The hardware (CPU, head-mounted display, batteries, etc.) should be light and should resist the environmental conditions. With the rapid development of computer hardware, mobile and wearable computers, PDAs and cellular phones will eventually evolve to new types of devices that combine powerful computing and communication functionalities with ease of use and mobility. Examples of devices combining cellular phones and PDAs are Handspring (Handspring, 2003) and PC-EPhone (PC-EPhone, 2003). Figure 10.4 shows three examples of thin mobile clients.

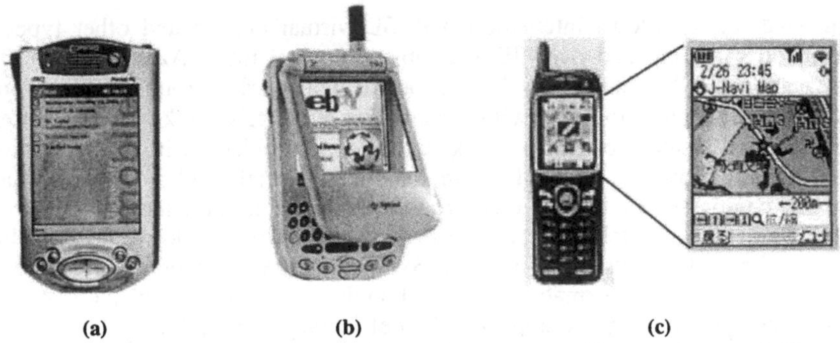

(a) (b) (c)

Figure 10.4 Examples of Thin Mobile Clients: (a) Compaq iPaq PDA (Source: Compaq, 2003), (b) Handspring Treo 300 (Cell phone and PDA. Source: Handspring, 2003), (c) J-Phone JP51 (The display shows navigation information on a map. Source: J-Navi, 2003)

As mentioned earlier, the design of the human–computer interface has to be unobtrusive. The attention needed for interaction with the system should be minimal allowing field workers to move freely from one location to another and to use their hands, as much as possible, for performing the main task in the field efficiently and safely. The main interaction modes that seem to be promising for mobile and wearable computers are:

(1) Graphical User Interfaces (GUIs)

GUIs are based on buttons, menus, lists, slide bars, etc. for selecting commands and data items. Interaction with a GUI needs a pointing device, such as a touch screen pen or a portable optical mouse.

(2) Direct manipulation mode

Direct manipulation of spatial models allows in general for better interaction with these models. For example, selecting an element of a bridge by clicking on that element in a 3D graphical model of the bridge is easier than selecting the same element from a list of identifiers in a GUI since this minimizes the syntax-memorization load of the user with respect to the design and contents of the GUI (Shneiderman, 1998). This mode is similar to GUIs in that it requires a pointing device.

(3) Speech-centered mode

Speech recognition and synthesis allow for hands- and eyes-free operations. Speech recognition systems can be categorized according to the flow of speech (discrete commands or continuous dictation), speaker dependency (training the system is needed), and vocabulary size (Sunkpho *et al.*, 2002). However, speech recognition may not be suitable in noisy environments, such as construction sites.

(4) Augmented reality mode

Augmented reality allows interaction with 3D virtual objects and other types of information superimposed over 3D real objects in real time (Azuma, 1997). The augmentation can be realized by looking at the real world through a see-through head-mounted display equipped with sensors that accurately track head movements (3 displacements and 3 rotations) to register the virtual objects with the real objects in real time as shown in Figure 10.5. In some commercial head-mounted displays, electronic compasses are used to measure the yaw angle, and tilt sensors are used to measure the roll and pitch angles. In this mode, field workers do not have to look back and forth at the structure and the computer screen to mentally achieve the spatial mapping of the information displayed on the screen. This helps them to better focus on their actual tasks and improves their efficiency and safety.

The main challenge for augmented reality is the requirement for very accurate 3D spatial databases and head tracking (further described in Sections 10.3.2 and 10.3.3, respectively). For example, if a bridge inspector is looking at a pier from a distance of 10 m, a 1 m position error will result in about 5 degrees error assuming that the inspector is still and not moving his or her head.

Augmented reality has many potential applications as a visualization aid in assembling, maintaining and repairing complex engineered systems. Examples of such augmented reality systems include a laser printer maintenance application (Feiner *et al.*, 1993), X-ray vision that allows seeing the location of reinforcement within a reinforced concrete wall (Webster *et al.*, 1996), and a construction support system for a space frame (Webster *et al.*, 1996). Mobile augmented reality systems are emerging as powerful support tools for field tasks. The "Touring Machine" is an example of such systems that allows users to view information linked to specific buildings of a university campus while walking (Höllerer *at al.*, 1999; also see Chapter 8). Thomas *et al.* (1998) have demonstrated the use of augmented reality to visualize architectural designs in an outdoor environment. Some practical systems in limited-range restricted environments exist today, such as Navigator 2 developed at

Carnegie Mellon University. Navigator 2 has been used by the Air Force to inspect the skin of Boeing aircraft for cracks and corrosion with 20% saving in inspection time (Smailagic and Siewiorek, 1999).

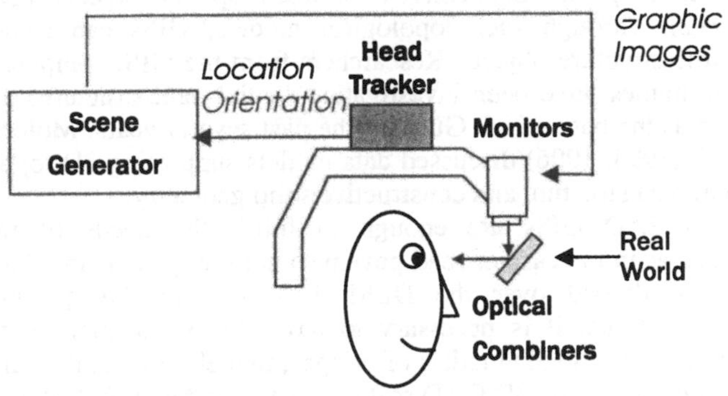

Figure 10.5 See-Through Head-Mounted Display (after Azuma, 1997)

The above interaction modes may be combined in a multi-modal adaptive user interface, which may use graphical, direct manipulation, speech, and augmented reality modes depending on the task, environment, and preferences of the user. For example, the user may have the choice to select the next command from a GUI menu or can use a speech recognition engine to input the same command. The system can be programmed to measure ambient noise levels of the environment and to turn on or off the speech recognition option depending on whether the noise level is lower or higher than a threshold value, respectively. These types of adaptive systems need to possess a level of intelligence to adjust to different situations without confusing the users.

Because of the complexity of spatial models used in LBC-Infra, direct manipulation and augmented reality modes are more promising, albeit more technically demanding, for interacting with these models. These modes can be combined with a speech interaction mode in a multi-modal interaction model to reduce the need for using a pointing device.

10.3.2 Spatial Databases

Spatial databases are an important component of LBC-Infra where a GIS is used for managing spatial databases, visualization, and spatial modeling and analysis. Current vector GISs can be categorized into one or more of the following: 2D GISs, 2.5D GISs, or 3D GISs. 2D GIS databases contain only the X and Y coordinates of the objects stored in them (points, lines, and polygons). When a GIS database contains the Z coordinate as an attribute of the planar points, the GIS is considered to be a 2.5D GIS. Digital Elevation Models (DEMs) are examples of 2.5D GIS models that can be represented using contour lines or a Triangular Irregular Network (TIN). 3D GIS databases contain 3D data structure representing both the geometry and topology of the 3D shapes, and allowing 3D spatial analysis.

A spatial database must contain two types of information about the represented objects: geometric data and topological data. Geometric data contain information about the shape of the objects, whereas topological data include the mathematically explicit rules defining the connectivity between spatial objects (Laurini and Thompson, 1992). Through such topological models, GISs can answer spatial queries about infrastructure objects. Researchers from the GIS, computer graphics, and CAD communities have been investigating spatial data structures and models that can be used as the base of 3D GISs for the past several years. Molenaar (1990) and Pilouk *et al.* (1994, 1996) discussed data models supporting 3D topology, such as the 3D formal data structure and constructive solid geometry.

The 2D or 2.5D GISs are enough to fulfill the needs of many LBC applications, such as in the case of road pavement survey systems in which the road geometry can be draped over the DEM. However, in the general case of infrastructure field tasks, it is necessary to have the 3D spatial models of the infrastructure itself, i.e., the 3D models of bridges, tunnels, etc., at the same level of detail as when represented by 3D CAD software, such as AutoCAD. Research in the area of product and process data models resulted in a number of standards for representing buildings and other types of structures, such as the STandard for the Exchange of Product (STEP) data model (STEP, 2003) and the Industry Foundation Classes (IFC) (Froese, 1996). Although these data models support topology representation, they do not provide the means for actually performing spatial analysis based on topology. The recent trend has been to integrate GIS and CAD functions into one environment, e.g., Autodesk Map (2003). However, this integration is based on geometry for visualization purposes and does not support topology and 3D spatial analysis (Nguyen and Oloufa, 2002). Many of the examples given in Section 10.1 require 3D spatial analysis. In the example of finding the locations of cracks, LBC-Infra equipped with accurate positioning can retrieve these locations from a 3D database and give the inspector 3D navigation guidance based on his or her present position.

In order to realize the full potential of the LBC Infra concept, a data structure and spatial analysis tools that integrate 3D GIS and CAD data need to be created and made able to answer geometric and topological queries. This integration will be of great benefit to all kinds of LBC applications related to infrastructure management, including construction, inspection, and monitoring. However, such 3D modeling and analysis require a large amount of storage and computing resources.

10.3.3 Positioning and Tracking Technologies

LBC-Infra requires that the position and orientation of the field worker are continuously tracked. Positioning technologies can be grouped into four categories: (1) active source systems; (2) passive source systems; (3) dead reckoning systems; and (4) hybrid systems (Azuma, 1997). Chapter 3 has more information on positioning technologies.

(1) Active Source Systems

Active source systems require powered signal emitters and sensors specially placed and calibrated. The signal can be magnetic, optical, radio, ultrasonic, or from the GPS satellites. These active source systems are based on the triangulation approach measuring signal arrival time (equivalent to the distances) between the mobile station and the multiple emitters. Traditionally, the need for special infrastructure and the short range of tracking make these systems more suitable indoors (except the GPS). Recently, some cell phone networks can be used as an active source positioning system that can provide cheap, low-accuracy positioning almost everywhere without the need for an additional infrastructure (Wirelessdevnet, 2003). In this case, the emitters are the cell phones and the infrastructure is the set of base stations in the network.

GPS is available anywhere within certain conditions, and it measures the horizontal and vertical positions of the receiver. Because of this availability and the relatively good accuracy and low cost of GPS, it is widely used for mobile mapping and other data collection tasks. Even with Selective Availability[1] turned off in 2000, other factors still affect GPS accuracy. These factors include ionospheric and tropospheric distortion of the radio signals from the satellites, orbital alignment and clock errors of the satellites, and signal multipath errors (reflections and bouncing of the signal near buildings). In addition, GPS is easily blocked in urban areas, near hills, or under highway bridges. The accuracy of a position is also a function of the geometry of the GPS constellation visible at that moment in time, i.e., when the visible satellites are well separated in the sky, GPS receivers compute positions more accurately. One method to increase the accuracy of GPS is by using Differential GPS (DGPS). DGPS is based on correcting the effects of the pseudo-range errors caused by the ionosphere, troposphere, and satellite orbital and clock errors by placing a GPS receiver at a precisely known location (base station). The pseudo-range errors are considered common to all GPS receivers within some range. Multipath errors and receiver noise differ from one GPS receiver to another and cannot be removed using differential corrections. DGPS has a typical 3D accuracy of better than 3 m and an update rate of 0.1-1 Hz (Kaplan, 1996). The DGPS corrections can be sent to the mobile GPS receivers in real time, or added later by post-processing of the collected data. Real-time kinematic GPS (RTK-GPS) receivers with carrier-phase ambiguity resolution can achieve accuracies better than 10 cm, but are computationally very intensive to run in real time (Kaplan, 1996).

(2) Passive Source Systems

These systems use naturally occurring signals. The main passive source systems are electronic compasses, sensing the earth magnetic field, and vision-based systems that depend on natural light. Electronic compasses are small, inexpensive, and accurate (accuracy of about 0.5 degrees). However, like magnetic sensors, they have the problem of magnetic distortion when in proximity to metals. Vision-based systems use video sensors to track specially placed marks called fiducials indoors, or natural features and landmarks outdoors. In augmented reality applications, these

---

[1] The random errors deliberately added to keep GPS a more powerful tool for the U.S. military.

systems allow for getting feedback from the scene and enforce better registration by matching reference features from the camera and graphics streams; this technique is called closed-loop tracking systems and can achieve tracking accuracy of few millimeters. However, this tracking requires that the tracked objects be always visible. In addition, achieving robust real-time video processing, especially for natural features tracking, is computationally intensive and requires special hardware that is not suitable for LBC-Infra.

(3) Dead Reckoning Systems

Dead reckoning systems do not depend on any external signal source. For example, an inertial system measures the linear accelerations and rotation rates resulting from gravity using linear accelerometers and rate gyroscopes, respectively. The measured signals from a gyroscope are integrated once and those from an accelerometer are integrated twice to produce orientation and position, respectively, causing linear and non-linear accumulation of drift errors with elapsed time. Although some rate gyroscopes may give sufficient accuracy of minutes or seconds for short time intervals, no accelerometers exist with enough accuracy (Azuma, 1997).

(4) Hybrid Systems

Hybrid systems use multiple measurements obtained from different sensors to compensate for the shortcomings of each technology when used alone. One promising hybrid system for LBC-Infra is to measure position by DGPS and inertial tracking, and orientation by a digital compass and tilt sensors. A Kalman filter can be used to integrate measurements from different positioning techniques and to provide the best tracking solution using the information obtained from the available sensors (You and Neumann, 2001). It also provides a positioning solution when measurements from one or more sensors become unavailable. In this case, each sensor in the hybrid system is considered as a backup for the others. For example, when the work to be done is under a canopy where GPS signals may be obstructed, the Kalman filter would use the information from the inertial tracking system to compute the position.

Table 10.1 shows a comparison of several positioning technologies and the associated accuracies.

Table 10.1 Positioning Technologies and Associated Accuracies

Technology	Accuracy (m)	Type of Position Data	Comment
GPS	20	Absolute	Stand-alone mode
Differential GPS	0.05-3	Absolute	Real time or post-processing
Laser rangefinders	0.05	Relative	Operators must aim device at a substantial target
Inertial navigation	0.05	Relative	Error accumulates with time

10.3.4 Wireless Communications

Field workers using LBC-Infra need to communicate with other members of the team who may be at different locations in the field or at the office. They also need to download and upload data from remote servers. The processing power of their mobile and wearable computers may not be enough to support the intensive processing needed for LBC. Wireless communications technology is becoming more available to field workers, allowing the design of distributed collaborative mobile systems that can satisfy the above communication requirements (Liu et al., 2002; Peña-Mora and Dwivedi, 2002). There are several standard protocols to achieve distributed computing that allow invoking methods of an object residing on a remote computer, e.g., Common Object Request Broker Architecture (CORBA) and the Java Remote Method Invocation (RMI).

Two types of wireless networks are conceivable for LBC-Infra: Wireless Wide Area Networks (WWANs) and Wireless Local Area Networks (WLANs). WWANs are used for low rate data transmission over cellular digital packet radios based on costly infrastructures. Advanced Mobile Phone System (AMPS) and Global System for Mobile Communication (GSM) are examples of second generation (2G) WWANs with data transmission rates of 4.8 Kbps and 9.6-14.4 Kbps, respectively. The 2.5G systems, such as the General Packet Radio Service (GPRS), and future 3G systems, such as the Universal Mobile Telecommunications System (UMTS), will allow digital packet data transmission rates of up to 160 Kbps and 2Mbps, respectively. On the other hand, WLANs are flexible data communication systems using electromagnetic waves (mostly radio) over a small area, such as a university campus. In a typical WLAN configuration, a transmitter/receiver (transceiver) device, called an *access point*, connects to the wired network from a fixed location using a standard Ethernet cable. End users access the WLAN through WLAN adapters, which are implemented as PCMCIA cards or fully integrated devices within mobile and wearable computers.

When fixed access points are established, the WLAN is called an infrastructure WLAN. All network traffic from mobile stations goes through an access point to reach the destination on either the wired or wireless networks. Another type of WLAN, called an ad-hoc (peer to peer) WLAN, has no central controller or access point. Each device communicates directly with other devices in the network rather than through a central controller. Bluetooth (Bluetooth, 2003) is expected to become popular in ad-hoc short range WLANs (about 10 m), also called Wireless Personal Area Networks (WPANs). Ad-hoc WLANs are useful in places where small groups of computers might congregate without the need for accessing another network. For example, in the case of an earthquake emergency, where a team of engineers needs to exchange information quickly, ad-hoc WLANs are more easily deployable than infrastructure WLANs.

WLANs have four main characteristics, i.e., range, throughput, interoperability and interference, that should be considered in LBC-Infra. **Range** (radius of coverage) is the distance over which radio waves can communicate and is a function of the transmitted power and receiver design. Most WLANs use radio waves because they can penetrate walls and surfaces, but interactions with buildings and other objects can affect the range. The range for typical WLANs varies from under 30 m to more than 200 m. **Throughput** depends on the type of WLAN used, airwave congestion (number of users), interference and multipath, as well as the latency and bottlenecks in the wired portions of the WLAN. Recent WLANs have high data rates. For example, IEEE 802.11b and 802.11a standards have data rates of 11 and 54 Mbps, respectively. **Interoperability** with wired and wireless infrastructure can be achieved by complying with WLANs industry standards and specifications, such as those of IEEE. **Interference** of WLAN waves with others in the same frequency spectrum transmitted by other devices may happen because of the unlicensed nature of radio-based WLANs. Table 10.2 summarizes the main characteristics of the major wireless communications technologies. Chapter 4 has a detailed description of wireless communications technologies.

Table 10.2 Summary of Wireless Communications Technologies

Technology		Average Range (m)	Data Rate (Mbps)
WLAN	Ultra-wideband	10	100
	IEEE 802.11a	50	54
	IEEE 802.11b	100	11
PLAN	Bluetooth	10	1
WWAN	GPRS	Wide region	0.16
	UMTS	Wide region	2

10.4 GENERAL REQUIREMENTS OF LBC-INFRA

LBC-Infra should satisfy a number of requirements in order to fulfill its functions. Some of these requirements are necessary at the component level of LBC-Infra, e.g., the requirements of WLAN, GPS, and mobile computing, while others are necessary at the system level, e.g., the requirements of an augmented reality system for bridge inspection. The following requirements can be considered as most essential for the successful deployment of LBC-Infra.

(1) **Availability**: Positioning and wireless communications systems, such as GPS and WLANs, may not be available in some situations.

(2) **Reliability**: Even when position information is available, the accuracy of this information and the data in the spatial database affect the reliability of LBC-Infra. The reliability is also affected by the quality of wireless communications.

(3) **Data storage**: The data volume of infrastructure-related 3D GIS/CAD databases is typically very extensive. The storage capacity of most of today's mobile and wearable computers is in most cases insufficient for storing these large databases required for field tasks.

(4) **Real time processing**: Spatial analysis using 3D GIS/CAD databases is computationally intensive. In LBC-Infra, the real time requirement is measured by the end-to-end system delay, which is the total time for tracking, spatial

analysis, communications, scene generation and display. An end-to-end delay of 100 ms is typical in existing augmented reality systems, which results in angular error of 5 degrees considering a head rotation speed of 50 degrees per sec. However, the angular accuracy that the human eye can detect is less than one minute (Azuma, 1997). This registration error causes visual conflicts that may not be acceptable. As GPS positioning is updated every 0.1 to 1 sec., position extrapolation may be used to compensate for the low update rate.

(5) **Wireless communications throughput and range**: In order to have efficient communications, the type of the deployed wireless technology should provide the necessary throughput and range for the specific system architecture as discussed in Section 10.5.

(6) **Interoperability**: LBC-Infra should be able to access heterogeneous data formats from different remote servers, and process and display the results on heterogeneous platforms, e.g., PDAs, wearable computers, etc.

(7) **Scalability**: When covering a wide geographical area, the volume of the spatial data of that area may become too large to be managed at a central server. In this case, the data should be distributed at several servers and dynamically partitioned based on the location of the field worker in order to answer a spatial query from the field. Also, the design of LBC-Infra should allow for increasing the number of field workers using the system without degrading the performance of the system.

(8) **Usability**: The usability of LBC-Infra is critical for its success and acceptance by field workers because of the small displays, limited input/output devices, and weight, energy and safety requirements.

It should be noted that most of the above requirements have strong dependency on the advancement in the technological components of LBC-Infra including spatial databases, mobile and wearable computers, tracking technologies, and wireless communications. Fortunately, these technologies have been progressing rapidly in the last few years, while becoming more available and less costly. In the following section, a framework for LBC-Infra is discussed that considers the above requirements and can accommodate the expected technological advancement in the near future.

10.5 INTERACTION PATTERNS AND FRAMEWORK OF LBC-INFRA

Based on the discussion of the components and requirements of LBC-Infra presented in Sections 10.3 and 10.4, respectively, a framework is discussed in this section that identifies a collection of high-level reusable software objects and their relationships. The generic structure of this framework embodies the general functionalities of LBC-Infra so that it can be extended and customized to create more specific applications, e.g., a bridge inspection application or a building construction progress monitoring application.

10.5.1 Interaction Patterns of LBC-Infra

The framework of LBC-Infra aims to facilitate data collection and access in the field by allowing field workers to interact with georeferenced infrastructure models to automatically retrieve the necessary information in real time based on their location and orientation, and the task context. The following interaction patterns are typical examples that have been identified based on common tasks that field workers usually perform and the type of information they collect (Bridge, 1995).

(1) Presenting navigation information

As an extension to conventional navigation systems based on 2D maps, LBC-Infra can also present navigation information in 3D. Within a specific field task, the system can guide a field worker by providing him/her with navigation information and focusing his or her attention on the next element to be inspected. Taking crack inspection as an example, the system displays arrows to navigate the inspector to the locations where cracks are most likely to be found or the locations found in previous inspections. The system intelligently asks the inspector for information about the data to be acquired based on similar previous cases in the database, thus reducing the data collection burden.

(2) Displaying graphical details

LBC-Infra displays to the field worker structural details retrieved from previous inspection reports. This can happen in a proactive way based on spatial events, such as the proximity of the inspector to specific elements. For example, once a cracked element is within an inspector field-of-view, the system displays the cracks on that element discovered during previous inspections. This will help focus the inspector's attention on specific locations. The user of the system can control the level of detail of representing objects depending on his or her needs.

(3) Displaying non-graphical information and instructions

The user interface can provide links to documents related to the project, such as reports, regulations and specifications. In addition, LBC-Infra allows for displaying context sensitive instructions on the steps involved in a specific task, such as instructions about the method of checking new cracks, and measuring crack size and crack propagation. These instructions are based on manuals, such as the Bridge Inspector's Training Manual (1995).

(4) Communication

LBC-Infra facilitates the wireless communications among a team of field workers, geographically separated at the project site, by establishing a common spatial reference about the site of the project. In some cases, the field workers may communicate with an expert engineer stationed at the office who monitors the same scene generated by the mobile unit in the field.

10.5.2 Interaction Framework

Because the LBC-Infra concept integrates several evolving technologies, its implementation should follow an open and extensible framework so that the application development can adapt to new requirements and new technologies while reducing the time and cost of the development. A framework is an abstract design that describes the interaction between objects to provide a skeleton of applications in a particular domain (Johnson, 1997; Fayad *et al.,* 1999). A framework also incorporates knowledge that is common to all applications in a specific domain.

The Model-View-Controller (MVC) software development model (Reenskaug, 1981; Potel, 1996) has been selected as the basic framework of LBC-Infra because of its simplicity and flexibility in manipulating complex and dynamic data structures requiring diverse representations. The MVC model has three main high-level objects: *Model*, *View*, and *Controller*. The *Model* represents the data underlying the application that are accessible only through the *View* object. The *Model* of LBC-Infra has three basic databases for managing the spatial data (e.g., the axial end coordinates of the members of a steel bridge), data about the attributes of the infrastructure (e.g., inspection data), and data about field tasks (e.g., information about inspection tasks, their order, and devices and methods used for performing them). The *View* object accesses the data from the *Model* database and specifies how these data are presented to the user, e.g., the information about the members of a bridge can be used to create 2D or 3D representations of the bridge. The *Controller* determines how user interactions with the *View*, in the form of events, cause the data in the *Model* to change, e.g., clicking on a "no defect" button causes the value of the "inspection" attribute of the selected member to be set to "no defect" in the attribute database. The *Model* closes the loop by notifying the *View* to update itself so that it reflects the changes that occurred in the data. Using the MVC model, new methods of interaction can easily be introduced to the system by developing new *View* objects.

Using the MVC model in distributed client-server applications involves deciding which parts of the model are implemented, in whole or in part, on the client or the server (Potel, 1996). The *Model* represents a typical server-side functionality while the *View* represents a typical client-side functionality. The *Controller* can be partitioned between the client and the server; most of the processing could be on the client side in a fat client or on the server side with only a simple GUI application on a thin client. In more elaborate architectures, other partitioning of the functionalities between the client and the server are possible, e.g., the client may have a proxy of the *Model* to lower the demand for communication with the server. In addition to the *Model, View* and *Controller,* LBC-Infra has a *Tracker* object that handles location-related functionalities as explained in the following paragraph. Figure 10.6 shows the relationship between the high-level objects of LBC-Infra using a Unified Modeling Language (UML) collaboration diagram, where objects interact with each other by sending messages (UML, 2003). The numbers in the diagram refer to the order of execution of the messages. Messages that are not numbered are threads that are executed at the beginning of the application and run continuously and concurrently with other messages, or they are event-driven messages that may occur at any time.

A *Field Worker* starts interacting with the system by sending a message (*start application*) to the *Controller* (message 1). As part of the initialization of the system, the *View* accesses the databases of the *Model* that reside on a remote server to retrieve the information necessary to create its contents (message 2) before it updates itself (message 3). Once the application is initialized, the *Tracker* starts continuously reading the location and orientation measurements of the *Field Worker* from the tracking devices and updating the *Controller* about the current coordinates of the *Field Worker*. The *Controller* retrieves the information about the next task to be performed from the *Task Database* that contains a plan defining the tasks (message 4). However, the *Field Worker* has the freedom to confirm this selection or override it by selecting another task if necessary (message 5). Based on the changing coordinates of the *Field Worker*, the *Controller* updates the viewpoint of the *View*, filters the contents of the selection menus, and updates the navigation information towards the best location and orientation to perform the present task. These updating and filtering steps insure that the presented information coincides with what the *Field Worker* can see in the real scene. The navigation is performed by presenting visual or audible guidance. As the *Field Worker* follows the navigation guidance towards the new location, the *Controller* provides him or her with information about the task (message 6). Another function of the *Controller* is to capture spatial events, such as the proximity to an element of interest. This flexibility allows switching the order of the tasks when adequate to inspect an element planned for inspection in a subsequent task. As in message 5 above, the *Field Worker* can accept or reject this change in the task order.

Once the *Field Worker* is at the right position and orientation to perform the task at hand, the *Controller* presents him or her with previously collected information (if available) that may help in performing the present task (message 7). At this point, the *Field Worker* performs the task, e.g., collecting inspection data visually of using some devices (message 8). He or she can input the collected data by interacting with the *View*, e.g., by clicking on an element displayed within the *View* for which data have to be updated. The input events are captured by the *View* (message 9), and used by the *Controller* to update the data in the *Model* (message 10). Finally, the changes in the databases are channeled to the *View* by distributed notification (message 11) so that the *View* can update itself (message 12). Client-server communications can happen whenever needed depending on the nature of the client (thin or fat).

10.5.3 Interaction Levels of LBC-Infra

Section 10.3 covered the whole spectrum of available technologies for implementing LBC-Infra. The framework of LBC-Infra presented in Section 10.5.2 provides the base for developing an open system architecture that is flexible enough to accommodate any combination of present and near-future technologies. However, the specific functional requirements of each version of LBC-Infra will largely vary depending on the specific technologies used in the implementation. For example, the following levels of interaction functionalities are identified based on the dimensionality of the spatial model (i.e., 2D, 2.5D or 3D), the tracking technologies that match this dimensionality, and the volume of information involved in the spatial model that need to be sent over the wireless network.

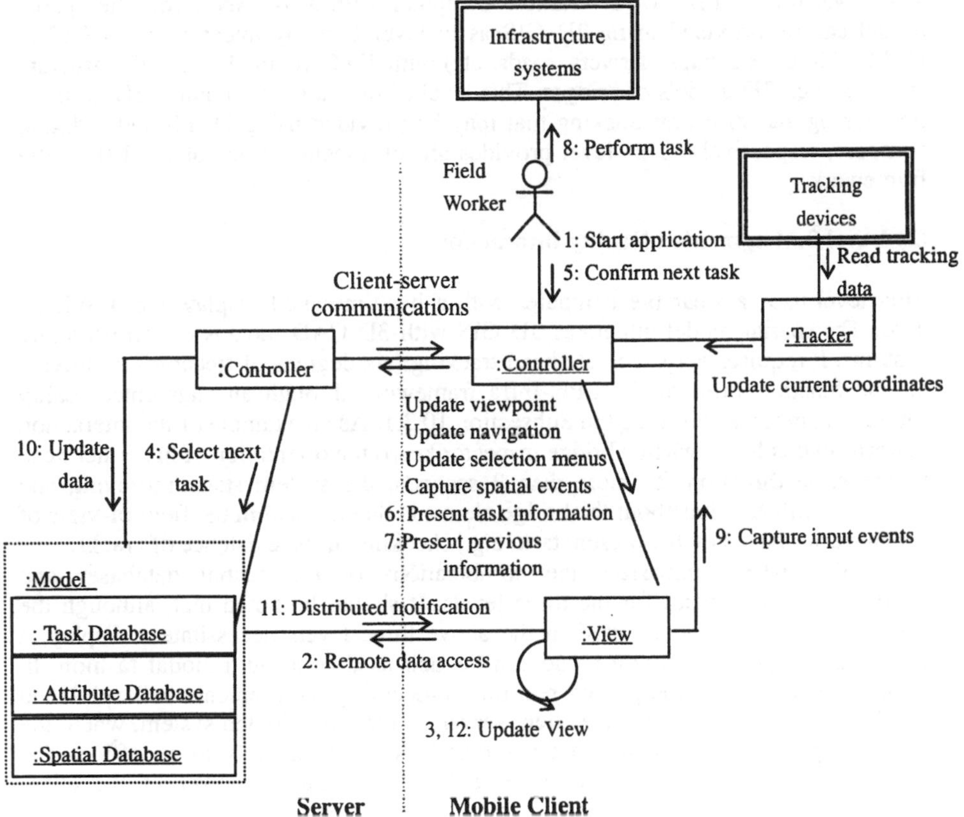

Figure 10.6 Collaboration Diagram Between the Entities of the Framework

(1) Level-1: Location-Aware Direct Manipulation of 2D GIS

This level uses a 3G cellular phone or a PDA equipped with a WLAN card as the mobile platform, and 2D maps of the infrastructure system. For example, in the case of a bridge inspection system, the piers and the decks of the bridges can be represented by a point layer and a polygon layer, respectively. Positioning can be achieved using GPS, DGPS or other methods depending on the needed accuracy. Because of the limited hardware and software capabilities, this level provides only simple navigation and data input/output functionalities using direct manipulation on the display of the cellular phone or the PDA.

(2) Level-2: Location-Aware Direct Manipulation of 3D GIS/CAD

This level uses a PDA or a wearable computer with a WLAN card. The spatial model can be provided using 2D GIS as in level-1, or by integrating 3D GIS (a DEM with draped maps of rivers, roads, etc) with 3D CAD models of infrastructure systems, e.g., 3D models of bridges. This level of functionality requires 3D accurate positioning and direction tracking that may be provided using DGPS and a digital compass, respectively. This level provides all the functionalities of the LBC-Infra framework.

(3) Level-3: Augmented Reality Interaction

This level uses a wearable computer with a head-mounted display and a WLAN card. The spatial model integrates 3D GIS with 3D CAD models of infrastructure systems. It requires very accurate head tracking (six degrees of freedom) to provide all the functionalities of the LBC-Infra framework through an augmented reality interaction mode as discussed in Subsection 10.3.1. As an example of the interaction patterns in this level, when a bridge inspector keeps the orientation of his or her head in the same direction for more than 3 seconds, the system starts retrieving and displaying information about the bridge elements that are within the field-of-view of the inspector based on the present task, e.g., checking of the existence of cracks.

Table 10.3 summarizes the combinations of the spatial databases and positioning technologies for the three levels. It should be noted that, although the primary interaction modes used in the above three levels necessitate a display, a secondary speech interaction mode can be also used in a multi-modal fashion. In order to address the storage and real-time processing requirements in level-2 and level-3, it is necessary to have a client-server wireless distributed system, where the field computer is the client and the server is located remotely in an office. The design of the system should consider both the computing power of the client computer and the data transfer rate over the wireless connection.

10.6 PROTOTYPE SYSTEM AND CASE STUDY

In order to demonstrate the potential of LBC-Infra in infrastructure projects, a prototype system of LBC-Infra following the framework discussed in Subsection 10.5.2 was developed and tested in a case study. The prototype system, whose architecture is based on level-2 functionalities, allows for location-aware direct manipulation of 3D GIS/CAD. In the case study, one of the authors simulating the action of a bridge inspector, walked around and above bridges and inspect their different elements visually or using some instruments. The Birmingham Bridge, erected in 1976 over the Monongahela River, is used in the case study (see Figure 10.7). The 185 m (607 ft) main span of this bridge is a steel arch bridge with wire rope suspenders.

Table 10.3 System Architectures for Combinations of Spatial Databases and Positioning Technologies

Type of spatial database / Type of tracking	2D GIS (Maps of roads, etc.)	3D GIS/CAD (Maps + DEM + 3D CAD models of infrastructure)
GPS/DGPS (2D positioning)	**Level 1: Location-Aware Direct Manipulation of 2D GIS** Uses 3G cellular phones or PDA with WLAN (IEEE 802.11b)	Not applicable
DGPS + digital compass (3D positioning and orientation)	**Level 2: Location Aware Direct Manipulation of 3D GIS/CAD** Uses PDA or wearable computer with WLAN (IEEE 802.11b)	
Accurate head tracking (Six degrees of freedom)	Not applicable	**Level 3: Augmented Reality Interaction** Uses wearable computer with head-mounted display and WLAN (IEEE 802.11a)

Figure 10.7 Birmingham Bridge

10.6.1 Software of the Prototype

The prototype system has the three databases of the *Model* object as discussed in Subsection 10.5.2, i.e., spatial database, attribute database including inspection data, and inspection tasks database. The databases were implemented in ORACLE 9i as relational databases. The spatial database includes the map of the rivers and major roads of the area around the bridge in addition to the 3D data of the members of the bridge. A detailed 3D model of the Birmingham Bridge was created and georeferenced based on the drawings of the bridge. The Virtual Reality Modeling Language (VRML) (VRML, 1997) was used for modeling and visualizing in the *View* object. VRML was selected as the 3D spatial modeling tool because of its powerful features in representing virtual environments and because it is supported by many CAD and graphics packages. VRML uses the concept of nodes to represent graphical objects. Nodes are also used to represent sensors, such as the TimeSensor, TouchSensor, and ProximitySensor nodes. For instance, the ProximitySensor node generates events when the user enters, exits, and moves within a region in a space defined by a box. Figure 10.8(a) shows the 2D rivers and roads maps converted into VRML. Figure 10.8(b) shows the 3D TIN representation of the elevation, which was created from the United States Geological Survey DEM data (30m x 30m resolution). In addition to the basic nodes of VRML, two of its extensions were used to facilitate the development of the prototype system, namely, GeoVRML (1997) and the steel element prototypes developed at the National Institute of Standard and Technology (NIST, 2003). GeoVRML enables the representation of geo-referenced data in standard coordinate systems, such as the Universal Transverse Mercator (UTM) grid, while the steel element prototypes provides a simple way to specify different types of beams with only the beam dimensions and position.

The inspection tasks database was partially implemented based on the information provided in The Bridge Inspector's Training Manual (1995). Examples of the inspection tasks in this manual are the tasks to investigate the arch ribs for signs of buckling, corrosion, and general deterioration. For each task, the manual explains the inspection procedures that should be followed. The information was elicited and organized in the database so that each element of the bridge has a relationship that links it to the tasks to be performed on it.

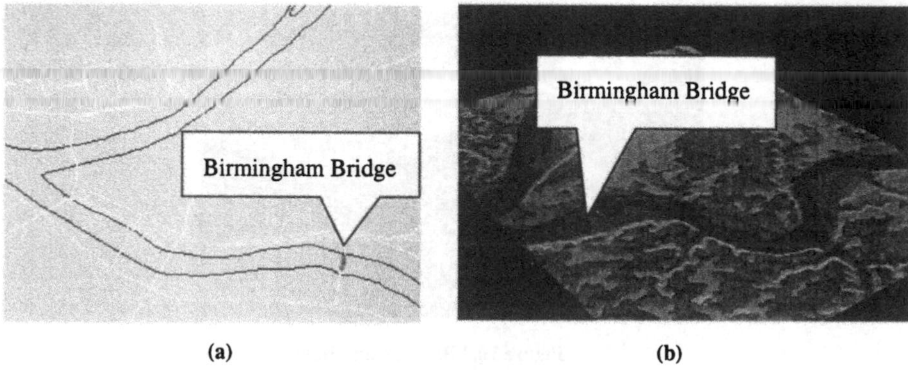

Figure 10.8 VRML Maps of Pittsburgh: (a) Rivers and Roads, (b) TIN Representation of the Elevation

The Java language was used for the implementation of the *Controller*. Figure 10.9 shows the components of the prototype system. The VRML External Authoring Interface (a Java API) was used to control the VRML browser and its contents through an applet. In the mobile client, a Java application reads the tracking data from the tracking devices and sends them to an applet that runs within the Web page containing the VRML file. The applet uses these data to control the location and orientation of the viewpoint so that the displayed scene matches the real scene in front of the user. The applet is also responsible for handling the user-interface for adding or removing objects (e.g., navigation objects) and interacting with menus that control the task flow (e.g., changing attributes). Another function of the Java application is to facilitate the communication between the mobile client and a remote server hosting the databases.

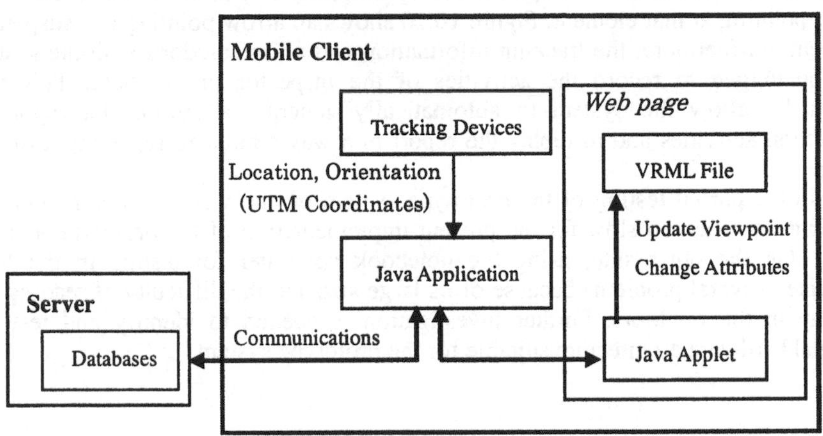

Figure 10.9 Components of the Prototype System

10.6.2 Hardware of the Prototype

The prototype system was developed and tested using a Sony notebook computer (Vaio, PCG-Z505CR/K) with Pentium III processor (750 MHz). The system was also partially tested using two wearable computers, a Xybernaut MA-IV with Intel Pentium III CPU (233 MHz), 128 MB RAM, and flat-panel all-light-readable display, and a Xybernaut Poma with Hitachi SuperH RISC CPU (128 MHz), 32 MB RAM, and Liquid Crystal on Silicon head-mounted display (see Figure 10.3).

A Trimble GeoExplorer 3 GPS receiver equipped with a magneto-resistive digital compass was used with a beacon for real-time DGPS correction. Although the software design is based on a client-server architecture, where the client and server are linked via a WLAN, because of the difficulty of establishing an access point at this stage of testing, both the client and the server reside on the same mobile notebook.

10.6.3 Preliminary Evaluation of the Prototype System

The purpose of the preliminary evaluation of the prototype system in this stage of the development is to demonstrate the feasibility of the proposed concept and framework of LBC-Infra, discussed in Section 10.5, by testing the basic functionalities of the prototype system. A more formal usability study of the system is planned in the near future after more detailed functionalities are implemented.

Figure 10.10 shows a snapshot of the user interface of the prototype system. At this stage, the prototype system successfully integrates the databases of the bridge model and inspection tasks and allows the user to interact with the 3D model to identify the elements of the bridge or to retrieve and update the related attributes in the database. User position and orientation are tracked and used to update the 3D view. In addition, a navigation function to a specific element of the bridge based on the location of the user was implemented by automatically inserting an arrow in the scene pointing at that element. Figure 10.10 shows an arrow pointing at a suspender element. Furthermore, the tracking information and all interactions with the system can be logged to record the activities of the inspector in the field. This will eventually allow the system to automatically generate a multimedia report of inspection activities and to replay the report in a way similar to replaying a digital movie.

In the partial testing of the prototype system on the wearable computers, the performance was too slow for the present implementation of the prototype as a fat client. On the other side, using the notebook computer for testing in the field presented several problems because of its large size and the difficulty of reading the display in the outdoors. Further investigation is needed to identify and test the optimal hardware architecture suitable for the prototype system.

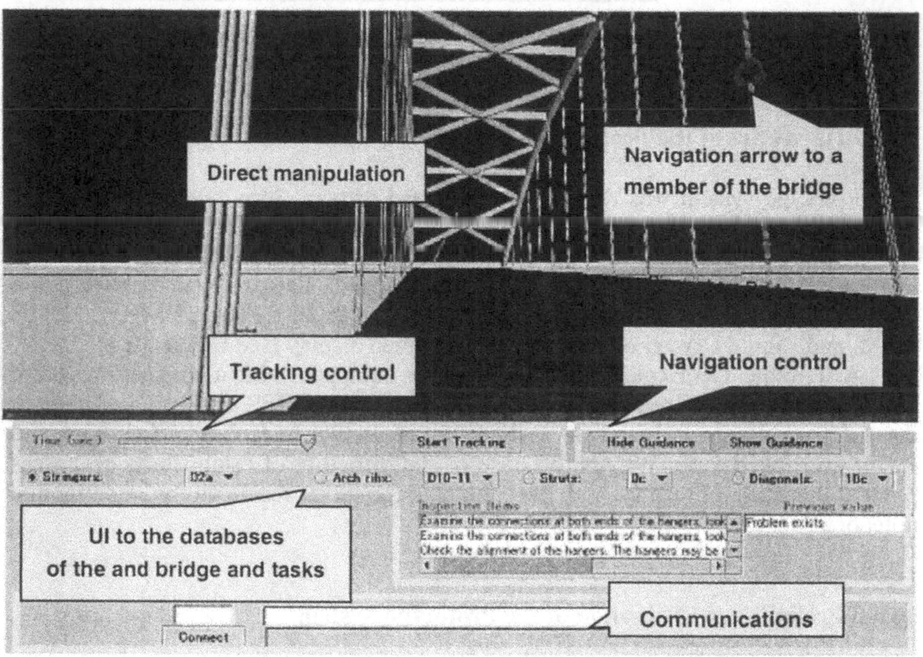

Figure 10.10 Snapshot of the User Interface of the Prototype System

10.7 CONCLUSIONS

In this chapter, we discussed the technologies, requirements, and the framework for **L**ocation-**B**ased **C**omputing for **Infra**structure field tasks (LBC-Infra). LBC-Infra facilitates collecting inspection data by allowing field workers to interact with georeferenced infrastructure models and automatically retrieve the necessary information in real time based on their location and orientation, and the task context. The framework of LBC-Infra integrates mobile and wearable computers, 3D GIS/CAD databases, positioning technologies and wireless communications to provide field workers with the specific information they need in a proactive manner. We investigated the promising system architectures that can be realized by integrating the wide spectrum of available technologies, and how these architectures can best support field tasks. We also demonstrated the potential and limitations of the framework and one system architecture through partially developing a prototype system and testing it in a case study. The prototype, using location-aware direct manipulation of 3D GIS/CAD, demonstrated the basic functionalities of LBC-Infra, such as tracking, navigation, and interacting with the 3D model to identify elements of the bridge or to retrieve and update the related attributes in the database. This is expected to improve the efficiency and safety of the field workers by allowing them to concentrate on their job. Future work will focus on further development and testing of the prototype system, considering both hardware and software issues, and on investigating the usage of LBC-Infra as a base for collaborative environment for field workers.

REFERENCES

ASCE, 2001, Report Card for America's Infrastructure, http://www.asce.org/reportcard/ (accessed March, 2003).

Autodesk web site, <http://usa.autodesk.com/> (accessed March, 2003).

Azuma, R.T., 1997, A Survey of Augmented Reality, *Presence: Teleoperators and Virtual Reality*, 6(4), pp. 355-386.

Beadle, H.W.P., Harper, B., Maguire, G.Q., Judge, J., 1997, Location Aware Mobile Computing, *Proc. IEEE/IEE International Conference on Telecommunications*, (ICTí97), Melbourne, April, pp. 1319-1324.

Bluetooth web site, <http://www.bluetooth.com> (accessed March, 2003).

Bridge Inspector's Training Manual, 1995, U.S. Department of Transportation, Federal Highway Administration.

Bürgy, C. and Garrett, Jr., J., 2002, Wearable Computers: An Interface between Humans and Smart Infrastructure Systems. *Bauen mit Computern*, Bonn, Germany. VDI Verlag GmbH, Duesseldorf, Germany.

Compaq web site: <http://athome.compaq.com> (accessed March, 2003).

de la Garza, J.M. and Howitt, I., 1998, Wireless Communication and Computing at the Construction Jobsite, *Automation in Construction*, Vol. 7, pp 327-347.

Dodge, M., Doyle, S., Smith, A., and Fleetwood, S., 1998, Towards the Virtual City: VR & Internet GIS for Urban Planning, paper presented at the *Virtual Reality and Geographical Information Systems Workshop*, Birkbeck College, London. Also available at the Centre for Advanced Spatial Analysis (CASA), University

College London <http://www.casa.ucl.ac.uk/newvenue/towards_virtual_city.pdf> (accessed March, 2003).

Fayad, M.E., Schmidt, D.C. and R. E. Johnson. 1999, Application Frameworks. In *Building Application Framework*, New York: John Wiley and Sons, pp. 3-27.

FCC web site: < http://www.fcc.gov/911/enhanced/> (accessed March, 2003).

Feiner, S., MacIntyre, B. and Sleegmann, D., 1993, Knowledge-Based Augmented Reality. *Communications of ACM*, 36:7, pp. 55-62.

Frangopol, D.M., Kong, J.S., Gharaibeh, E.S., 2001, Reliability-based Life-Cycle Management of Highway Bridges, *Journal of Computing in Civil Engineering*, 15(1), pp. 27-47.

Froese, T., 1996, STEP Data Standards And The Construction Industry, *CSCE Annual Conference*, Vol. 1, Edmonton, pp. 404-415.

Fujitsu web site: <http://www.fujitsupc.com> (accessed March, 2003).

Garrett, J.H. Jr., Sieworiek, D.P. and Smailagic, A., 1998, Wearable Computers for Bridge Inspection, *Proceedings of the International Symposium on Wearable Computers (ISWC)*, Pittsburgh, PA,USA, October, 1998, pp.160-161.

Garrett, J.H. Jr., Sunkpho, J., 2000, Issues in Delivering IT Systems for Field Users, *International Conference on the Application of Computer Science and Mathematics in Architecture and Civil Engineering* (IKM2000). Weimar, Germany.

Garrett, J.H. Jr., Bürgy, C., Reinhardt, J. and Sunkpho, J., 2002, An Overview of the Research in Mobile/Wearable Computer-Aided Engineering Systems in the Advanced Infrastructure Systems Laboratory at Carnegie Mellon University. *Bauen mit Computern*, Bonn, Germany. VDI Verlag GmbH, Duesseldorf, Germany.

GeoVRML web site: <http://www.geovrml.org/> (accessed March, 2003).

Handspring web site: <http://www.handspring.com> (accessed March, 2003).

Höllerer, T., Feiner, S., Terauchi, T., Rashid, G. and Hallaway, D., 1999, Exploring MARS: Developing Indoor and Outdoor User Interfaces to a Mobile Augmented Reality System, *Computers and Graphics*, Elsevier Publishers, 23(6), pp. 779-785.

J-Navi web site: < http://www.j-phone.com/english/products/kisyu/j_p51/index.html > (accessed March, 2003).

Johnson, R.E., 1997, Frameworks = (Components + Patterns). *Communications of the ACM* 40 (October), pp. 39-42.

Kaplan, E.D., 1996, *Understanding GPS: Principles and Applications*, Artech House.

Kuo, S. S., Clark, D. A. and Kerr, R., 1994, Complete Package for Computer-Automation Bridge Inspection Process. *Transportation Research Record*, No. 1442, pp. 115-127.

Laurini, R. and Thompson, A.D., 1992, *Fundamentals of Spatial Information Systems*. A.P.I.C. Series, Academic Press, New York, NY.

Leung, A., 1996, Perfecting Bridge Inspecting. *Civil Engineering Magazine*, pp. 59-61.

Liu, D., Cheng, J., Law, K.H. and Wiederhold, G., 2002, An Engineering Information Service Infrastructure for Ubiquitous Computing, *CIFE Technical Report #141*, Stanford University.

Lynch, J.P., Kiremidjian, A.S., Law, K.H., Kenny, T.W. and Carryer, E., 2002, Issues in Wireless Structural Damage Monitoring Technologies, *Proceedings of the 3rd World Conference on Structural Control (3WCSC),* Como, Italy, April 7-12.

Microsoft Windows XP Tablet PC Edition web site: <http://www.microsoft.com/windowsxp/tabletpc/> (accessed March, 2003).

MIT Wearable Computing Group web site: <http://www.media.mit.edu/wearables> (accessed March, 2003).

Molenaar, M., 1990, A formal data structure for 3D vector maps, *Proceedings of EGIS'90*, Vol. 2, Amsterdam, The Netherlands, pp. 770-781.

Nguyen, T.H. and Oloufa, A.A., 2002, Automation of Building Design with Spatial Information, *the 19th International Symposium on Automation and Robotics in Construction,* NIST, Washington, D.C., pp. 179-184.

NIST web site: <http://cic.nist.gov/vrml/cis2.html> (accessed March, 2003).

PC-Ephone web site: <http://www.pc-ephone.com> (accessed March, 2003).

Peña-Mora, F. and Dwivedi, G.H., 2002, Multiple Device Collaborative and Real time Analysis System for Project Management in Civil Engineering, *ASCE Journal of Computing in Civil Engineering,* 16(1), pp.23-38.

Pilouk, M., 1996, Integrated modelling for 3D GIS, PhD thesis, ITC, The Netherlands.

Pilouk, M. and Tempfli, K., 1994, An object-oriented approach to the Unified Data Structure of DTM and GIS, *Proceedings of ISPRS*, Commission IV, Vol. 30, Part 4, Athens, USA.

Potel, M., 1996, MVP: Model-View-Presenter, The Taligent Programming Model for C++ and Java, <ftp://www6.software.ibm.com/software/developer/library/mvp.pdf>

Reenskaug, T., 1981, *Smalltalk Issue Of Byte*, 6(8), August.

Shneiderman, B., 1998, *Designing the User Interface*, Addison Wesley Longman, Inc., Reading, Massachusetts.

Smailagic, A., Siewiorek, D., 1999, System Level Design as Applied to CMU Wearable Computers, *Journal of VLSI Signal Processing Systems*, Kluwer Academic Publishers, 21(3).

STEP web site: <http://cic.vtt.fi/links/step.html> (accessed March, 2003).

Sunkpho, J., Garrett, J., and McNeil, S., 2002, A Framework for Field Inspection Support Systems Applied to Bridge Inspection, *Proceedings of the 7th International Conference on the Applications of Advanced Technologies in Transportation,* Cambridge MA, August 5-7, pp. 417-424.

Terminal Server web site: <http://www.microsoft.com/ntserver/ProductInfo/terminal/thinclient.asp> (accessed March, 2003).

Thomas, B., Piekarski, W. and Gunther, B., 1998, Using Augmented Reality to Visualize Architectural Designs in an Outdoor Environment, *DCNet'98 Online Conference*, <http://www.arch.usyd.edu.au/kcdc/journal/vol2/dcnet/sub8> (accessed March, 2003).

Trilon, Inc. web site: <http://www.trilon.com/> (accessed March, 2003).

UML Resource Center web site: <http://www.rational.com/uml/> (accessed March, 2003).

ViA web site: <http://www.via.com.tw> (accessed March, 2003).

VRML Specifications, 1997, <http://www.vrml.org/Specifications/VRML97> (accessed March, 2003).

Wang, K.C.P. and Gong, W., 2002, Real time Automated Survey of Pavement Surface Distress, *Proceedings of the 7th International Conference on the Applications of Advanced Technology in Transportation Conference (AATT' 2002),* Cambridge, pp. 456-472.

WearableGroup at Carnegie Mellon University web site: <http://www.ce.cmu.edu/~wearables> (accessed March, 2003).

Webster, A., Feiner, S., MacIntyre, B., Massie, W., and Krueger, T., 1996, Augmented reality in architectural construction, inspection and renovation, *Proc. of ASCE Third Congress on Computing in Civil Engineering*, Anaheim, CA, pp. 913-919.

Wirelessdevnet web site: <http://www.wirelessdevnet.com/channels/lbs/features/mobilepositioning.html> (accessed March, 2003).

Xybernaut web site: <http://www.xybernaut.com> (accessed March, 2003).

You, S. and Neumann, U., 2001, Fusion of Vision and GyroTracking for Robust Augmented Reality Registration, *IEEE Virtual Reality 2001*, Yokahama, Japan, March, pp.71-78.

CHAPTER ELEVEN

The Role of Telegeoinformatics in ITS

Chris Rizos and Chris Drane

11.1 INTRODUCTION TO INTELLIGENT TRANSPORTATION SYSTEMS

11.1.1 The ITS Vision and Functional Areas

Intelligent Transportation Systems (ITSs) are not so much a technology as they are a vision. In the 1980s, traffic planners and researchers saw that the application of "high technology" offered the prospect of revolutionizing their national transportation sector. They foresaw a transportation system of the future that would be more efficient, less polluting and safer. The key technologies that could enable this vision were the group of technologies we call today, *Telegeoinformatics*.

Initially different countries commenced their own programs to achieve this vision. As a result there were separate initiatives in the U.S., Europe, Japan, as well as other developed countries. By the 1990s it became apparent that there were many common elements to these programs, and that there were potentially considerable benefits associated with a global collaborative approach. The ITS movement was born (Catling, 1994; Whelan, 1995; ITS America, 1992).

Today, many countries around the world have an organization dedicated to the fulfillment of the ITS vision. Examples include ITS America, ITS Japan, ITS Korea, ITS Australia, and ITS Brazil (Ertico, 2002a). In Europe, the major organization is ERTICO (Ertico, 2002b). An important characteristic of each of these organizations is that they represent a collaboration of government officials, road transportation user groups, research institutes, representatives from industry, and academics. This is in recognition that the breadth of the vision requires the active participation of researchers from many disciplines, as well as the cooperation from all sectors of society. ITSs can be divided into five functional areas (Drane and Rizos, 1998), each briefly described in the following paragraphs.

Advanced Traffic Management Systems (ATMSs) involve the use of sophisticated technologies to manage the traffic on the transport network. An important element of ATMS is advanced traffic control systems that will, e.g., phase all the traffic lights in a particular area, provide such functionality as a "green wave" to vehicles. ATMS also includes electronic road tolling, adaptive signposting, traffic congestion monitoring, and incident management systems.

Advanced Traveller Information Systems (ATISs) are systems that provide information directly to the driver of a vehicle (McQueen *et al.*, 2002). An important service is route guidance, where the driver is informed of the best route to travel in order to reach a particular destination, taking into account road congestion conditions. ATIS also includes concierge services such as the location of nearby restaurants, parking space availability, and other geographically relevant information.

Advanced Vehicle Control Systems is the most ambitious of the ITS functional areas, and ultimately will involve having the vehicle controlled by computer, so that it can travel along the highway with no (or minimum) human intervention. In the short term this functional area deals with collision warning systems and intelligent cruise control.

Commercial Vehicle Operations involve the use of Automatic Vehicle Location Systems (AVLSs) linked with computer-aided dispatch systems. These systems enable more efficient dispatch and scheduling as well as increased driver safety. Many examples of such systems have already been deployed. These systems also have applicability to fleets of emergency vehicles, such as ambulances, fire engines, and police vehicles, as well as public transport.

Advanced Public Transport Systems involve the development of special purpose public transport information and control systems. These will provide passengers with information on the arrival times of buses, trams or trains, allow smart card payment of fares, and a much higher level of operational efficiency. In addition, it will be possible to use ATMSs and ATISs to provide higher priority to buses and trams. This area is also likely to see the development of personalized public transport that will provide a service that is intermediate in terms of cost, timeliness, and proximity between those provided by buses and taxis.

These five functional areas give only a broad overview of ITSs. However, an important feature of ITSs is that many of their constituent systems emerge from the interaction of a host of related applications. Each of these applications will fulfil a particular need within the transportation system. Many of these applications will not be autonomous, but will use information drawn from other ITS applications. Some applications that use Telegeoinformatics are discussed in Section 11.2.2.

11.1.2 The ITS Architecture

The five functional areas of ITSs provide the broad vision. The applications listed above provide specific examples of how different aspects of Telegeoinformatics fit into the vision. In order to better understand the relationship of Telegeoinformatics to ITS, it is useful to consider the logical architecture of ITSs (McQueen and McQueen, 1999). Different countries are developing their own logical architectures. Here, we will refer only to the U.S. architecture. The full logical architecture requires literally hundreds of pages of description (US DOT, 2002a). The U.S. was the first country to develop a national ITS architecture, and its approach has been highly influential in other countries. The starting point for the ITS architecture is a context diagram (see Figure 11.1).

In this diagram, the square boxes are known as *terminators*. Terminators are sources or sinks of data. From the viewpoint of Telegeoinformatics, an important terminator is the Location Data Source, defined as (Drane and Rizos, 1998):

> *This terminator represents an external entity which provides accurate position information. External systems which use GPS, terrestrial trilateriation, or driver inputs are examples. This*

> *terminator contains sensors such as radio-position receivers (for example GPS), and (or) dead reckoning sensors (for example odometer, differential odometer, magnetic compass, gyro, etc.), and (or) map-matching techniques.*

It is interesting that in the U.S.A. architecture a key Telegeoinformatics component is placed at such a high level in the architectural description.

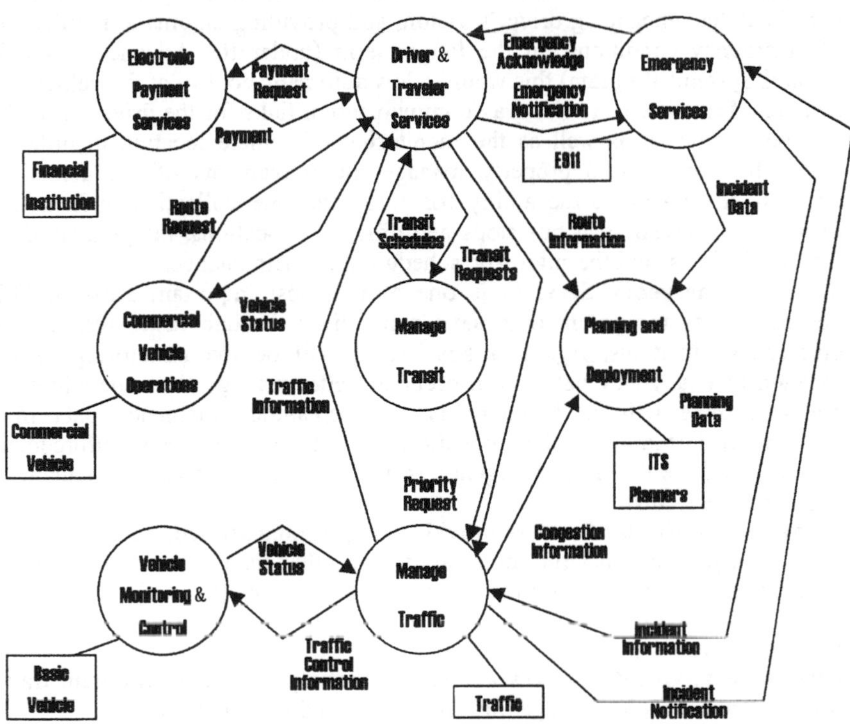

Figure 11.1 Simplified Top-Level Architecture (Source: US DOT, 2002b, reprinted with permission)

The ITS logical architecture is described by a series of *decompositions*. The decomposition of the main context diagram is done in terms of eight processes: (1) Manage Traffic, (2) Manage Commercial Vehicles, (3) Vehicle Monitoring and Control, (4) Manage Transit, (5) Manage Emergency Services, (6) Provide Driver and Traveller Services, (7) Provide Electronic Payment Services, and (8) Plan System, Deployment and Implementation. Each of these is briefly described below:

- **Manage Traffic** provides the ability to monitor traffic conditions and then use this information to control the traffic, using tools such as traffic lights or variable message signs. This facility can be used to manage traffic flow, react to incidents (such as traffic accidents), mediate travel demand, and monitor pollution levels.

- **Manage Commercial Vehicles** can be divided into two sub-functions: the need for the owner of the fleet to manage their vehicles, and the need for government to administer regulations related to commercial vehicle operation. The former requires AVLSs and efficient communications for computer-aided dispatch. The latter requires automated roadside checkpoint facilities, electronic interchange of documents, and an overall administration system.
- **Vehicle Monitoring and Control** describes the functionality for monitoring vehicle status, enhancing driver's vision, and providing automatic notification of emergency conditions. In the longer term (under the Advanced Vehicle Control Systems Program) this would allow automatic control of the vehicle.
- **Manage Transit** covers two-way communication links for the driver of public transport vehicles, as well as the monitoring of vehicle location in order to allow the operators to properly manage the movements of bus fleets. In addition, there will be the ability for automatic fare collection, information provision to travellers at bus stops and other key locations, the generation of driver schedules, and the automatic scheduling of maintenance.
- **Manage Emergency Services** is one of the most important areas of ITS, because it has the potential to save many lives through faster response to emergency situations. Again, a key aspect will be the monitoring of the location of emergency vehicles in order to better manage fleet operations. In addition, there will be functionality for displaying emergency data in a meaningful fashion and the allocation of tasks between law enforcement agencies and the different emergency services (e.g., fire brigade, ambulances, police rescue teams).
- **Driver and Traveller Services** provides trip planning services, route guidance, "yellow pages" or concierge information, and other travel related information. This information can be provided to people in vehicles (drivers), or to people accessing information from their home, office, hotel, or some other fixed location (travellers).
- **Electronic Payment Services** allows automatic payment for different travel related services such as bridge tolls, road tolls, and parking fees.
- **Plan System Deployment and Implementation** provides for the collection of large amounts of data relating to the transport system, so as to allow the transport planner to properly deploy and implement ITSs.

It is worth emphasizing that so far in this chapter, we have described ITSs from two different viewpoints. The first was in terms of the functional areas; these identify the broad vision. Then, we considered the top layers of the ITS logical architecture. Figure 11.1 illustrated the importance of Telegeoinformatics to ITS, and the eight processes described above provide a complete framework for the operation of an ITS.

11.2 TELEGEOINFORMATICS WITHIN ITS

To some extent, Telegeoinformatics applications are a subset out of the total number of possible ITS applications or processes. In this chapter we have focused on those aspects of ITSs that fit the definition of Telegeoinformatics given in Chapter 1. The term *Telematics* is often used within the Information Technology (IT)/wireless communications community to refer to products and services for a "mobile society." The automotive industry quickly adopted the term to describe *any* system that combines wireless technology with Location-Based Services (LBSs) or systems (e.g., Zhao, 2002). However, they do not typically accord position determination technologies and geospatial information the same prominence as mobile computing devices and wireless telecommunications. This is understandable, as IT professionals and communication engineers see their role as central to the development of ITS systems, and their experience or knowledge of geospatial technologies and applications is generally limited.

However, there is a certain ambiguity in the term "Telematics," as there are other application areas of increasing importance that also rely on IT and communication technologies. One of these is in the area of home-based Health Services, and the term *Health Telematics* is often used to refer to such applications as the remote monitoring of a person's physiological parameters, and response by health services when requested, telemedicine, and so on. A more accurate descriptive term for Telematics in the context of ITS therefore is *Transport Telematics*. However, in this chapter we prefer to use the term *ITS-Telegeoinformatics* in order to emphasize the importance of *Geoinformatics*. Note that although most of our examples will be of ITS-Telegeoinformatics systems, we will provide examples of more general ITS applications to enable the reader to gain a better understanding of the overall context of ITS.

11.2.1 ITS-Telegeoinformatics Technologies

The beneficiary or customer of an ITS-Telegeoinformatics system may be the vehicle driver, a passenger, a fleet manager, a dispatcher, a road traffic monitor, or a government regulator. In general, the task of an ITS-Telegeoinformatics system or service will be to answer a question that has a spatial aspect to it, such as: "Where am I?"; "Where are you?"; "Where is the nearest parking station?"; "When is the next bus traveling south along Z Street?"; "What is the traffic density at Location X?"; "How many miles has the vehicle traveled since its last service?"; "Which vehicle is nearest to Location Y?"; and so on. Each of these requirements or applications will involve a different configuration of the constituent technologies (see Section 11.6). The constituent technologies of ITS-Telegeoinformatics are IT devices and software, wireless telecommunications systems, and position determination systems and spatially referenced data.

Information Technologies in this context refer to the range of embedded processors, mobile computing devices and software that integrate a wide variety of discrete components, control the transmission of data within and from the vehicle-mounted system, and manage or analyze the disparate datasets and data streams involved in ITS-Telegeoinformatics. An example is the Vehicle Navigation System (VNS), typically consisting of a dash-mounted Liquid Crystal Display (LCD), controller and onboard computer. This device can handle the tasks of accepting input commands or requests from the driver, interrogating onboard digital map datasets, handling data or service requests to an outside server, and displaying information in the appropriate form. Clearly its functions can be *emulated* by a desktop PC-based system, but it is the fact that the system design is optimised for vehicle operations that makes it application-specific and worthy of special study. On the other hand, the widespread adoption of the Personal Digital Assistant (PDA) computing device has facilitated true mobility.

Wireless telecommunications are a crucial technology supporting a wide range of data transmission and networking functions, such as linking together devices within a vehicle, sending service requests or data to central servers, and receiving data over broadcast or one-to-one communication links. In ITS-Telegeoinformatics, a variety of communication links may be used: Bluetooth, Wireless Local Area Network (WLAN), mobile telephony, trunk or packet radio networks, and digital and analogue radio broadcast (Section 11.5). For example, Bluetooth will increasingly be used to connect a mobile phone to the "vehicle navigation system", or a CD player to a backseat LCD. WLAN is considered one option for mobile broadband Internet link that can be used to download massive amounts of data (such as video streams and music). Yet probably the explosion in use of mobile phones will have the greatest impact on ITS-Telegeoinformatics as more than 70% of the U.S. population owns one, and the infrastructure gives almost 100% coverage. Hence, mobile telephony is without equal as a convenient and reliable channel for the transmission of voice requests or short text messages to ITS service providers.

Geoinformatics is a convenient umbrella term for all aspects of geospatial information. In particular, the position determination technologies (Section 11.3), as well as the "soft" elements such as map data and other spatially referenced information (Section 11.4). There is no doubt that the Global Positioning System (GPS) (see Chapter 4) has had a revolutionary impact on many navigation and telematic applications. It would be fair to say that without the capabilities of GPS to provide on-demand position information, Telematics would refer to little more than the transmission, analysis and display of information within (and beyond) a vehicle. In its most trivial form it would be wireless games and "downloads", accessing emails, stock market prices and sports results while in the car! *It could therefore be argued that without a location determination device, many Telematics products and services would not be attractive to consumers*. However, it is important to emphasize that Geoinformatics is not just GPS, it includes all geospatial data and services as well. Hence the humble map (and its sophisticated variants) is a crucial component of ITS-Telegeoinformatics applications. In addition, newer location technologies, including mobile telephony-based location technologies (Tekinay, 2000; Drane *et al.*, 1998; Cursor, 2002), will supplant or supplement GPS for certain Telegeoinformatics applications.

11.2.2 ITS-Telegeoinformatics Applications: General Comments

In addition to considering the constituent technologies, and the integration architecture, it is also useful to examine ITS-Telegeoinformatics systems with respect to which of the following categories they may belong:

1. **Driver Assistance**: Systems that provide vehicle route guidance and navigation, enhance security and safety (such as calls for breakdown services

and alerting authorities when an airbag is inflated), respond to requests for traffic information, concierge or LBSs, and so on.

2. **Passenger Information**: Products and services that target the passenger in an automobile (e.g. LBS, in-vehicle games and entertainment), or public transport (e.g. providing information on times of services, multi-modal connections, etc.).
3. **Vehicle Management**: Including such applications or systems as automated vehicle monitoring and fleet management, in order to improve the efficiency of commercial fleet operations, permit stolen vehicle recovery, monitoring of young drivers (by parents), remote opening of doors or disabling of a vehicle, the provision of real-time vehicle diagnostics services, and so on.
4. **Road Network Monitoring**: Such as the tracking of dangerous cargoes, or the tracking of probe vehicles in order to monitor the traffic congestion, electronic road tolling, parking spaces allocation, intelligent speed adaptation, and so on.

The first three categories are discussed in more detail in Section 11.6.

11.2.3 The ITS-Telegeoinformatics Development Drivers

The above categorization cannot be rigidly applied. The first three categories refer to "behind the wheel, in the backseat, or under the bonnet" type applications. These are the ones that have been identified as being of particular interest to the following industry segments:

- **Vehicle instrumentation suppliers** address the factory, dealer or after-market installation of Telegeoinformatics devices, potentially worth billions of dollars in the coming years – *it is estimated that the value of the electronics in automobiles will be over 40% of the total value within a few years.*
- **Vehicle manufacturers** are tempted to get more involved in ITS-Telegeoinformatics because of the promise of massive revenue streams – *however, it is questionable whether they concern themselves with the service aspects of ITS-Telegeoinformatics.*
- **Telecommunication carriers** clearly will play a central role in the provision of communication services that enable a wide range of ITS-Telegeoinformatics applications – *such as driver assistance and concierge services, remote vehicle diagnostics, fleet management, and vehicle tracking.*
- **Service providers** in general are crucial as they are expected to develop the "content" for LBSs and concierge services – *they may be relatively small companies or large service organizations, that address niche markets or offer a broad portfolio of services that might be ITS-specific or also accessed by any mobile device.*
- **Dealers** are intending to also participate in ITS-Telegeoinformatics in order to put into practice the principles of Customer Relations Management – *e.g. by providing remote vehicle diagnostic and other after-sales services.*

On the other hand, many vehicle tracking applications may be viewed as benefiting the community as a whole. For example, as a result of the tragic events of 11 September 2001 in the U.S., the tracking of vehicles carrying hazardous materials is seen by many as essential for "homeland security." The tracking of probe vehicles to monitor traffic congestion is another example of an application with wide ranging environmental, commercial and road safety benefits.

What sets ITS-Telegeoinformatics apart from other road, traffic or vehicle innovations is the generation of *continuous* revenue streams through the provision of driver and passenger information or response services. There have been many attempts to estimate the size of the ITS-Telegeoinformatics market. Typically it is predicted to grow vigorously to be of the order of tens of billions of dollars per annum worldwide by the year 2010. In general, such predictions assume that some proportion X% of all vehicles will be fitted with a Telegeoinformatics system that will generate $Y per month service fees. Summing the capital cost of hardware installed in Z vehicles, and the estimated annual service fees $Z*12*Y, one arrives at a prediction of the ITS-Telegeoinformatics market value. Only the total number of new vehicles sold per year can be known with any certainty. This is currently of the order of 40 million vehicles. The aim obviously is to maximize the installed base of ITS-Telegeoinformatics systems, and to encourage regular use of fee-paying services. However, it is not clear which segments of industry will ultimately gain the dominant market share for products and services. Consider the following:

- It is estimated that less than 10% of new cars currently have a Telegeoinformatics device installed, and that this number is expected to inextricably rise to over 50% by the latter half of the decade.
- A vehicle's lifetime is much longer than those of most electronics devices, posing a special challenge to those promoting ITS-Telegeoinformatics applications.
- Interoperability is a particular concern, as industry standards need to be agreed upon for data bus interfaces, operating systems, wireless communication portals, map and geospatial data formats, etc.
- There is confusion concerning ITS-Telegeoinformatics products and services being offered by different parties. ITS service providers, telecommunication carriers and vehicle manufacturers.
- The public appears to be averse to paying monthly charges for what many seen as "unnecessary" vehicle options such as breakdown or emergency services, navigation and route guidance, and the search for a "killer application" has been a frustrating one.
- There is a blurring of the distinction between what are clearly ITS-Telegeoinformatics products and services, and those offered also to people via their mobile devices (phones and PDAs).

ITS-Telegeoinformatics are a very diverse set of applications, products and services. For example, the Telegeoinformatics "product package" for a Driver Information System will be very different from an AVLS. The most important design feature of the former is the appropriate User Interface, allowing for input of requests and the delivery of information to the driver in a form that is not distracting. On the other hand, the latter application will usually involve a hidden device that must communicate reliably with a central server. Yet both make use of a position determination technology and an appropriate embedded processor to manage the task. Despite the diversity, all ITS-Telegeoinformatics systems are characterized by the use of positioning systems, geospatial data, and communication systems. Each of these will be discussed in the following sections.

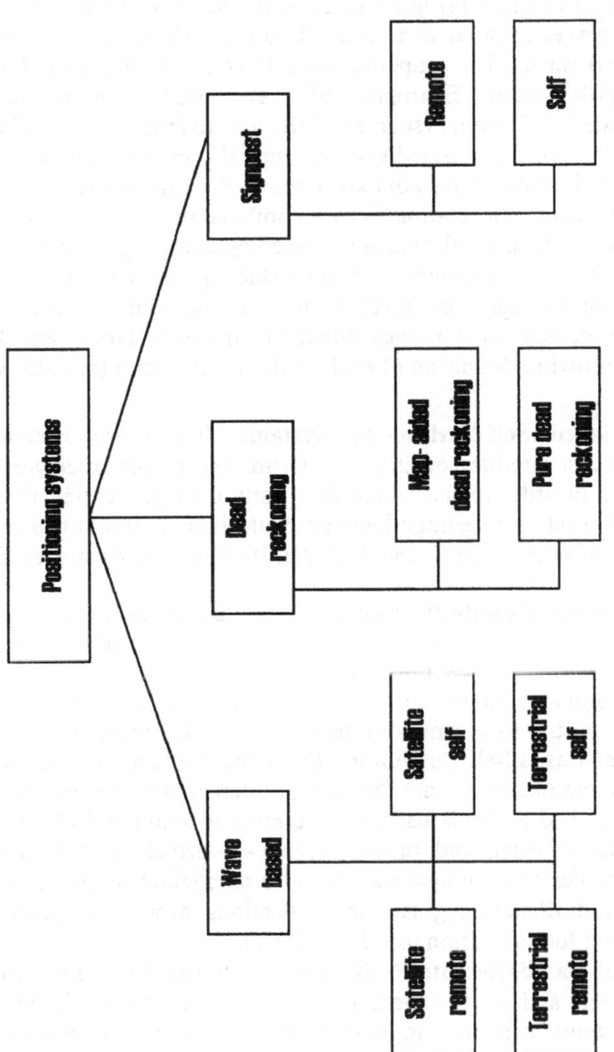

Figure 11.2 Taxonomy of Positioning Systems (Source: Figure 8.8, Drane and Rizos, 1998, reprinted with permission)

11.3 THE ROLE OF POSITIONING SYSTEMS IN ITS

11.3.1 Taxonomy of Positioning Systems

Positioning systems can be classified according to whether they are wave-based, dead reckoning, or signpost (Figure 11.2). *Wave-based* systems use the propagation properties of waves in order to measure location. These can use electromagnetic waves or sound waves. The properties include velocity, direction of arrival, power level, and polarization. Examples of wave-based systems include Global Navigation Satellite Systems (such as GPS) and mobile telephony-based systems (see Section 11.3.3). Wave-based systems normally involve separated transmission and reception sub-systems. By contrast, *dead reckoning systems* are self-contained, deriving the location information from a combination of internal sensors, such as gyroscopes, odometers, accelerometers and compasses. S*ignpost* systems are based on proximity detection that detects when a vehicle passes close to a known location or signpost. For example, toll readers on a freeway can be used to monitor the progress of a vehicle as it moves down the freeway. Drane and Rizos (1998) provide the following definition of each of these categories (see also Chapter 4 for more details):

1. **Satellite-Based Self-Positioning Systems:** These are systems that have transmitters in satellites orbiting the Earth, and require receivers mounted on the vehicle in order to determine the position by processing the signals. The techniques used to determine location could include round-trip time and time-difference-of-arrival. GPS and GLONASS (Chapter 4) are examples of the latter.
2. **Satellite-Based Remote-Positioning Systems:** These systems would operate by placing the receivers on the satellites and the transmitters on the vehicles. These systems could be implemented as "add-ons" to Low Earth Orbiting satellite communications systems. Techniques to determine location could include round-trip time, time-difference-of-arrival, and angle-of-arrival.
3. **Terrestrial-Based Self-Positioning Systems:** These systems consist of a series of transmitters placed around the area of interest, with the receiver mounted in the vehicle. The methods used to determine location include round-trip time, phase measurements, and time-difference-of-arrival. In the past, there were many examples of such systems for radio navigation purposes, principally for maritime and offshore engineering applications. Most of these have since been discontinued for navigation use (FRP, 2001).
4. **Terrestrial-Based Remote-Positioning Systems:** These have the transmitter in the vehicle and the receivers at fixed locations around the area of interest. These systems can use round-trip time, time-difference-of-arrival, phase measurement, and angle-of-arrival. Quiktrak is one example of such a system that has been specifically developed for ITS applications (Quiktrak, 2002). Mobile telephony-based systems fall into this category, as several base stations make measurements on the signals being transmitted by a mobile phone to derive the device's location.
5. **Pure Dead Reckoning:** These are systems that use dead reckoning sensors such as gyroscopes, accelerometers, magnetic compasses and odometers. They

are typically used in combination, such as gyroscope and odometer, but make no use of mapping information. There is no pure dead reckoning system which has seen widespread implementation in ITS applications. With the advent of low-cost GPS hardware, VNSs nowadays typically use an instrument package of dead reckoning sensors and GPS.

6. **Map-Aided Dead Reckoning Systems:** These are dead reckoning systems that also make use of maps to correct for position, using the principles of map-matching (Section 11.4.4). The most sophisticated VNSs use a combination of dead reckoning (gyroscope and odometer), GPS and map-matching.
7. **Signpost Self-Positioning:** These are signpost systems where the beacons emit signals which allow the vehicle to work out if it is in the proximity of a particular beacon, but there is no two-way data exchange. *Cell-ID* information on a mobile phone is an example of a signpost self-positioning system. In the future, WLAN access points will be used in a similar manner.
8. **Signpost Remote-Positioning:** In these systems, the signposts sense when a particular vehicle is in the proximity of the signpost. Mobile telephony and WLAN services are able to carry out such a positioning function.

11.3.2 Attributes of Positioning Systems

In order to understand the capability of each of these types of systems, it is necessary to understand some of the important attributes of positioning systems. We consider the important attributes to be: autonomy, absolute positioning, number of simultaneous users, cost, accuracy, coverage, susceptibility to jamming, capability for data transmission, and privacy (Drane and Rizos, 1998):

- **Autonomy** is the capability to provide position information in a vehicle without reference to external services or signals.
- **Absolute positioning** means that the vehicle location is referenced to a global reference system or datum (such as used by maps), rather than to an arbitrarily defined local system.
- An important attribute of a particular positioning system is the **number of simultaneous users** that the system supports. This can range from one to infinity.
- The **cost** of a system: including the in-vehicle cost and the infrastructure cost. For example, in the case of the GPS, the in-vehicle cost is the purchase price and installation cost of the onboard receiver. The infrastructure cost refers to the satellites and ground stations (which in the case of GPS are not passed on to the user).
- **Accuracy** refers to the "correctness" of the position information derived from the measurements. This accuracy may be enhanced in the case of special augmentation techniques. For example, the GPS positioning accuracy can be improved using differential techniques. Mobile telephony-based positioning techniques can also be enhanced using assistance data.
- **Coverage** is a critical attribute and refers mainly to: the signal availability and the geographical distribution of base stations or the infrastructure needed by the system. An example of the former is the severe attenuation of GPS signals by buildings, tunnels and even trees. However, when the GPS signals are available

(e.g. in the open), the coverage is otherwise global. On the other hand, beacon systems only function in areas where the necessary infrastructure is provided.

- **Susceptibility to jamming** can be important for ITS applications relevant to homeland security. GPS in particular suffers in this respect.
- **Uplink data capability** allows data to be sent from the vehicle to the central control and monitoring facilities. This combines the two operations of positioning and wireless communications into one system. The best example of this is mobile telephony-based positioning systems and WLAN systems.
- **Downlink data capability** allows data to be sent from central control and monitoring facilities to the vehicle. Mobile telephony and WLAN are the primary general purpose systems, though a variety of other beacon systems share this attribute.
- **Privacy risk** is the susceptibility of the positioning system to allow invasion of privacy.

Table 11.1 compares the different types of positioning systems with respect to the above attributes.

It can be seen that different systems exhibit quite different characteristics. In addition, no single system provides a high level attainment for each one of the attributes. One way around this problem is to combine different systems, hence providing enhanced capabilities. For example, the combination of dead reckoning and GPS provides absolute positioning together with autonomy, so that the combined system does not need to be told the starting location, and can work satisfactorily in tunnels and car parks (Zhao, 1997).

11.3.3 E911 and Positioning System Development

The major impetus for the development of positioning technologies has traditionally been *navigation*, and in particular the technologies needed to support military operations. Examples include wave-based systems such as GPS and GLONASS (and the future Galileo system), and many of the miniaturized inertial sensors. However, the U.S. Federal Communication Commission's mandate to telecommunications carriers to deploy an *enhanced* 911 emergency response system for mobile phones from 2002 onwards (known as E911), has also been an important driver for the development of positioning technologies capable of being implanted within mobile phones.

The E911 mandate does not specify which technology to use. It only defines the general specifications, including that the accuracy be of the order of 50 meters 67% of the time, 150 meters 95% of the time for "handset solutions", or 100m (67%) and 300m (95%) if a "network solution" is used. Initially it appeared that GPS could not satisfy such requirements, as GPS signals could not be received indoors. Hence much of the attention in the late 1990s was on mobile telephony-based systems. The cellular positioning techniques include cell-ID, time-of-arrival (TOA), time-difference-of-arrival (TDOA), angle-of-arrival (AOA), signal-strength, and several variations of the TOA technique (Hjelm, 2002). The enhanced-observed-time-difference (E-OTD) technique, as used in the cellular location technology Cursor™, appears to be the most accurate of the mobile telephony-based positioning systems.

Table 11.1 Comparison of Different Types of Positioning Systems with Respect to a Number of Characteristics (adapted from Drane and Rizos, 1998)

	Satellite-Self	Satellite-Remote	Terrestrial-Self	Terrestrial-Remote	Pure Dead Reckoning	Map-Aided	Signpost-Self	Signpost-Remote
Autonomy					✓	✓		
Absolute Positioning	✓	✓	✓	✓			✓	✓
Number of Users	∞	finite	∞	finite	∞	∞	∞	finite
Infrastructure Cost	very large	very large	large	large	zero	large	large	large
In-Vehicle Cost	low-medium	low-medium	low-medium	low-medium	low	medium	low	low
Potential Coverage	global	global	wide area	wide area	wide area	wide area	road ways	road ways
Susceptibility to Jamming	high	low	high	low	low	low	medium	medium
Uplink Data Capability		✓		✓				✓
Downlink Data Capability	✓		✓				✓	
Privacy Risk	low	medium	low	medium	low	low	low	medium

During the last few years, there has been renewed interest in improving the GPS receiver technology in order to track very low signal strengths, as would be encountered inside buildings. Wireless Assisted-GPS (A-GPS) together with so-called *high sensitivity GPS receiver designs* have been developed (Djuknic and Richton, 2000), and we believe that such systems may be deployed within mobile phones, competing with (or complementing) the mobile telephony techniques. The high sensitivity GPS system manufacturers include Snaptrack (2002), Enuvis (2002), Global Locate (2002), SiRF (2002), and Sigtec (2002).

11.4 GEOSPATIAL DATA FOR ITS

After position determination technology, geospatial data in the form of maps and databases is the next most important geoinformatics element of ITS-Telegeoinformatics. The road map is a compact, graphical representation of the essential geospatial information that a driver needs to negotiate a journey to a new location, and to be able to answer such questions as "where am I?" and "how do you get to location A?" Hardly any vehicle is without its street directory. However, a compact, informative road map is only the end product of a long and complex map-making process whose principles and procedures have been refined over centuries. A discussion of map-making principles is outside the scope of this book, and the reader is referred to, e.g., Bossler *et al.* (2002). The end result of the map-making process is, however, the collection of two basic forms of information about every physical feature of interest (at the appropriate scale): its *position* information, and the *attribute* information associated with that feature. The former is generally given in a mathematical form – coordinates expressed within a well-defined geodetic datum or reference system. Position may be expressed in one of a number of different coordinate systems, however the most common are *latitude and longitude* or *easting and northing* within a plane map projection system (height information is of limited utility for most ITS-Telegeoinformatics applications). One's own position may also be expressed in the same coordinate and reference system as the map data, using one of the self-positioning technologies mentioned in Section 11.3.

11.4.1 The Digital Map

Nowadays map data are generally stored in electronic form so that they may be displayed on a computer screen. The map database contains the geometric and attribute information of all physical features that are likely to be of interest to a map user. These maps are then created on-the-fly, as would any other form of computer graphics. However, many digital or electronic maps are derived merely by scanning existing printed maps. These are *raster* images that do not explicitly contain topological information that describes how the different points and lines are related to each other in a geometric sense. Such images are therefore similar to a photograph or image file, ordered as an array of pixels (picture elements). The scanned image may, however, be *vectorized* – essentially a process of automatically determining the coordinates of lines and point features on the map, and the necessary topological information, from their pixel space information.

Whatever the source of the map coordinate data, be it the scanning-vectorization of existing paper maps mentioned above, or the "capture" of geospatial data during the map-making process, such map data can be stored and managed by separating them into different layers of geospatial information *themes*. For example, one layer will contain the road centerline coordinate information. Another layer may have the contour lines depicting the topography, while another will have the land parcel boundaries, and so on. Hence a map may be generated by overlaying one layer on another, as can be done within a Geographic Information System (GIS) (see Chapter 2). Such a customized map includes only those layers of geospatial information which are relevant to its application, e.g., as in the case of topographic maps, tourist maps, air navigation maps, geological maps, census and political boundary maps, and so on.

The maps, which are used for ITS-Telegeoinformatics applications, are generated from the raw (layered) geospatial datasets that already exist. Additional geospatial data capture may have had to be carried out in order to locate points-of-interest such as restaurants, parking areas, theaters and cinemas, banks, post offices, police stations, etc. As far as the user is concerned, the result is the electronic or computer-drawn version of the humble *street directory*, and many of them have been developed to have almost the same *look and feel* of the paper equivalent.

11.4.2 Map Attribute Data

Geometric data is not the only information that may be contained within an electronic map. As discussed earlier, additional address information may be linked to the coordinate data so that a position may be transformed into a street name, a point of interest, etc. This extra information is typically in the form of attribute data associated with a point, line or polygon feature within a GIS database (Chapter 2). For example, the name of the restaurant is included, the car park entrance is described, the brand of gasoline sold at the gas station is included, etc. If the basic map data (coordinates and attributes) have been carefully collected, and it is both current and comprehensive, then this process is not too onerous. However, there may be attribute data which is useful to drivers that is more difficult to obtain from traditional map databases, such as which streets carry one-way traffic (and in which direction), the weight restrictions on roads, speed restrictions, where right turns (or left turns) are prohibited, which streets have median strips, and so on.

The requirement for "intelligent" road maps is even more stringent for ITS-Telegeoinformatics applications such as fleet management and emergency vehicle dispatch. In such applications the software which advises the driver on how to travel from location *A* (present position) to location *B* (point of cargo pickup, site of emergency, etc.) would require not only information on the road *geometry*, but also traffic restrictions and conditions, etc., in order to perform optimum route guidance computations. The ultimate attribute data would be up-to-the-minute information on traffic conditions, so that route guidance will be sensitive to congestion and other ephemeral traffic restrictions. A variation of this is *map-aided navigation* (Section 11.4.4).

11.4.3 Map Display

The issue of map readability, particularly on in-vehicle computer displays, is an important one, and currently an area of active research. This is inextricably linked to the map *scale*. Map scales may be confusing to those who are not trained in the basics of cartography. Scale is usually expressed in the form of a statement such as "1 in xxx xxx," e.g. 1 in 100 000, or simply 1:100 000. This implies that 1 unit on the map represents 100 000 units "on the ground," hence at this scale 1mm on the map is equivalent to 100m in the real world. This would be generally classed as a "small scale" map. One's first reaction is that 100 000 is a *large* number! Or that a small-scale map must depict *small* details. This is, of course, the opposite of what really happens. The words "small" or "large," as they relate to map scale, refer to the *fraction* 1/100000. Hence, a 1:1 000 map is at a large scale, while 1:1 000 000 is a very small scale map.

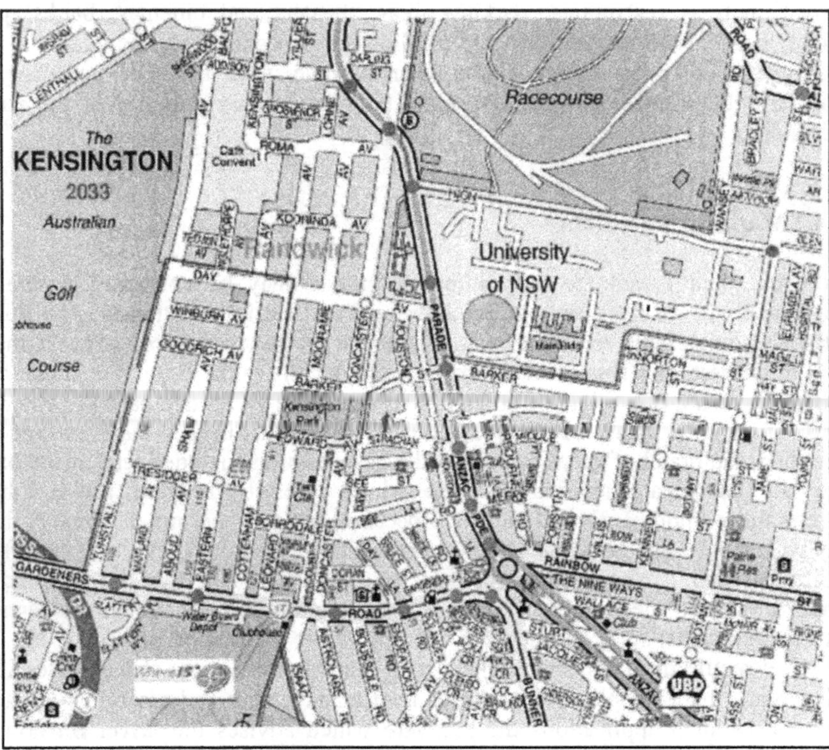

Figure 11.3 "Street Directory" Type Map Example (© Telstra Corp Ltd 2002, where WhereIS™ and Sensis™ are registered trade marks of Telstra Corp Ltd; © Universal Press Pty Ltd, where UBD™ is a registered trade mark of Universal Press Pty Ltd)

Due to inaccuracies in the map source data, many road map databases may not be accurate to the meter-level, and in fact inaccuracies of several tens of meters are not unusual in many countries. It is therefore possible for a vehicle's location device to provide positioning information at a higher accuracy than the map data. Often this is not a critical issue as a navigation technique known as *map-matching* (Section 11.4.4) may be applied, in which the vehicle is assumed to be traveling along the road and is displayed as doing just that – even if the vehicle's coordinate implies that it is in fact not on the road segment as defined by the map database!

Another factor that has an impact on the accuracy with which a feature is shown on a map is the size or bulk of the feature, and whether, relative to the scale of the map, it should be drawn at an *exaggerated* scale. The best example of this is the depiction of a road reserve by a pair of lines the apparent separation of which may be much larger than the actual width of the road (Figure 11.3).

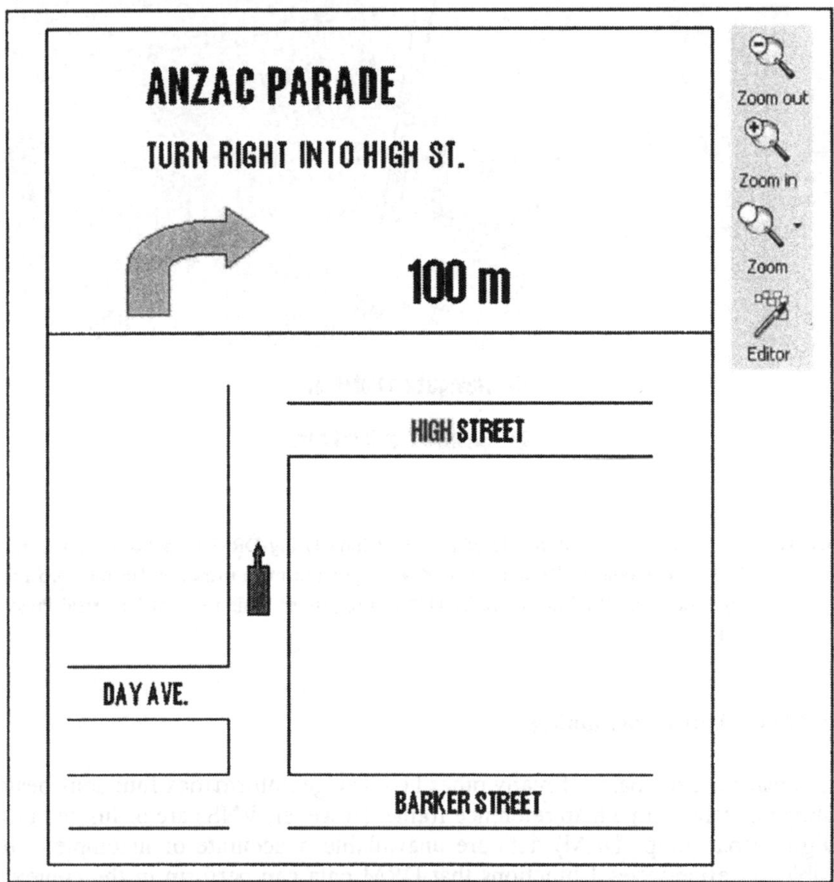

Figure 11.4 Schematic of Simplified Road Map with Turn-by-Turn Directions

As a consequence, small-scale maps show only generalized features such as roads and major junctions, while large-scale maps may show considerably more detail – if it does not impact on *readability*. For example, a street directory at a scale of 1:10 000 may have the appearance of Figure 11.3, however, this may be simplified considerably on an in-vehicle computer screen to show only the vehicle's location, the street it is traveling on, and the streets intersecting (Figure 11.4).

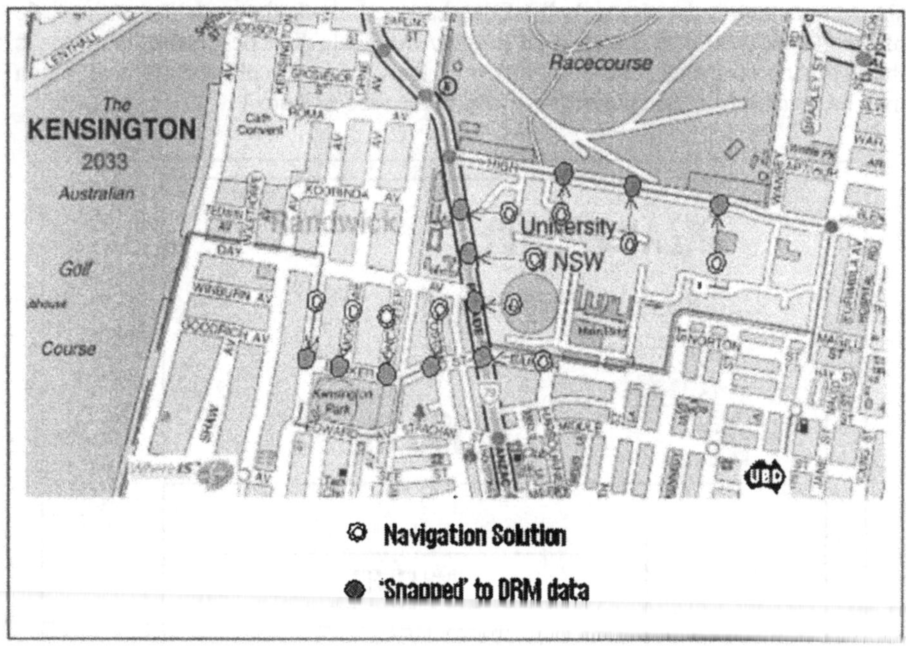

Figure 11.5 Schematic Showing Principle of Map-Matching Using DRM Data (© Telstra Corp Ltd 2002, where WhereIS™ and Sensis™ are registered trade marks of Telstra Corp Ltd; © Universal Press Pty Ltd, where UBD™ is a registered trade mark of Universal Press Pty Ltd)

11.4.4 Map-Aided Positioning

Digital maps are the basis of many other ITS-Telegeoinformatics functions besides locating the vehicle in a map reference frame. However, VNSs are of limited utility if Digital Road Map (DRM) data are unavailable, inaccurate or incomplete. It is possible to define several functions that DRM data can perform in the context of ITS-Telegeoinformatics applications:

1. **Address Matching** transforms a coordinate (e.g., from GPS) into a street address, or vice versa. People know the address of their destination rather than its coordinates. Hence to *navigate* to an address requires that the positioning

system be able to reliably convert the address to a coordinate.

2. **Map Matching** is based on the premise that the vehicle is on a road. Hence when a positioning system outputs coordinates that are not on a road segment as defined by the DRM, the map-matching algorithm finds the nearest road segment and "snaps" the vehicle onto that road segment (Figure 11.5).
3. **Best-Route Calculation** supports driver planning by providing assistance on selecting the optimal travel route. A DRM coupled with a best-route calculation algorithm provides an optimal route based on travel time, travel distance, or some other specified criterion (Figure 11.6). Best-route calculation requires a high level of map information, e.g., the DRM must include traffic and turn restrictions so that the route selected is not illegal or dangerous.
4. **Route Guidance** supports the driver as he or she navigates along a route (selected by the driver, or the best-route algorithm). Route guidance includes turn-by-turn instructions, street names, distances, intersections, and landmarks (Figure 11.4). This is particularly challenging in real-time because the algorithm must process position information, perform address and map-matching, and display the DRM to the driver. If the driver misses a turn, the system must be able to compute a new best-route "on-the-fly," and provide new guidance information.

11.4.5 Navigable Road Map Databases

Each of the four functions of map-aided positioning discussed in the previous section relies on specific features in the DRM database, however the most demanding are the last two: best-route calculation and route guidance. If a DRM supports these functions it is said to be *navigable* (Drane and Rizos, 1998). Navigable DRM databases are significantly more expensive to produce and maintain. The most sophisticated ITS-Telegeoinformatics map functions can therefore be supported only when accurate, complete and seamless DRMs are available. Hence, the maintenance of a navigable DRM database is a time-consuming and expensive process.

The standards for geospatial data for ITS-Telegeoinformatics are such that they go beyond conventional map databases. Although there are a variety of DRM data formats, including the ESRI's Shapefiles, U.S. Census TIGER files, Etak's MapBase files, increasingly the Geographic Data File (GDF) format is emerging as the global standard. GDF is a relatively new spatial data format standard, which was globally promoted in 1995 by the Central European Normalization (CEN), and is the basis for European map data production. GDF is much more than a generic GIS standard because it gives rules about how to capture the data, and how the features, attributes and relations have been defined. GDF files contain a vast amount of information, and therefore vehicle navigation system manufacturers generally only use a subset of the available GDF data for their own systems.

Currently there are two major companies supplying navigable DRMs around the world. TeleAtlas is a European company which claims to be the leading supplier of digital map data for ITS applications, having started over 15 years ago (Teleatlas, 2002). Recently, Teleatlas acquired the U.S.-based Etak company. Navigation Technologies, based in Sunnyvale, California, is the other company

(NavTech, 2002). Both have established strategic alliances with companies that have national DRM databases, and hence Teleatlas and Navigation Technologies can be considered the lead companies in two international consortia of DRM data providers. Increasingly they are also developing products such as route guidance, incident reporting, and traffic congestion reporting services for specific regions and cities.

Not all DRM data are offered by these two companies. In 1988 the Japanese Ministry of Construction supported the creation of the Japan DRM Association (JDRMA) of approximately 80 companies involved in vehicle navigation, and then provided the original DRM databases. JDRMA sells the basic 1:50 000 and 1:25 000 maps to commercial map suppliers, who value-add features to these basic databases.

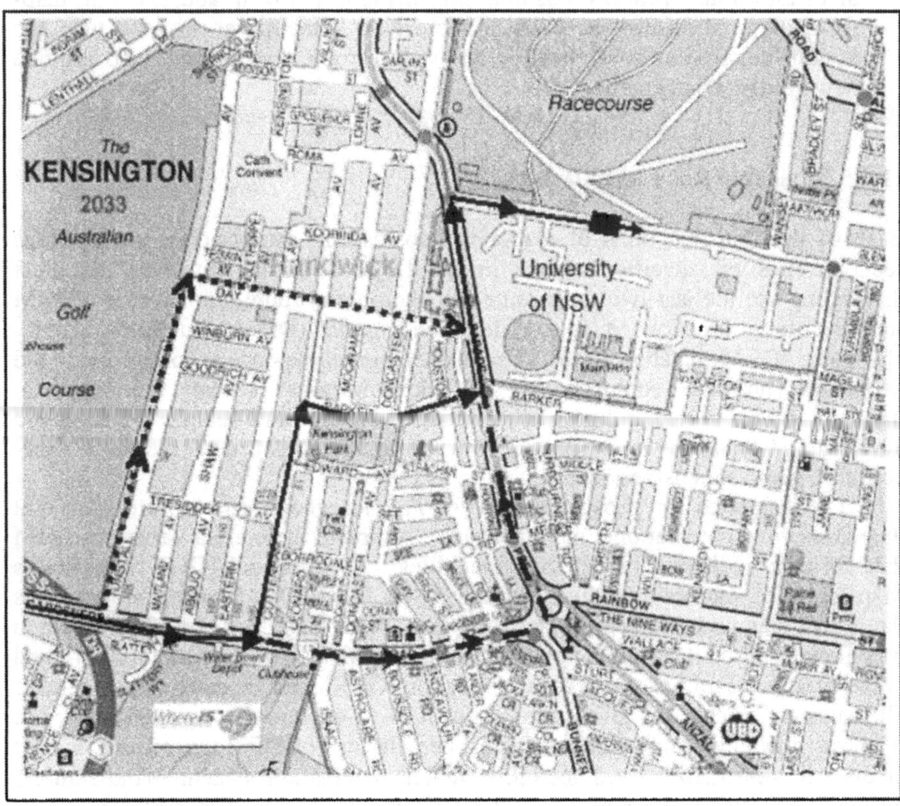

Figure 11.6 Schematic Showing Alternative Routes to Destination (© Telstra Corp Ltd 2002, where WhereIS™ and Sensis™ are registered trade marks of Telstra Corp Ltd; © Universal Press Pty Ltd, where UBD™ is a registered trade mark of Universal Press Pty Ltd)

11.5 COMMUNICATION SYSTEMS IN ITS

Wireless communication technologies have been discussed in Chapter 5. Many organizations may use their own private communication systems/networks. This can be through special allocation of broadcast spectrum (e.g., police radio), or for business usage (e.g., voice radio systems as used by taxis and commercial vehicles). However, in general, ITS applications targeted at the private user are likely to use existing modes of communication. The most important of these communications systems from the viewpoint of ITS are Broadcast Networks, Public Land Mobile Networks, Private Networks, Public Switched Telephone Networks, and Packet Switched Data Networks (Elliott and Dailey, 1995):

- **Broadcast Networks** include television and radio stations that are able to transmit information to a large number of people.
- **Public Land Mobile Networks** refer to the mobile telephony networks that allow users to communicate almost anywhere.
- **Private Networks** are dedicated networks that allow an organization to exchange data and voice in a secure and cost-effective fashion.
- **Public Switched Telephone Network** refers to the standard landline telephone service.
- The most important example of a **Packet Switched Data Networks** is the Internet.

Another way to view communication systems is in terms of the services that are provided by such networks. Figure 11.7 illustrates a communication services hierarchy.

Interactive services are ones that allow a back and forth interaction between the mobile user and another node. A *node* could be a fixed site, such as a traffic control center, or another mobile user. In conversational mode, there is a two-way connection, in that it is expected that there will be frequent interchange between the two parties. An ITS application involving interaction could be an Emergency Response Center handling a call from a driver needing assistance. This could be done with a voice call. Messaging is a one-way point-to-point communication, where one node sends information to another node. An example would be the transmission of a message to a driver informing him/her that he/she is approaching his/her favorite type of restaurant. Distribution involves the sending of information from one node to many nodes. In broadcast mode, every user in a particular region is sent the same message. For example, all travellers within an area might be informed of the approach of a hailstorm. In Multi-cast mode only a subset of users are targeted, such as the transmission of a warning of an approaching rainstorm to drivers of convertible automobiles.

11.5.1 Mobile Telephony Systems: GSM and SMS

Of these networks, probably the one that will find the greatest use in ITS-Telegeoinformatics applications will be the Public Land Mobile Networks, or simply *mobile telephony*. As already mentioned, these networks can be used for positioning, as well as the transmission of data and voice messages to and from vehicles.

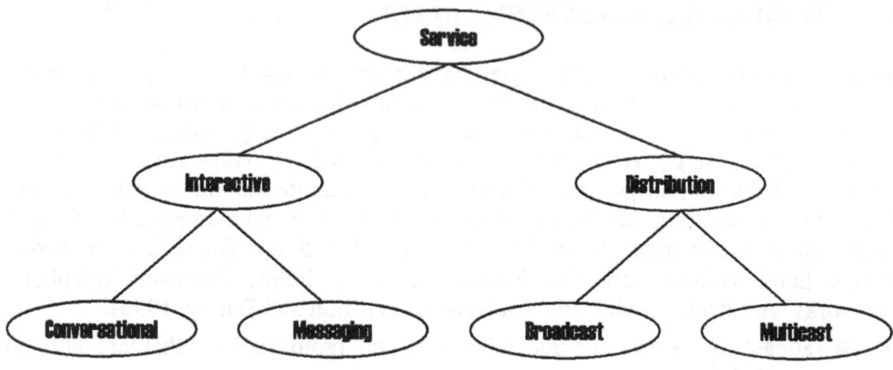

Figure 11.7 Communications Services Hierarchy (Source: US DOT, 1997, reprinted with permission)

An important development in mobile telephony is the seamless capability for data transmission. If we concentrate on the most successful mobile standard, Global System for Mobile (GSM) communication, there already exists the capability of sending small packets of data using the Short Message Service (SMS). It is already very popular with teenagers in many countries, but can be used for ITS-Telegeoinformatics applications as well. SMS allows the transmission of up to 160 characters, either from the vehicle back to a Central Control Facility, or from a Central Control Facility or an Information Service Provider to a vehicle. Because of the limited number of characters in an SMS packet, such messages will be limited to telemetry or brief updates. Cellular location systems such as Cursor™ (Cursor, 2002) provide a means of determining the location from where an SMS packet was sent (provided the user allows this facility).

SMS has two significant limitations for ITS-Telegeoinformatics applications. There is an indeterminate latency so that it is not possible to predict how long a message will take to be delivered. This makes SMS unsuitable for *time critical applications*, such as the real-time coordination of vehicle movements. Another limitation is that a significant number of messages can be lost. This makes SMS unsuitable for *safety critical applications*.

The basic GSM network is known as a 2G or "second generation" network. Some network operators have already deployed 2.5G networks, and there are even some early implementations of 3G networks. The 2.5G and 3G networks have capabilities that make them suitable for many ITS-Telegeoinformatics applications.

11.5.2 Mobile Telephony Systems: GPRS and 3G

An important characteristic of 2.5G networks is the General Packet Radio Service (GPRS). This allows data transmission speeds of up to 171.2 kbps, making it possible to transmit images and video files. One possible ITS-Telegeoinformatics application is the transmission of pictures from slowly scanning video cameras that could be used to track vehicle progress down a freeway. General ITS applications include the monitoring of parking lots, transmission of maps to drivers showing traffic congestion, and sending bus timetables to a potential passenger. Another feature of 2.5G networks is High Speed Circuit Switched Data (HSCSD). Unlike GPRS, HSCSD establishes a continual connection between the nodes. Accordingly, the latency is generally lower than GPRS. Hence, HSCSD is suitable for safety critical ITS applications requiring low latency, such as remotely operating safety barriers.

3G systems will have many of the capabilities of 2.5G systems, but offer higher data rates. It seems that, from an ITS perspective, there will be little difference between 2.5G and 3G networks as far as the functionality that can be achieved. It is likely that there will be a mixed rollout of 2.5G and 3G systems, with some regions going directly to 3G. Other regions might go to 2.5G, and then later migrate to 3G, or perhaps even directly deploy 4G networks.

11.6 ITS-TELEGEOINFORMATICS APPLICATIONS

In Section 11.2.2, we categorized the majority of ITS-Telegeoinformatics applications according to four classes: Driver Assistance, Passenger Information, Vehicle Management, and Road Network Monitoring. The first three categories represent those that could be distinguished from other ITS applications by virtue of them all being examples of LBSs.

In this section, these LBSs are analyzed with respect to the *technologies* involved – such as the in-vehicle devices, the central server, the wireless communications and the *service providers* – the information sources and associated infrastructure. It must be emphasized that the development and deployment of ITS-Telegeoinformatics technology does not, on its own, imply that the full range of ITS-LBSs can be offered. Often the main impediment to the widespread adoption of these applications is the lack of vital infrastructure to underpin the provision of the necessary services that make these applications commercially viable. On the other hand, physical communications infrastructure is required in order to address all the data flow requirements within the vehicle, and between the vehicle and the outside world.

11.6.1 Driver Assistance

The VNS is the quintessential in-vehicle ITS terminal device. However, the sophistication of the VNS can vary considerably:

1. The **Electronic Street Directory** (ESD) is its most rudimentary form and consists of a fixed in-vehicle LCD screen or removable PDAs, upon which map data is displayed. The map database is contained within a memory module (RAM, CD-ROM, compact flash, etc.). The system's functionality is very limited, and generally restricted to *pan* and *zoom* of the digital map display.
2. The **Electronic Vehicle Locator** (EVL) is an enhancement of the above, and permits the current vehicle's location determined by a positioning system to be displayed on the LCD screen. Such a combination of mobile map display and positioning system need not be permanently installed within the vehicle, and in

fact the archetypal configuration is the PDA to which a GPS receiver is attached via the compact flash or serial port. The map database is identical to the one used in the ESD.

3. The **Electronic Navigation Assistant** (ENA) makes use of navigable DRM data to aid the driver. The DRM data can support either enhanced positioning via map-matching, or best-route calculation and route guidance (Section 11.4.4). Typical installations involve an integrated GPS plus dead reckoning positioning system (Section 11.3.1), an in-vehicle computer and LCD screen, a means of command input (mobile pointer/controller, or voice recognition), and the DRM database (usually in the form of a CD-ROM or DVD-ROM). The DRM data include all geospatial information on roads and points-of-interest, and are able to support map queries such as "how do you get from my current location to location X?," "where is the nearest gas station?," etc.
4. The **Server-assisted ENA** (SENA) is the logical extension of the ENA. The in-vehicle equipment is essentially the same as for the ENA, but by means of a wireless communications link, the in-vehicle system can receive additional dynamic data, such as traffic congestion information, to improve best-route calculation and route guidance. In addition, this wireless link permits access to a wide range of LBSs and concierge services, significantly enhancing the driving experience.

Because of the combination of the vehicle position determination capability and the extensive database included in ENA and SENA-type VNSs, the driver has access to a wealth of information:

- Turn-by-turn directions from his/her current location to any specified destination.
- Information on (including directions to) points of interest, such as restaurants, banks, cinemas, ATMs, gas stations, parking stations, hotels, tourist attractions, etc.
- Digital street maps of varying scales, from large-scale city road maps to small-scale cross-country route maps.
- Miscellaneous data, such as public transport routes, road restrictions (e.g., roads with weight and height restrictions, one-way thoroughfares, etc.), tolls, and so on.

However, it is the SENA that is perhaps the defining example of the Driver Assistance implementation that fits the ITS-Telegeoinformatics definition of Chapter 1. The wireless connectivity permits access to a range of services:

- **Broadcast traffic information** such as the level of current congestion on roads, reports of breakdowns or accidents likely to cause delays, highway construction information, traffic lights that are not functioning, locations of speed cameras, and so on.
- **Concierge services** that *find* a particular type of restaurant, shop or point of interest (by accessing a database at a server), that *reserve* an item or service (e.g., a parking spot, a hotel room, etc.), or that otherwise enable a *transaction* to take place.
- **Request for roadside assistance** in the event of a vehicle breakdown, the driver being locked out, or if an accident has occurred.
- **Alert emergency services**, e.g., if there has been a car crash, or the vehicle has

been stolen.

- **Route guidance** in which driver instructions are generated by a central server and delivered in the form of voice commands.

In fact the SENA concept, if carried to its natural conclusion, means that all of the services can be accessed from a very "thin" mobile device, even a PDA. Therefore, the SENA device need not even be permanently installed within a vehicle, and the various concierge and LBSs can be offered to all mobile users, whether in a vehicle or not. All they require is a mobile device with the necessary positioning and wireless communication link accessories. However, in the context of ITS-Telegeoinformatics, the focus will be on in-vehicle devices connected to external information sources via an appropriate communication link. It is possible to distinguish between two types of wireless connectivity: the continuous *broadcast* of information to mobile users, and information sent upon *request* by a user.

The main types of broadcast information relate in some way or the other to *traffic conditions*, e.g., breakdown incident alerts, weather and road traction conditions, and information on traffic congestion, road restrictions and parking availability. Such information has been broadcast to drivers by AM/FM radio for decades. However, the trend is to broadcast such information in a standardized digital format that computers can readily understand and use in, e.g., best-route calculations and route guidance.

In Europe, the Radio Data System (RDS) (Kopitz and Marks, 1998) is the delivery channel for real-time traffic and weather information via the Traffic Message Channel (TMC) (TMC, 2002). The TMC is a set of message formats that can be decoded by a TMC-equipped car radio or navigation system. Although the current channel is based on RDS, TMC can also be delivered via digital radio, Internet, or GSM/GPRS systems.

The Vehicle Information & Communication System (VICS) was established by the Japanese government, coming online in 1996, and currently provides real-time traffic and accident information at no charge to the drivers. Drivers can receive VICS information through a variety of means, principally via special beacon receivers installed in vehicles or FM broadcasting. The service is extremely popular, with over four million installed units, and is available for all expressways and major general roads in Japan. The situation in the U.S. as a whole is less well developed. To support such continuous broadcast services, a significant investment in road network monitoring technology is required, something the industry is reluctant to undertake. Hence, much of the basic infrastructure to support broadcast services has been, to date, provided by government agencies.

It is important to emphasize that the difference in levels of interest in broadcast road network information is not just a function of economic prosperity, but reflects the different priorities of the vehicle driving population. In the U.S., Driver Assistance Services to enhance safety and security are considered a higher priority than services that aid navigation on congested or unfamiliar road networks, as is the case in Europe and Japan.

On the other hand, information or services requested by a driver are best provided by a point-to-point wireless link, such as mobile telephony. The mobile phone (or an embedded modem that accesses the mobile phone network) is therefore the primary enabling technology for the request to, and delivery of information from central servers or call centers. For example, a driver may initiate a "mayday" call, either by a voice call to a human operator or by the press of a button (placing the emergency call via a computer-to-computer hook-up). Alternatively, the "mayday" call could be generated automatically in the event of airbag activation (and the coordinates of the vehicle sent in a small message). This example shows that some driver assistance services will rely on a voice call, while others involve only SMS.

There are essentially two means of accessing ITS-Telegeoinformatics Driver Assistance services, either on an *ad hoc* (e.g., once-off) basis, or via *subscription* to a service provider. The former effectively provides similar traffic information that the European and Japanese broadcast systems currently do, but does so over a telephone link. In the U.S., since mid-2000, the telephone number 511 has been reserved for callers who are seeking information on traffic and weather conditions in their state (US DOT, 2002c). This free service delivers the information to the caller via voice, but cannot provide input to the onboard navigation computer. The 511 service provides similar information to web-based services currently available for many cities, counties and states around the world.

In 1996, General Motors (GM) launched its OnStar system (Onstar, 2002). Initially, the devices to access navigation and Driver Assistance Services were installed in top-of-the-line vehicles. However, GM has expanded the range of vehicles it sells that have OnStar equipment factory-installed. Furthermore, several Japanese and German automakers have adopted the OnStar architecture. With over 2.5 million subscribers, OnStar is the most successful of the ITS-Telegeoinformatics services offered to vehicle owners. Subscription costs vary from $199 per annum for the basic service, rising to over $300 for premium services. However, the setting up and maintenance of such services by automakers is a very expensive undertaking, and very few (if any) are profitable (the Ford-backed venture, known as Wingcast, was abandoned in mid-2002). The range of Driver Assistance Services include: air bag deployment notification, emergency services, roadside and accident assistance, stolen vehicle tracking, remote door unlock, remote diagnostics, concierge services, route support, and general information services. These services are accessed via a mobile phone link, transmitting voice requests and responses, or data, from the vehicle to the OnStar call centers/servers, and visa versa.

Many commentators believe that the successful service providers will not be automakers like GM, but rather they will be wireless carriers, road service organizations, or Internet content providers. They already have the expertise in developing wireless services; they have the billing software systems and are best able to leverage the mobile phone user base to promote the new ITS-Telegeoinformatics Driver Assistance Services. In fact, with Bluetooth enabling devices to connect without wires, the mobile phone handset can function as the "terminal" that communicates with all the embedded devices within a vehicle, and therefore can be used to access and receive all the Driver Assistance Services that GM's OnStar system currently offers. However, it is a poor substitute for an in-vehicle ENA or SENA device.

11.6.2 Passenger Information

Essentially, under this category are all services requested by travellers in the vehicle other than the driver, as these were considered in the previous section. The breadth of services and the variety of system configurations that support them is very wide. The Passenger Information Services range from the frivolous to the critical, and cover a wide range of users: passengers within private vehicles, those engaged in pre-trip planning, and public transport commuters.

The archetypal class of services are those that can be accessed by the passengers via wireless communications through the in-vehicle mobile terminal or via the mobile phone. Increasingly, as the amount of time spent each day in traffic grows, efforts will be made to improve the quality or "productivity" of this vehicle time. Although some services are available to the vehicle driver, given the potential for driver distraction, most of these services are offered to the other occupants. These services include access to electronic games, attending to email, receiving reports on weather, stock prices and sports scores, and ultimately (if there is enough bandwidth in the wireless communications link) downloads of music and video clips. Not all of these use the vehicle's position as a "filter," and hence not all can be considered examples of LBSs. However, the same IT and wireless communications components are used as in the case of Driver Assistance Services. Furthermore, extra infrastructure does not need to be deployed to allow for the provision of such services. In fact the only issue may be how to charge for such services, whether as part of general mobile phone charge billing, or as an ITS-Telegeoinformatics service subscription.

The trend for the automobile becoming "wired to the Internet," of being an entertainment center, and even of evolving into a mobile office (or home) is unmistakable. Many parties are seeing the vehicle (and its occupants) as a new source of revenue for services and products. All this is made possible by the revolution in auto electronics and wireless communication networks. The vehicle's location simply enables the process by providing a means of filtering unwanted or irrelevant services.

The Driver Assistance Services can be also accessed during pre-trip planning. ESD, ENA or SENA-type VNSs can be used to simulate the intended trip and to obtain, e.g., best route information before setting out. This would be done in the vehicle if the terminal were permanently installed within the vehicle. Hence, details, e.g., of the nearest gas station to the trip's destination could be obtained, and programmed into the in-vehicle device. Alternatively, if the mobile device is a PDA, then this pre-trip planning can occur anywhere, at home, in the office, and even on the move. In addition, web-based services could be accessed at home, or from any convenient PC.

Passengers planning to use public transport could obtain advice on their itinerary, with step-by-step instructions on which buses or trains to use. They could be informed of which transport services are running late, and the estimated arrival times of the service at stops of interest to the passenger. Currently, such public transport information can be obtained via the web, via information kiosks, LCD screens in the bus or train, or computer displays at selected bus or train stops. Increasingly such services will be available to mobile devices such as PDAs, but in a more "intelligent" manner than is currently the case. PDAs can be used to access the web via wireless links (such as mobile telephony, or WLAN-IEEE802.11 systems), but they are indistinguishable from other (wired) web users. If position information was automatically part of the query process, then the public transport information provided to the user could be tailored to the PDA's location. Hence, only information on transport options relevant to that place and time needs to be sent.

The infrastructure needed to support public transport Passenger Information Services is currently being deployed as part of programs to improve the efficiency and attractiveness of public transport. Hence, buses, trains, trams, etc., will need to be tracked using some form of remote-positioning technology (Section 11.3), and this information used to predict the time the services arrive at different stops on their route. In many respects such operations are similar to commercial or emergency fleet management, with one major difference – public transport routes are fixed. The challenge therefore is primarily in distributing this information to passengers, or would-be passengers. Accessing such information from mobile personal devices, via wireless communication networks, is a natural extension of current practice.

11.6.3 Vehicle Management

Vehicle Management Services are equally as broad as Passenger Information Services, but with the primary difference being that the services are not, in the first instance, targeted to the driver or passenger of the vehicle, but to a central agency or center.

Commercial Vehicle Operations: Fleets of trucks, couriers, taxis and other commercial vehicles can be managed using ITS-Telegeoinformatics applications such as the Fleet Management System (FMS). The "back-office" FMS application is generally a form of a GIS, containing the road network data and incorporating data about the vehicle fleet, customer addresses and other information. The FMS has information transmitted to it of each vehicle's position, and often other data such as engine performance, as well as cargo-specific information such as the temperature of refrigerated goods, etc. If the goods are of a *hazardous* nature (e.g., waste, chemicals, dangerous materials, etc.), or *valuable* (e.g., cash, electronic goods, cigarettes or alcohol, etc.), then the FMS is effectively a means of tracking the progress of consignments. On the other hand, the FMS can be used to dispatch the appropriate vehicle to a customer, on the basis of the location of the vehicles in relation to the customer. Yet another use of the FMS is to help schedule vehicle maintenance.

Emergency Vehicle Dispatch: In many respects the operations of police, ambulance and fire fighting vehicles are similar to those of commercial vehicle fleets. Therefore, they can be "managed" by an FMS, and in general the FMSs for emergency vehicles are identical to those used for commercial vehicle operations. However, the most important functions of such FMSs are: *dispatch* of vehicles (including route guidance), *monitoring* of the vehicles and occupants (to ensure the safety of emergency personnel), *managing situations* (e.g., monitoring the nature of the emergency, and the response), and *handling requests* by emergency personnel (e.g., for additional or specialist services). Emergency vehicles typically have sophisticated communication systems, and the trend is to incorporate broadband wireless links.

Personal Vehicle Monitoring: A private vehicle may be tracked or monitored remotely for different reasons. The driver may request this monitoring (e.g., as a Driver Assistance Service if they are in an unfamiliar area and need guidance), or it may be initiated by the vehicle owner (e.g., in the case of a parent lending a car to a teenager driver), or it may be precipitated in the event of car theft or accident. Clearly, unlike Commercial Vehicle Operations, or Emergency Vehicle Dispatch, the monitoring of private vehicles is an "invasion of privacy", and hence such an action must be agreed to (explicitly or implicitly) by the owner or driver. Such monitoring may be *passive*, involving tracking only, or it may also lead to an *action* such as the switching off of the car's motor remotely, as in the case of vehicle theft.

Remote Vehicle Diagnostics: This is perhaps one of the more interesting of this class of ITS-Telegeoinformatics applications. Imagine that the vehicle's Engine Management System was monitoring the parameters of the vehicle, such as engine and brake performance, electronics, cooling and lubrication subsystems, and transmitted this information to a service center if something was on the brink of malfunctioning. The service center could "communicate" with the vehicle, diagnosing the possible problem, alerting the driver to this, and even make an appointment to have it repaired! Such Remote Vehicle Diagnostic Services are already being offered to some luxury cars, and the trend will see such services extended to more and more vehicle owners.

All of these services require a data link from the vehicle to the dispatch or control center, by which position information, driver requests, vehicle data, etc., are transmitted. This link may be radio based, as in the case of emergency and commercial vehicles, or provided by the mobile telephony network in the case of private vehicle services. The tracking of the vehicle by an organization is invariably required, raising the issue of privacy (Section 11.7). Often a communications channel to the vehicle is also required, e.g., to provide up-to-date information concerning a parcel pickup, or to send details to the vehicle attending to the emergency.

Many commentators believe that this class of ITS-Telegeoinformatics applications will be the most effective, as they impact directly on industrial productivity, as well as making the greatest contribution to the community's safety and sense of security. In addition, the in-vehicle costs are the lowest because the complexity of such systems is largely at the dispatch or control center.

11.7 NON-TECHNICAL ISSUES IMPACTING ON ITS

There is a range of issues that go beyond the merely technical. For example, ITS-Telegeoinformatics applications or services can be characterized according to whether they are:

- **Essential**, impacting on the safety and security of drivers, passengers and cargoes, and addressing the goals of ITS, i.e., improving the efficiency and safety of the road transport network and mitigating the negative effects on the environment.

- **Useful**, providing assistance, navigation services and road network information in order to enhance the journey experience.
- **Trivial**, such as those services that are entirely self-indulgent, e.g., access to email, sports news, games and entertainment.

This raise concerns about *inequity*. Should society be concerned if a privileged segment of the community has access to information that others do not? Should people who can pay more be provided with better travel information and so enjoy shorter travel times? Should requests for publicly available road traffic congestion information be prioritized so that only essential applications are serviced if there is the chance of system overload? What incentives are there for offering LBSs, which might be viewed as "non-essential," over the same channels and devices as services intended to enhance safety and security? Does a local community have the right to monitor the profile of commercial traffic through its neighbourhood? ITS-Telegeoinformatics will raise many such questions. As with many such ethical questions, the best approach is an open public debate, allowing the various stakeholders to develop an informed consensus.

The transmission to, and analysis or storage by, an external party of vehicle location information, especially in an AVLS application but also in certain configurations of Driver Assistance Services, raises legitimate concerns about *privacy*. When can the need for privacy be overridden by community concerns about safety and security, involving the monitoring of traffic movements? What provisions for "opting out" of tracking schemes should there be? There are a number of things that could be said about this issue:

- Firstly, it seems appropriate that all callers to the E911-type emergency services should forfeit their right to privacy. A legitimate call to an emergency will always request a rapid response by emergency crews to a particular location. Any false call endangers public safety; hence rapid location of "prank" callers is in the public interest.
- Otherwise, an "opt in" system appears to be best. Knowledge of a user's position is a critical piece of information for LBS applications. However, users should have the power to decide if they are to be tracked or not, or whether they should broadcast their location to others. An "opt out" system, on the other hand, is likely to encounter fierce resistance by users.
- There appears to be times when the police and other authorities have a legitimate need to track the location of persons or vehicles. However, this power should only be exercised under the supervision of the courts, as happens with other invasive surveillance techniques such as wiretaps. In many jurisdictions, legislation will be needed to govern the use (or abuse) of such Telegeoinformatics systems.

What about car rental companies tracking their customers and sending them fines when they drive faster than the speed limit? There have been a number of such cases, but when challenged in court the ruling has generally gone in favor of the driver. Civil liberty groups have also raised concerns. This is an issue that is still far from resolved. In the context of "Homeland Security," instances of tracking trucks, buses and cars will increase. However, it is likely that there will be a heated debate about "where to draw the line" so that civil liberties are not systematically abused.

A further issue with many ITS-Telegeoinformatics systems is the current difficulty in developing a successful business model. This is because many of these systems require widespread deployment of some form of infrastructure before it is possible to operate them. This means that very large investments and long timeframes are needed before it is even clear which of the applications will "make money." Hence, this argues the need for public-private partnerships to foster innovation in this area, and also highlights the role of such groups as ITS America, ITS Japan, ERTICO, etc. (Section 11.1) in developing such partnerships. We believe that this dependence on public-private partnerships is only a feature of the early stages of ITS. Once the various infrastructure platforms are in place, it should be much easier, and cheaper, to develop and market new applications.

This raises another interesting issue about ITS applications. That is, the potential to transform the transport sector. At present, compared to the IT and telecommunication sectors, the transport sector is marked by a heavy dependence on public funding and government oversight. ITSs have the potential to transform this situation, by establishing a common platform that allows new and innovative applications to be developed. This will bring many small-to-medium sized firms into the industry, so transforming it to a form much more like the IT sector.

11.8 CONCLUDING REMARKS

Despite problems with business models, the lack of government investment in infrastructure, and the initial reluctance of drivers to pay for LBSs, it seems that the deployment of ITSs, and the corresponding ITS-Telegeoinformatics applications, are proceeding at a rapid rate. It is one of the ironies of ITSs that it is virtually invisible to the general public. The average consumer today hears of navigation systems in luxury vehicles, is vaguely aware that the traffic lights have some level of coordination (or grumbles when none is apparent), and is blissfully unaware of the existence of traffic control centers that aid their journey and are prepared to assist them in an emergency. We believe that within the next twenty years, the full panoply of the ITS vision is likely to be deployed, with many of the ITS-Telegeoinformatics applications identified in this chapter being the first to be "market tested." However, the general public will tend to see each component in isolation, rather than appreciating that it is just a part of a broader vision.

REFERENCES

Bossler, J.D., Jenson, J.R., McMaster, R.B. and Rizos, C. (Eds), 2002, *Manual of Geospatial Science and Technology*, (London, New York: Taylor & Francis).

Cursor, 2002, <http://www.cursor-system.com> (accessed 20 September 2002).

Djuknic, G.M. and Richton, R.E., 2000, Geolocation and Assisted-GPS. Online. Available HTTP: <http://www.lucent.com/livelink/090094038000e51f_White_paper.pdf> (accessed 15 September 2002).

Drane, C.R. and Rizos, C., 1998, *Positioning Systems in Intelligent Transportation Systems*, (Boston, London: Artech House).

Drane, C.R., Macnaugtan, M. and Scott, C.A., 1998, Positioning GSM telephones. *IEEE Communications Magazine*, 36(4), pp. 46-59.

Catling, I., 1994, *Advanced Technology for Road Transport: IVHS and ATT,* (Boston: Artech House).

Elliott, S.D. and Dailey, D., 1995, *Wireless Communications for Intelligent Transportation Systems*, (Boston, London: Artech House).

Enuvis, 2002, <http://www.enuvis.com> (accessed 20 September 2002).

Ertico, 2002a, <http://www.ertico.com/links/links.htm> (accessed 20 September 2002).

Ertico, 2002b, <http://www.ertico.com> (accessed 20 September 2002).

FRP, 2001, Federal Radionavigation Plan, U.S. Dept. of Transport. Online. Available HTTP: <http://www.navcen.uscg.gov/pubs/frp2001> (accessed 20 September 2002).

Global Locate, 2002, <http://www.globallocate.com> (accessed 20 September 2002).

Hjelm, J., 2002, *Creating Location Services for the Wireless Web*, (New York: John Wiley & Sons Professional Developers Guide Series).

ITS America, 1992, *Strategic Plan for Intelligent Vehicle Highway Systems in the United States*, (ITS America, Washington, DC).

Kopitzer, D. and Marks, B., 1998, *RDS: The Radio Data System*, (Boston, London: Artech House).

McQueen, J. and McQueen, B., 1999, *Intelligent Transportation Systems Architectures*, (Boston, London: Artech House).

McQueen, B., Schuman, R. and Chen, K., 2002, *Advanced Traveler Information Systems*, (Boston, London: Artech House).

NavTech, 2002, <http://www.navtech.com> (accessed 20 September 2002).

Onstar, 2002, <http://www.onstar.com> (accessed 20 September 2002).

Quiktrak, 2002, <http://www.quiktrak.com.au> (accessed 20 September 2002).

Sigtec, 2002, <http://www.signav.com.au> (accessed 20 September 2002).

SiRF, 2002, <http://www.sirf.com> (accessed 20 September 2002).

Snaptrack, 2002, <http://www.snaptrack.com> (accessed 20 September 2002).

Tekinay, S., 2000, *Next Generation Wireless Networks*, (The Kluwer International Series in Engineering and Computer Science, Volume 598).

Teleatlas, 2002, <http://www.teleatlas.com> (accessed 20 September 2002).

TMC, 2002, Traffic Management Channel Forum. Online. Available HTTP: <http://www.tmcforum.com> (accessed 20 September 2002).

US DOT, 1997, *Communications Document, ITS Architecture Report,* Joint Architecture Team (Loral Federal Systems and Rockwell International) ITS Architecture, Figure 3.1-2, Physical Architecture, (Washington DC: US Department of Transportation, Federal Highways Administration).

US DOT, 2002a, *ITS Architecture Report,* Joint Architecture Team (Loral Federal Systems and Rockwell International) ITS Architecture - Physical Architecture, (Washington DC: US Department of Transportation, Federal Highways Administration).

US DOT, 2002b, *Executive Summary, ITS Architecture Report.* Joint Architecture Team (Loral Federal Systems and Rockwell International) ITS Architecture, Figure 1, Physical Architecture, (Washington DC: US Department of Transportation, Federal Highways Administration).

US DOT, 2002c, U.S. Dept. of Transport 511 Service Information. Online. Available HTTP: <http://www.its.dot.gov/511/511.htm> (accessed 20 September 2002).

Whelan, R., 1995, *Smart Highways, Smart Cars*, (Boston: Artech House).

Zhao, Y., 1997, *Vehicle Location and Navigation Systems*, (Boston, London: Artech House).

Zhao, Y., 2002, Telematics: safe and fun driving. *IEEE Intelligent Systems*

CHAPTER TWELVE

The Impact and Penetration of Location-Based Services

Narushige Shiode, Chao Li, Michael Batty,
Paul Longley, and David Maguire

12.1 THE DEFINITION OF TECHNOLOGIES

Since the invention of digital technology, its development has followed an entrenched path of miniaturization and decentralization with increasing focus on individual and niche applications. Computer hardware has moved from locations remote from the individual to desktop and handheld devices while becoming embedded in various material infrastructures. Software has followed the same course. The entire process has converged on a path where various analogue devices are becoming digital, increasingly being embedded in machines at the smallest scale. In a parallel development, there has been a convergence of computers with communications ensuring that the delivery and interaction mechanisms for computer software is now focused on networks of individuals, not simply through the desktop, but in mobile contexts. Various inert media such as fixed television is becoming more flexible as computers and visual media are becoming one.

With such massive convergence and miniaturization, new software and new applications continually define the cutting edge. As computers are being increasingly tailored to individual niches, then new digital services are emerging, many of which represent applications which hitherto did not exist or, at best, were rarely focused on a mass market. Location-Based Services (LBSs) form one such application and in this chapter, we will both speculate on and make some initial predictions of the geographical extent to which such services will penetrate different markets. We define such services in detail below but suffice it to say at this stage that such functions involve the delivery of traditional services using digital media and telecommunications. High profile applications are now being focused on handheld devices, typically involving information on product location and entertainment, but wider applications involve fixed installations on the desktop where services are delivered through traditional fixed infrastructures. Both wired and wireless applications define this domain. The market for such services is inevitably volatile and unpredictable at this early stage but here we will attempt to provide some rudimentary estimates of what might happen in the next five to ten years.

The "network society" which has developed through this convergence is, according to Castells (1989, 2000), changing and re-structuring the material basis of society such that information has come to dominate wealth creation both as a raw material of production and tradable commodity. This has been fueled by the way technology has expanded at rates of doubling every 18 months or so as implied in Moore's Law and by fundamental changes in the way telecommunications, finance, insurance, utilities and so on are regulated. LBSs are becoming an integral part of this fabric, and these reflect yet another convergence between Geographic Information Systems (GISs), Global Positioning System (GPS), and satellite remote sensing. The first geographical information system, the Canada Geographic Information System (CGIS), was developed as part of the Canada Land Inventory in 1965 and the acronym 'GIS' was introduced as early as 1969. 1971 saw the first commercial satellite, LANDSAT-1. The 1970s also saw prototypes of the Integrated Services Digital Network (ISDN), mobile telephony, and the introduction of TCP/IP as the dominant network protocol. The 1980s witnessed the introduction of the IBM PC (1982) and the beginning of deregulation in the U.S., Europe and Japan of key sectors, particularly telecoms, within the service economy. Finally in the 1990s, we saw the introduction of the World Wide Web and the ubiquitous pervasion of business through networked PCs, the Internet, mobile communications, GPS for positioning, and GISs for the organization and visualization of spatial data. By the end of the 20th century, the number of mobile telephone users had reached 700 million worldwide. The increasing mobility of individuals, the anticipated availability of broadband communications for mobile devices, and the growing volumes of location-specific information available in databases will inevitably lead to the demand for services that will deliver location-related information to individuals on the move. Such LBSs, although in a very early stage of development, are likely to play an increasingly important part in the development of the social structure and business in the coming decades.

In this chapter, we begin by defining LBSs within the context just sketched. We then develop a simple model of the market for LBSs developing the standard non-linear saturation model of market penetration. We illustrate this for mobile devices, namely mobile phones and then we develop an analysis of different geographical regimes which are characterized by different growth rates and income levels worldwide. This leads us to speculate on the extent to which LBSs are beginning to take off and penetrate the market. We conclude with scenarios for future growth using an analogy between the way GISs have developed and the diffusion and penetration of mobile devices.

12.2 LBSs: DEFINITIONS, SOFTWARE, AND USAGE

There are many different definitions of LBSs presented in this book and in one sense, the reader will already have a good idea of these from simply scanning the previous chapters. However, here we must be specific again in our definitions for these relate now to a peculiar combination of hardware, software and usage which we are able to measure in terms of their penetration and diffusion. To us, LBSs are geographically-orientated data and information services to users across mobile telecommunication networks. LBSs can thus be seen as a convergence of New Information & Communication Technologies (NICTs) such as the mobile telecommunication system, location-aware technologies, and handheld devices, with the Internet, GISs and spatial databases as illustrated in Figure 12.1.

Mobile telecommunication networks have developed dramatically in Europe in the late 1990s, from the second generation based on the Global System for Mobile Communication (GSM) to 2½ generation called the General Packet Radio Service (GPRS) with an expected third generation termed the Universal Mobile Telecommunication System (UMTS). Bandwidth is moving from 9.6 Kbps to 115 Kbps, and will support 2 Mbps with UMTS. The Wireless Application Protocol (WAP), started in 1997, is being developed into a new standard, aiming for "an architecture and a language which will facilitate the convergence of fixed Internet, ... toward one and the same mobile Internet solution." (Ericsson, 2001). The development of mobile telecommunications has also transformed the services from focusing on transmission of voice data to various applications transmitting multimedia information. Handheld, mobile and small size wireless devices such as Personal Digital Assistants (PDAs) and mobile phones have been enhanced and can be connected via infrared, GSM modems, or radio signals to wireless networks. With the development of location-aware technologies and the increasing number of mobile device users, services providing location-related information are likely to become major applications of these new technologies.

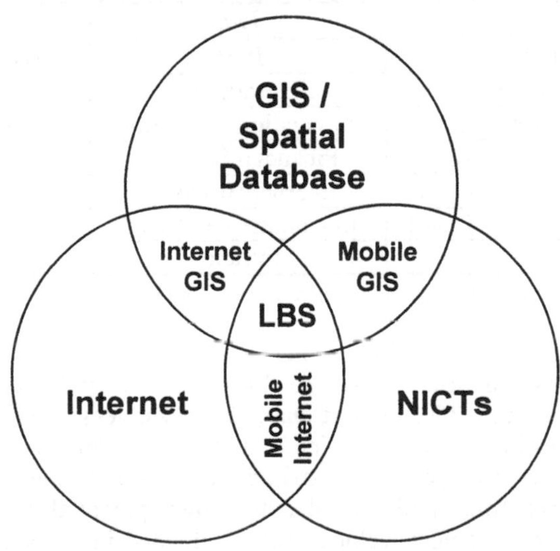

Figure 12.1 The Intersection of Technologies Creating LBSs (Brimicombe, 2002)

Given the location-aware nature of LBSs, locational information is vital for communications system set-up, navigation and system management, while the handling and processing of such information (spatial and temporal) is essential for providing LBSs in the first place. Many industry analysts see a central role for GISs which form the foundation for many of these LBS developments as a core area in managing, processing and delivering spatial information in meaningful ways (Bishr, 2000). The META Group reported that "... about 40 percent of all mobile data applications in utilities are linked to GIS" (quoted in Wilson, 2001). Maguire (2001) noted that "... geographic data and processing are provided as a type of service over a wireless network connection," which suggests the terms "mobile geographic services" and "mobile GIS." A number of statements have alluded to GIS solutions being able to analyze objects in particular locations in terms of their relationships to each other, thus providing tools for integrating spatial information. Such developments will transform the average LBSs into valuable information sources for supplying value-added services (Coleman, 2001; Wilson, 2001).

The incentive in the E911 (emergency services in the U.S.) mandate is that it provides a wide range of additional, value-added LBSs. The concern for locating such wireless emergency calls has only been tackled since the year 2000 in Europe. Nevertheless, there is currently fast development and deployment of both wireless and location-aware networks. Also in 2000, mobile network operators, after immense investment in 3G licenses, began to look for content-driven services. LBSs in Europe are increasingly being driven to provide differentiated and value-added services in a competitive marketplace.

Table 12.1 Range of LBS Applications and Their Perceived Ranking by Mobile Users over Time

Rank	***2000***	***2003***	***2005***
1	PIM*	PIM	Navigation/Location
2	Entertainment	Entertainment	PIM
3	Financial Services	Navigation / Location	Entertainment
4	Internet Browsing	Financial Services	Financial Services
5	Navigation / Location	Internet Browsing	Internet Browsing
6	m-Commerce / Retail	m-Commerce / Retail	m-Commerce / Retail
7	Intranet	Intranet	Intranet

**PIM: Personal Information Management* (source: Beinat, 2001).

Gartner, Inc. (Gartner, 2002) predicts that by 2005, almost 40% of European mobile subscribers will use LBSs. The 2001 surveys by Jupiter and McKinsey (Wireless World Forum, 2002) report that 38% of wireless users would like to see navigation services being provided. The ARC Group's figure shown in Table 12.1 demonstrates the wide range of potential applications and their perceived ranking by mobile users over time. With considerable growth potential in LBSs, the various applications currently being developed can be divided into seven categories with some of these applications commercially available (Hunter, 2001). The list is still growing and consists of:

- Pushed online advertising;
- User-solicited information, such as local traffic information, weather forecasts and local services;
- Instant messaging for communication with people within the same or nearby localities;
- Real-time tracking,
- Mapping/route guidance, directing people to reach their destination;
- Emergency services for stationary location; and
- Location-based tariffs.

An open industry specification for delivering navigation, telematics and related geographic information services across multiple networks, platforms and devices is being promoted by a consortium that includes Panasonic, Microsoft, and TeleAtlas. They are also developing a suite of APIs and protocols designed to create a development environment for LBS applications (Bastiaansen, 2001). Some companies have developed geo-spatial engines optimized for LBS applications, while others have chosen to embed existing GISs into their services.

Personal profiling has been regarded as a way to enrich and enhance LBSs by supplying people with the information that they are likely to need. 'Tailor made' LBSs as facilitated by such profiling are considered to be a significant development in the mobile data market, which will deliver information specifically compiled for the individual according to his/her interests, history and location. An example of future LBS personalized intelligent services tuned to individual's lifestyle and demands would be theater tickets, automatically reserved during a business trip away because of a user's cultural profile. Information related to individuals and their locations has been suggested as classifiable into short-, middle- and long-term information in its impact on behavior. Recent spatial and temporal trends are considered as short-term information, personal preferences as middle-term, and the person's identity as long-term (Mountain and Raper, 2000; 2001). Personal profiling would clearly be attractive to marketers in their pursuit of 'one to one' marketing of products and services as it would be more accurate than current geodemographic profiles.

A range of input and output methods have been proposed and offered in some of the LBS applications, thus trying to identify the desirable and effective way for delivering information. Voice has been suggested as a user-friendly method for inputting requests and giving instructions. Using speech via mobile phones as navigating instructions during driving is one example. Phonetic transcriptions of city names, street names, and 'points of interest' in digital maps are related examples inputting information through voice (van Es, 2001). Despite many exciting applications of wireless services, LBSs are still in the earliest stages of deploying mobile solutions. Many of the assumptions concerning usage and behavior are untested and should not be taken for granted. There is thus considerable uncertainty still in predicting how these applications and markets will pan out, but we consider that the analogy with all kinds of hardware and software in the computer industry during the last 50 years is the one that provides the best basis for a speculation of the impact and penetration of LBSs during the next decade.

12.3 THE MARKET FOR LBSs: A MODEL OF THE DEVELOPMENT OF LBSs

Although the concept of LBSs has emerged only recently, it is expected to follow the remarkable growth demonstrated earlier by other IT technologies. The demand for various geographical information-oriented services is also likely to provide a large share for LBS industry in the IT market within the next ten years. This wave of market growth is influenced by a number of elements and factors that are themselves growing rapidly and are increasingly becoming available in the market. These include hardware vehicles such as the WAP-enabled mobiles and other wireless Internet terminals, software services ranging from secure online transaction protocols to various entertainment functions, and data resource on which LBSs rely.

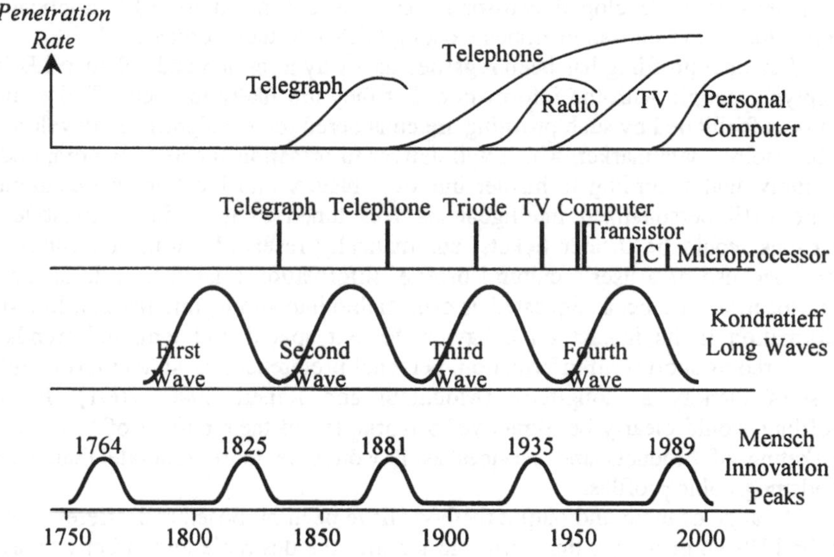

Figure 12.2 Schematic Diagram of Radical New IT Innovations and the Kondratieff Waves from Hall and Preston (1988) (reproduced with permission)

The series of successive waves of new IT inventions and their diffusion appear to follow a certain pattern known as the Kondratieff cycle. This cycle is based on a theory that inter-relates diverse events such as war, knowledge discovery, public opinion, weather, price behavior, and so on as being integral parts of a long-term economic life-cycle (Kondratieff, 1935). Within a capitalistic society, Kondratieff proposed that economic trends tend to repeat themselves every 50-60 years. This alternation of *the long wave* from prosperity to depression, complemented by many shorter cycles, lends a dynamic trend to the economy that to a large degree becomes predictable. The cycle is strongly associated with the discovery and ingestion hence dissemination and implementation of new technologies, digital computing being the most recent but electricity being the more fundamental. Hall and Preston (1988) argue that the array of innovations of IT-related technologies and services can be regarded as part of the fifth wave of innovations in this particular context (Figure 12.2). The figure illustrates the transition of innovations and technologies where the new generation of innovations are continuously emerging under a certain frequency, occasionally replacing their predecessors.

Taking this analogy further, we can argue that the LBS-related technologies also follow the wave of innovations as a group of inventions forming the terminal period of the fifth Kondratieff wave, where these technologies will rapidly become available in the market and will promote the LBS industry in making transitions in service types within the wireless market.

Whether or not the Kondratieff wave is theoretically robust remains still somewhat controversial, but there is a widespread evidence which empirically supports the idea that the economic market and innovations curve reveal common ground in their periodic behavior. We are also aware of numerous other waves or cyclic models suggested in the past. For instance, Schumpeter's (1939) *Business Cycles* proposed a three-cycle model of economic fluctuations or waves:

- the Kitchin inventory cycle (3-5 years),
- the Juglar investment cycle (7-11 years), and
- the Kondratieff long cycle (50-60 years).

However, these are fundamentally inter-related with one another, essentially pointing in the same direction. For instance, three Kitchins make up one Juglar and six Juglars make up one Kondratieff. There is also the mid-long wave of infrastructural investment suggested by Kuznets – *the Kuznets infrastructural investment cycle* (15-25 years) – where, again, we can fit three Juglars to one Kuznets and two to three Kuznets to one Kondratieff.

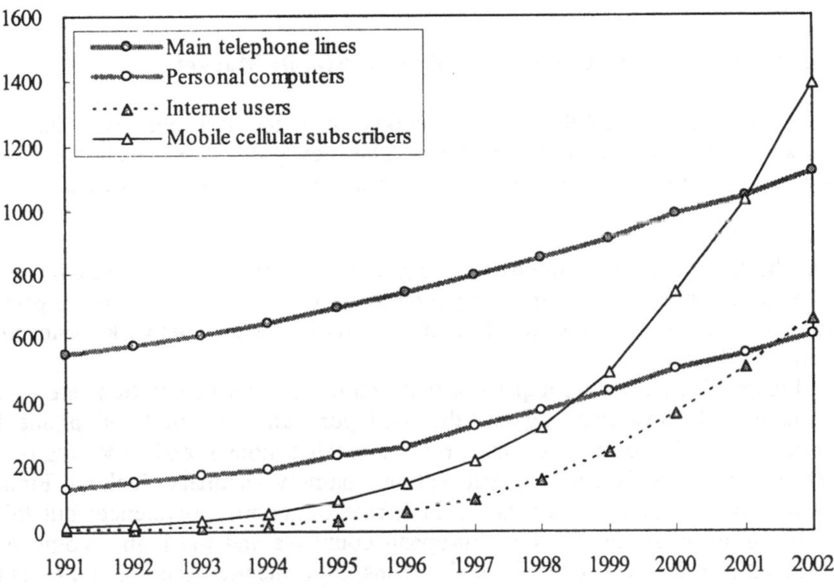

Figure 12.3 Growth of Some IT-Related Services and Commodities (in millions) (Data source: International Telecommunication Union, 2002)

Here we follow the analogy of IT deployment which forms the so-called fifth wave of innovations, applying the Kondratieff cycle model to depict the market growth of LBSs and related industries. Figure 12.3 shows the growth of some of the IT-related commodities over the last ten years. The four indices can be categorized into two different groups each of which is increasing at a near-exponential growth rate but with a different scale and slope. The first group contains the basic hardware commodities that were invented prior to the convergence of the fifth wave, and these include fixed phone lines and personal computers which are well deployed, their market having reached a state of maturity over the years. The second group comprises users of various IT services provided through such devices where the curve is still steep but is likely to reach saturation as it exceeds the overall growth of hardware from the first group. They are growing at a remarkable rate, rapidly catching up with the number of users, suggesting that the LBS market follows the same exponential growth observed in many other IT services.

In the next section, we observe the growth of the mobile industry in terms of its penetration rate in different parts of the world, estimating its growth in the next five to ten years. The analysis will be primarily focused on handheld devices in general, but as the number of WAP-enabled mobile devices is increasing, it can be regarded as an indicator of the increase in the number of hardware vehicles for LBSs.

12.4 PENETRATION OF MOBILE DEVICES: PREDICTIONS OF FUTURE MARKETS

12.4.1 Summary of the Growth Trend in the Mobile Market

The growth of the mobile industry market over the last ten years has been remarkable. However, as it applies to any other newly developed technologies, the growth has been very variable across the world's nation states and within different consumer groups. The North European nations, for instance, are known as the most wired and they have constantly enjoyed a high penetration rate of mobile devices, while the less-developed countries have yet to take off as a large mobile phone market. This may change dramatically over the next couple of years, as the price of mobile devices becomes even cheaper and more mobile network connections become available.

Figure 12.4 shows a comparison between the current penetration rate — i.e., the number of subscribers against the total population — of fixed phone line connection and that of mobile subscription in each continent and wider region. As of December 2000, North European nations, namely Denmark, Iceland, Finland, Norway, and Sweden, already had a high rate of mobile prevalence; but this is closely followed by the Western European countries and the high income Asia-Pacific nations. In terms of individual nations, high income Asia-Pacific countries such as Taiwan (80%), Hong Kong (77%) and Singapore (62%) are more visible among the most mobile saturated nations. The rank of the actual number of mobile subscription has a strong positive correlation with the GDP of strong economic powers such as the United States, Japan, Britain, and Germany dominating the list; with the notable exception of China which ranks at the very top due to the sheer size of its population. This also applies to several other countries with large populations such as Brazil, Turkey, and Mexico, all of which have a low penetration rate for mobile phones. This alone clearly indicates a potentially explosive growth in the number of mobile devices over the next few years as the price becomes more affordable. The divide among those who have and those who do not have is more visible in Figures 12.5 and 12.6 where groups of nations with similar economic and demographic characteristics naturally form loose clusters.

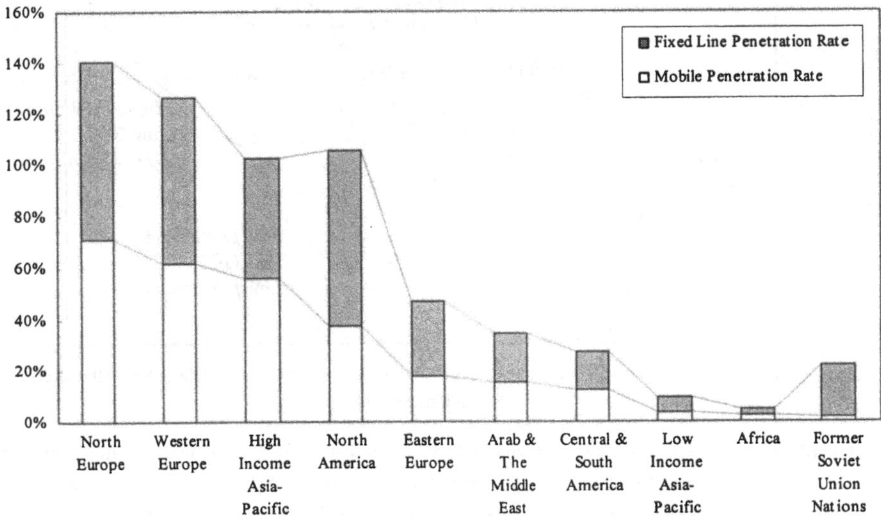

Figure 12.4 The Global Fixed Line and Mobile Penetration Rate as of 2000 (Data source: International Telecommunication Union, 2002)

In Figures 12.5 and 12.6, the more developed countries form an almost log-linear distribution towards the upper right, suggesting that they are already in a steady state. These are indicated as "nations with high GDP" in Figure 12.5 and "West and North Europe" in Figure 12.6, both of which are predicted by a regression with an r^2 value of over 97%. These countries may still see a healthy growth in the number of devices, but this would exceed the total number of population who are likely to use mobile phones (e.g., excluding children under 10) suggesting that the surplus will be used as a secondary device where one user would have multiple handheld devices, possibly because of the different functions they offer.

Countries with medium GDP per capita (e.g., East European countries) which are rapidly growing in economic presence, currently form a cluster between the superpowers and the smaller yet developed nations. This group may diminish as they gradually catch up with developed countries and come to form a combined cluster of high mobile subscription rate, although the current r^2 value of 74% for its regression implies that this cluster is wider than the former and may take more time before convergence. The most controversial element would be the large cluster in the lower-right in Figure 12.5 formed by less (and the least) developed countries which have low GDP figures and naturally low mobile penetration rates. Potentially, these nations may boost the mobile industry, especially in the case of nations with large populations where even a small percentage increase would significantly affect the market.

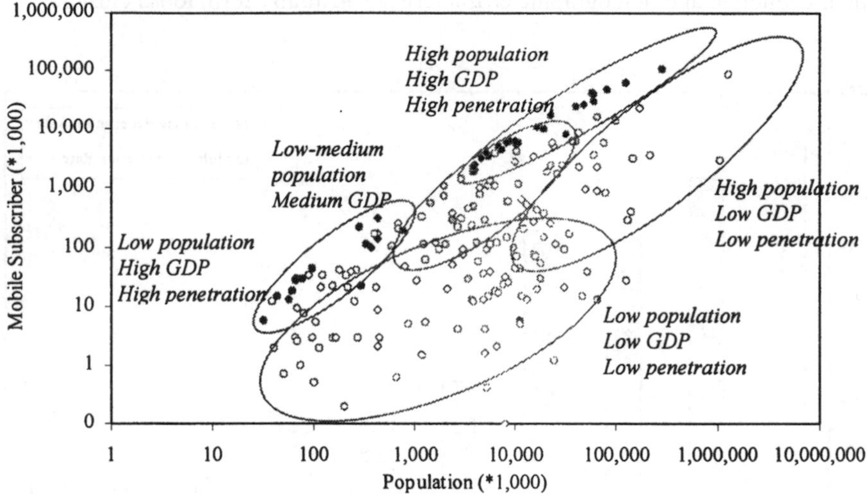

Figure 12.5 Log-log Plot of Mobile Subscribers Against Population for Each Country in the Year 2000 (Data source: International Telecommunication Union, 2002). The white, gray, and black dots represent countries with low GDP (per capita) value, intermediate value, and high value, respectively

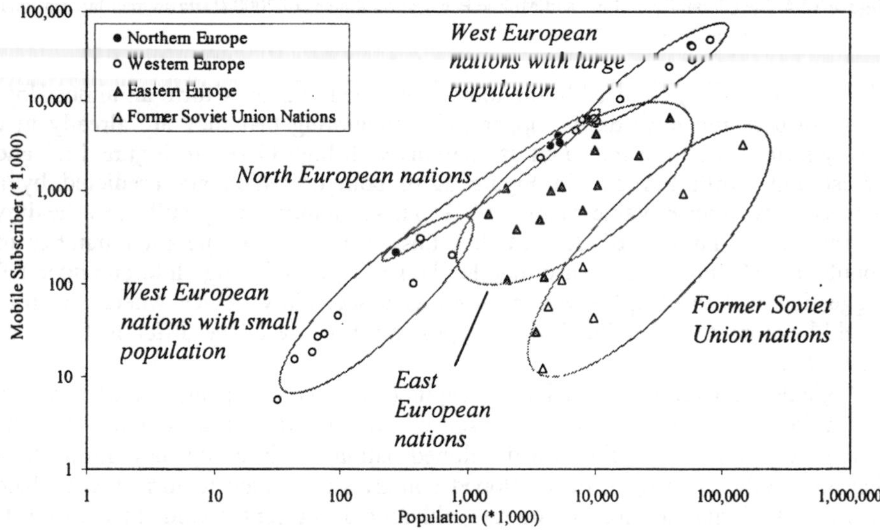

Figure 12.6 Log-log Plot of Mobile Subscribers Against Population for Countries in Europe in 2000 (Data source: International Telecommunication Union, 2002)

These observations are rather superficial and speculative but are nonetheless confirmed in Figure 12.7, which shows that the accumulative growth of mobile penetration rate within each continent or region can be categorized into three groups. The Northern European countries lead the first group where the Western Europe and the high income sector of Asia-Pacific nations are also catching up rapidly. North America is yet to reach the steepest penetration increase but is constantly increasing, and so are the other regions with a mixture of developed and less developed countries such as the Middle East, Eastern Europe and Central and South America, forming the second category of penetration growth. The last category consists of the least developed regions such as Africa, lower income countries of Asia-Pacific, as well as the former Soviet Union nations such as Armenia, Belarus and Russia. The figure also shows the increasingly steep rise of overall penetration rate which can be divided roughly into short periods of 4-5 years between which there is a sudden increase in the penetration rate (possibly following the Kitchin's inventory cycle).

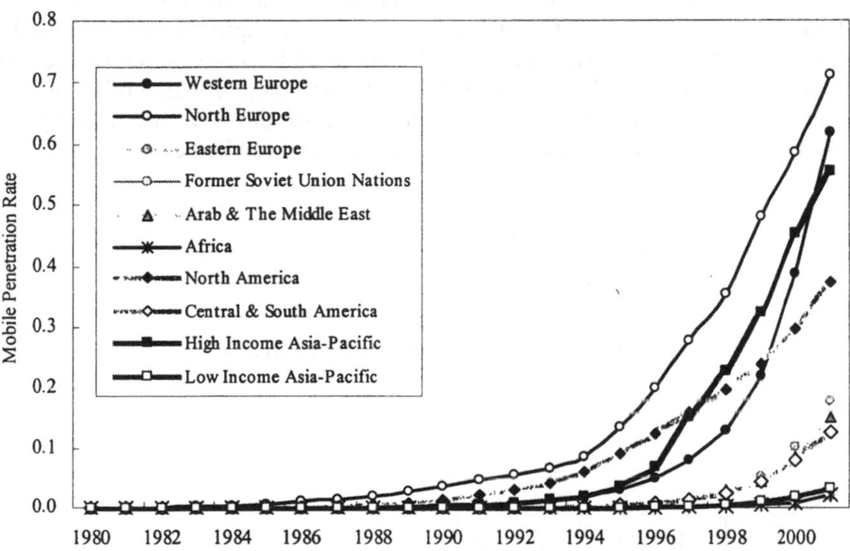

Figure 12.7 Accumulative Growth of Mobile Penetration Rate between 1980 and 2000 (Data source: International Telecommunication Union, 2002)

12.4.2 Prediction of Growth Trend in the Mobile Market

In order to predict the growth of the mobile market, we apply a logistic regression model to the annual data of mobile penetration rate between 1980 and 2000 (i.e., we use the data illustrated in Figure 12.7 as our input value to make our prediction). A basic logistic model can be written as follows

$$P_t = \frac{L}{1 + b\exp(-at)} \tag{12.1}$$

where L is the upper boundary, and a and b ($a<0$, $0<b<1$) are variable parameters. This can be rewritten as

$$\frac{1}{P_t} = \frac{1}{L} + \frac{b}{L}\exp(-at).$$

Thus

$$Y = A + B\exp(-at) \text{ where } Y = \frac{1}{P_t}, A = \frac{1}{L}, B = \frac{b}{L}. \tag{12.2}$$

Since Equation (12.1) is non-linear in its parameter a, and since we know there is sufficient number of elements n, we apply a successive approximation method to estimate a (note that, because of the high degree of freedom in equation (12.4), there are cases where δ will never converge). Suppose that P consists of finite set of elements n. Using Newton-Cotes formula (to be precise the m=2 case of Simpson's rule), the initial value for a, a_0 ($a = a_0 + \delta$), can be approximated as

$$a_0 = \frac{1}{m}\log\left(\frac{S_1 - S_2}{S_2 - S_3}\right) \quad \text{where}$$

$$S_1 = \sum_{i=1}^{m}\log(y_i),\ S_2 = \sum_{i=m+1}^{2m}\log(y_i),\ S_3 = \sum_{i=2m+1}^{n}\log(y_i) \text{ and } m = \frac{n}{3}.$$

Hence

$$P_t = \frac{L}{1 + b\left[\exp(-a_0 t) - \delta t\exp(-a_0 t)\right]} \quad \text{or}$$

$$Y = A + B\left[\exp(-a_0 t) - \delta t\exp(-a_0 t)\right]. \tag{12.3}$$

Let $X_1 = \exp(-a_0 t), X_2 = t\exp(-a_0 t), C = B\delta$. Then

$$Y = A + BX_1 - CX_2. \tag{12.4}$$

Since Equation (12.4) is a multiple regression with two independent variables X_1, X_2, we can obtain A, B, and C. We continue iterating on Equation (12.4) until δ converges to a certain threshold.

Due to the dynamic nature and the sheer magnitude of data, the source data itself would, to some extent, inevitably consist of estimated values, and as this is combined with high dimension in the parameter variables, the predictions made are only illustrative. We have nonetheless estimated the approximate trend of how the mobile devices are likely to penetrate markets in different parts of the world over the next ten years.

We fit the logistic curve to the initial values in Figure 12.7. However, the model is based on the assumption that there is an upper threshold of 100%, which effectively means that once the penetration rate reaches the ceiling, then the number of handheld devices would balance with overall increases through the population growth. This would still account for an estimated growth rate of 120-170% of the actual users of handheld devices (e.g., excluding children under 10), although the fluctuations would clearly depend on the demography of each region.

Figure 12.8 shows one particular case of such a prediction where the ceiling threshold is 100% and the growth rate also reflects the GDP increase within each region. The results indicate that the mobile market will come to near saturation in the developed countries within the next ten years or in even lesser time, and that even the least developed countries will see penetration rates of around 30%. Again, it should be noted that this prediction is based on the assumption that the mobile market will continue to grow under the current penetration rate. That said, this figure indicates the rise of mobile and wireless society that will provide an ideal ground for the market penetration of LBSs which will doubtlessly become the most popular utility for mobile users, provided that LBS-oriented technologies mature and deploy on time. But will these geographical information-based methods truly prevail amongst other services? In the next section, we will speculate on this point through the analogy with the diffusion model of GISs.

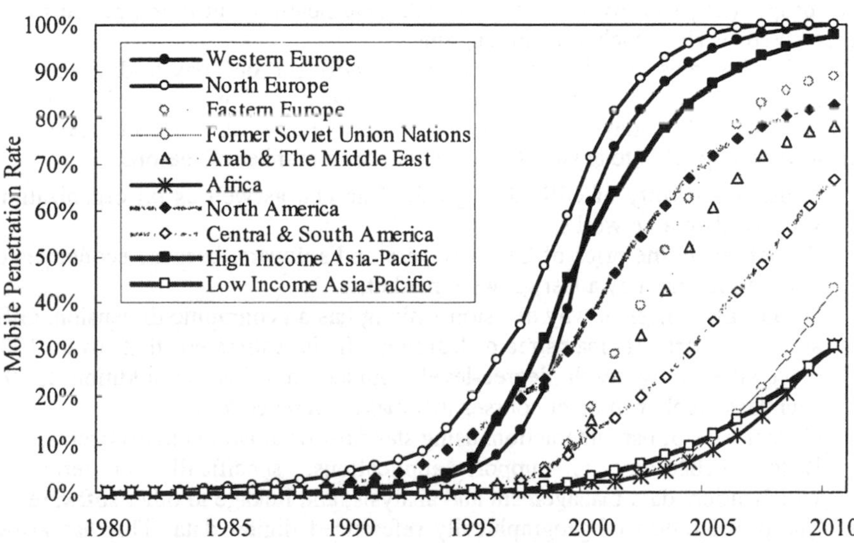

Figure 12.8 Prediction of the Mobile Penetration Rate over the Next Decade Using a Logistic Model

12.5 IMPACTS OF LBSS ON GEOGRAPHICAL LOCATIONS

The growth of the GIS industry can be illustrated by considering the recent history of ESRI – perhaps the largest GIS company worldwide. Longley *et al.* (2001) suggest that this company has many more than three-quarters of a million licensees for its different products, and that this translates to about half a million active users at any one time. ESRI income has risen at a compound rate of between 15% and 20% over the last decade, with the financial year 2000 revenues being $400 million. If the revenues of the partly owned ESRI "franchises" in other countries are added in, the total revenue comfortably exceeds half a billion US dollars. Another indicator of growth has been the number of attendees of their annual conference from 23 attendees in 1981 to over 10,000 in 2000 – in many ways an allegory for GISs as a whole. The company also estimates that their activities leverage at least 15 times the revenues they get in terms of staff, hardware, training and other expenditures on the part of the users. All of this suggests that ESRI alone generates expenditures of between $5 and $6 billion annually.

If we add to this the known market share of other vendors and their own leverage, the total expenditure on GISs and related activities worldwide cannot be much less than $15 to 20 billion. Depending on how wide is the definition of GISs, it could be much higher still (e.g., if the finances of both commercial and military satellites are included). Given the number of ESRI licenses, those of other commercial firms, use of 'free software' provided by governments and other parties plus the illegal, unlicensed copying and use of commercial software which still occurs, the likely number of active GIS users must be well over one million people world-wide – perhaps 1.5 to 2 million. At least double that number of individuals will have had some direct experience of GISs and perhaps an order of magnitude more people (i.e., well over 10 million) will have heard about it and perhaps used passive Web services such as local mapping.

This illustrates the degree to which support for our day-to-day activities requires a vast array of organizations to answer the fundamental locational question, *where?* Today, more and more individuals and organizations find themselves using GISs to answer this question, for a number of reasons:

- Wider availability of GISs through the Internet, as well as on organization-wide local area networks,
- Reductions in the price of GIS hardware and software, because economies of scale are realized by a fast-growing market,
- Greater awareness of why decision-making has a geographic dimension. GISs are now part of mainstream learning. It is estimated that over 1000 universities now teach degree-level courses in GISs, in addition to the countless number of other courses that make reference to it,
- Greater ease of user interaction, using standard windowing environments,
- Better technology to support applications, specifically in terms of visualization, data management and analysis, and linkage to other software,
- The proliferation of geographically referenced digital data. This has arisen through the routine use of the GPS technology at high resolutions, the proliferation of value-added resellers (VARs) who update, edit, and otherwise increase the value of existing data, and the accumulation and maintenance of

data by mapping organizations, census agencies, and environmental organizations,

- Availability of packaged applications, which are available commercially off-the-shelf (COTS) or "ready to run out of the box," and
- Development of research procedures and operational workflows that are built around GISs in fields as diverse as local-government land-parcel development, business location, and climate modeling.

GIS applications may be classified as traditional, developing, and new. Traditional GIS application fields include military, government, education, and utilities. The mid-1990s saw the development of contemporary business uses, such as banking and financial services, transportation logistics, real estate, and market analysis. The early years of the 21st century are seeing new forward-looking application areas in small office/home office (SOHO) and personal or consumer applications.

A further way to examine trends in GIS applications is to examine the diffusion of GIS use. The classic model of diffusion originally developed by Everett Rogers (1993) is directly applicable. Rogers' model divides the adopters of an innovation into five categories:

- Venturesome Innovators – willing to accept risks and sometimes regarded as oddballs.
- Respectable Early Adopters – regarded as opinion formers or "role models."
- Deliberate Early Majority – willing to consider adoption only after peers have adopted.
- Skeptical Late Majority – overwhelming pressure from peers is needed before adoption occurs.
- Traditional Laggards – people oriented to the past.

GISs seem to be in the transition between the Early Majority and the Late Majority stages. The Innovators who dominated the field in the 1970s were typically based in universities and research organizations. The Early Adopters were the users of the 1980s, many of whom were in government and military establishments. The Early Majority, typically in private businesses, came to the fore in the mid-1990s.

A wide range of motivations underpins the use of GISs, although it is possible to identify a number of common themes (Longley *et al.*, 2001). Applications dealing with day-to-day issues typically focus on very practical concerns such as cost-effectiveness, service provision, system performance, competitive advantage, and database creation, access, and use. Other, more strategic applications are more concerned with creating and evaluating scenarios under a range of circumstances.

12.6 CONCLUSIONS

In this chapter, we have studied the impact and penetration of LBSs through an analogy between their diffusion and the Kondratieff cycles. We reviewed the growth and penetration of mobile industry and then applied logistic-type forecasting which is based on classic diffusion-innovation trends to predict the impact of technologies. Assuming that the LBS-related technologies will grow in a similar fashion following the same innovation wave as the handheld devices and other mobile hardware, we drew on the growth of the mobile industry to predict the impact of LBSs. We also observed some of the growing elements in the field of GISs which is a key element supporting the evolution of LBSs.

The success of these technologies naturally depends on the intersection of different growth profiles associated with hardware, networks, software, and data. In fact, the market for LBSs and other related services is wide and diverse, involving many kinds of fixed devices and networks as well as wireless applications. It has been popularly envisaged that such services will rapidly take-off in the next 5 years as networks, handheld devices, GIS technology in the form of software and data begin to converge. Our preliminary prediction supports this view in that LBS-enabled devices will diffuse massively within the next 10 years following the penetration of WAP-enabled mobile devices and the deployment of online GISs and other geographical information-oriented services.

The study provides some preliminary predictions and speculations, and would clearly benefit from further reworking and extension. For instance, while the difference in economic conditions in different parts of the world will affect the penetration of LBSs and the growth of their market, there are also other elements such as culture and language as well as demography which are essential in determining such growth. We also need to track different kinds of growth and development in various LBS software. The differentiation of the market as illustrated in many of the devices and services summarized in other chapters of this book, is an essential factor in developing a clear view of this future. Thus, our preliminary predictions will hopefully provide a context for a wider study.

We have not spent much time in this chapter on an analysis of the social and organizational impacts of LBSs, and thus it is fitting that we conclude with some speculation on these issues. The social impact of the Internet, in particular, and the emergence of a network society, in general, is something which has far reaching social implications with respect to the way we work and socialize. Already it is clear that far from spreading spatial activities out, networks grow and change by concentrating resources at significant hubs and thus the image of our future cities is not necessarily one where everything will be spread out. In fact, recent evidence shows that location activity is concentrating even further within large cities and that information, in particular, seems to gravitate to the largest cities. We have known for a long time that cities are places which transmit information. The dissemination of information devices to the person and hence their usage in mobile situations is something that is extremely difficult to gauge in terms of whether this will concentrate activity further or spread it out. Like so much of the IT revolution, the impacts of such change are likely to occur in both directions.

In fact, we consider that mobile devices may well concentrate things even further because their successful use seems to be where there are dense agglomerations of people and where walking rather than driving is still significant. This suggests that the largest and densest places will be those with more than average use of mobile devices. Yet to counter this, evidence from less dense peripheral western economies such as Scandinavia suggests that such devices are attractive to low density societies where fixed communications infrastructure is more costly. How LBSs will play out in terms of their spatial impacts is still quite uncertain. The key issue is whether or not such services will be absorbed in a routine context without affecting very much major locational decision-making or otherwise. So far, the evidence shows, like most other features of the IT revolution, that such services will have a pervasive but low-level impact reinforcing wider trends in location rather than providing a new basis for dramatically different ways of living in cities.

REFERENCES

Bastiaansen, C., 2001, LBS industry needs MAGIC to grow. *GI News*, April/May Issue, pp. 6.

Beinat, E., 2001, Location-based services - market and business drivers, *GeoInformatics*, April 2001 Issue, **4**(3), pp.6-9.

Bishr, Y., 2000, Positioning position technology in the new mobile marketplace. Sun Microsystems. Online. Available HTTP: <http://www.jlocationservices.com/company/Bishr/PositioningPositionTechnology.html> (accessed 5 May 2001).

Castells, M., 1989, *The Informational City*, (Oxford: Blackwell).

Castells, M., 2000, *The Rise of The Network Society (2nd edition)*, (Oxford: Blackwell).

Coleman, M., 2001, Role reversal: can GIS breathe life into LBS. *GI News*, September Issue, pp. 47-48.

Ericsson, 2001, Ericsson mobile world. Online. Available HTTP: <http://www.ericsson.com/mobilityworld/> (accessed 5 May 2001).

Federal Communications Commission, 2001, FCC wireless 911 requirements. Online. Available HTTP: <http:// www.fcc.gov/e911/> (accessed 5 May 2001).

Gartner, Inc. (2002) *Mobile Location Services for Governments*, (Stamford, CT: Gartner Inc.).

Hall, P. and Preston, P. 1988, *The Carrier Wave: New Information Technology and The Geography of Innovation 1846-2003*, (London: Unwin Hyman).

Hunter, P. 2001, Location, location, location. *Computer Weekly*, May Issue, p.64.

International Telecommunication Union (ICT), 2002, Basic Global Indicators. Online. Available HTTP: <http://www.itu.int/ITU-D/ict/statistics/> (accessed 7 October 2002).

Kondratieff, N. D., 1935, The long waves in economic life. *The Review of Economic Statistics,* **17**, pp. 105-115.

Longley, P., Goodchild, M. F., Maguire, D.J. and Rhind, D.W. (Editors), 2001, *Geographical Information Systems*, 2nd edition, (Chichester: John Wiley & Sons).

Maguire, D., 2001, Mobile geographic services come of age. *GeoInformatics*, March Issue, pp. 6-9.

Mountain, D. and Raper, J., 2000, Designing geolocation services for next generation mobile phone systems. *Proceedings for GIS 2000*, (London: Earls Court) (Workshop 2.4, on CD-ROM).

Mountain, D. and Raper, J., 2001, Spatio-temporal representations of individual human movement for personalising location-based services. *Proceedings for GISRUK 2001*, (Pontypridd, UK: University of Glamorgan), pp. 579-582.

National Research Council, 1997, *The Future of Spatial Data and Society*, (Washington DC: National Research Council).

Rogers E. H. 1993, The diffusion of innovations model. In *Diffusion and Use of Geographic Information Technology*, edited by Masser I. and Onsurd H. J. (Dordrecht: Kluwer), pp. 9-24.

Schumpeter, J.A., 1939, *Business Cycles*, (New York: McGraw-Hill).

van Es, P. 2001, Where is the LBS industry heading to? *GI News*, April/May Issue, pp. 3-5.

Wilson, J., 2001, The next frontier. *GeoEurope*, July Issue, pp. 40-45.

Winter, S., Pontikakis, E. and Raubal, M., 2001, LBS for real-time navigation – a scenario. *GeoInformatics*, May 2001 Issue, 4(4), pp.6-9.

Wireless World Forum (WWF), 2002, Location-based services – long term optimism prevails. Online. Available HTTP: <http://www.w2forum.com/news/w2fnews10209.html> (accessed 1 October 2002).

About the Authors

Michael Batty is professor of spatial analysis and planning, and director of the Centre for Advanced Spatial Analysis (CASA) at University College London (UCL). He was previously Director of the NCGIA at SUNY-Buffalo and professor of city planning at Cardiff University in the 1980s. He has published several books on urban simulation such as *Urban Modelling* (Cambridge University Press, 1976), and *Fractal Cities* (Academic Press, 1994), and is editor of the journal *Environment and Planning B*.

Michael Chapman is a professor in the Department of Civil Engineering at Ryerson University, Toronto, Canada. He received his Ph.D. degree in Photogrammetry from Laval University, Quebec, Canada. He was previously a professor in the Department of Geomatics Engineering at the University of Calgary for 18 years. His research interests are in the areas of photogrammetry, digital mapping, digital terrain modeling, image metrology and biometrology. He has supervised and graduated over twenty five graduate students at the M.Sc. and Ph.D. levels. He is a reviewer for scientific journals including *Geomatica, Photogrammetric Engineering and Remote Sensing*, and the *Journal of Geospatial Engineering*. He is the author or co-author of over 130 technical articles. Dr. Chapman is the photogrammetry editor of *Geomatica*. He has also served on several conference organizing and scientific committees. He is a professional engineer and a registered member of the Association of Ontario Land Surveyors.

Panos K. Chrysanthis is an associate professor of computer science at the University of Pittsburgh and an Adjunct Professor at Carnegie Mellon University. He received his B.S. from the University of Athens, Greece, in 1982 and his M.S. and Ph.D. from the University of Massachusetts at Amherst, in 1986 and 1991 respectively. His current research focus is on mobile and pervasive data management. In 1995, he was a recipient of the National Science Foundation CAREER Award for his investigation on the management of data for mobile and wireless computing. Besides journal and conference articles, his publications include a book and book chapters on advances in transaction processing and on consistency in distributed databases and multidatabases. He is currently an editor of the *VLDB Journal*, and was program chair of several workshops and conferences related to mobile computing. He is the ICDE 2004 vice chair for the area of distributed, parallel and mobile databases and the general chair of MobiDE 2003.

Chris Drane is vice president, R&D Australia, for Cambridge Positioning Systems (CPS). His specific area of research is positioning systems. He is the author of two books, two patents, and numerous papers dealing with positioning systems. Prior to joining CPS, he worked in variety of positions. He has been a full professor at the University of Technology, Sydney, has worked with an aerospace electronics company in the United States, a computing consultancy company, as technical manager in a high technology startup, and as a principal research fellow at the University of Sydney. He has spent sabbatical leaves at Cambridge University in the United Kingdom and at Intelligent Transportation Systems in Washington DC.

Steven Feiner is a professor of computer science at Columbia University, New York, where he directs the Computer Graphics and User Interfaces Laboratory. He received a Ph.D. in computer science from Brown University in 1987. Professor Feiner's research interests include virtual environments and augmented reality, knowledge-based design of multimedia, wearable computing, information visualization, and hypermedia. He is coauthor of *Computer Graphics: Principles and Practice* (Addison-Wesley) is an associate editor of ACM Transactions on Graphics, and is a member of the steering committees for the IEEE Symposium on Information Visualization, the IEEE and ACM International Symposium on Mixed and Augmented Reality, and the IEEE Computer Society Technical Committee on Wearable Information Systems. Professor Feiner is program co-chair for the 2003 IEEE International Symposium on Wearable Computers, and was symposium co-chair of the 2001 IEEE and ACM International Symposium on Augmented Reality. In 1991 he received an Office of Naval Research Young Investigator Award.

Nobuo Fukuwa is professor of earthquake engineering and architecture at Nagoya University, Japan. He has experience in the seismic design of nuclear power plants, super high-rise buildings and base isolated buildings. He is now involved in disaster mitigation activities and development of emergency response systems for the national and local governments. He is also interested in the development of educational materials about earthquake disaster mitigation.

James H. Garrett, Jr. is a professor of civil and environmental engineering at Carnegie Mellon University in Pittsburgh, Pennsylvanis. His research interests are oriented toward computer-aided civil engineering system development. Specifically, he is investigating applications of MEMS to civil infrastructure condition assessment; mobile hardware/software systems for field applications; and representations and processing strategies to support the usage of engineering codes, standards, and specifications.

Dorota A. Grejner-Brzezinska is an assistant professor at the Department of Civil and Environmental Engineering and Geodetic Science, The Ohio State University. She holds an M.S. degree in surveying and land management from the University of Agriculture and Technology, Olsztyn, Poland, and an M.S. and a Ph.D. in geodetic science from The Ohio State University. Her research interests include Global Positioning System (GPS) algorithms and application, integration of GPS and inertial navigation system (INS), positioning and tracking techniques, multi-sensor mobile mapping systems, estimation techniques, GPS orbit modeling and GPS-based positioning in space.

Amin Hammad Amin Hammad obtained his MSc and PhD degrees in computer-aided engineering at Nagoya University, Japan, in 1990 and 1993, respectively. His research interests are in the area of computer-aided engineering with special focus on applications in the fields of infrastructure and urban management systems. Prior to his present position as associate professor at Concordia institute of information systems engineering, he held faculty and visiting positions at Nagoya University, Carnegie Mellon University, the University of Pittsburgh and the University of Tokyo.

Tobias Höllerer is assistant professor of computer science at the University of California, Santa Barbara, where he leads a research group on imaging, interaction, and innovative interfaces. With undergraduate and graduate degrees in computer science from the Technical University of Berlin in Germany, Höllerer did his U.S. graduate work at Columbia University. His dissertation for a Ph.D. is on user interfaces for mobile augmented reality systems. His work has been published in international journals and presented at more than 10 international conferences and workshops. He co-organized several workshops on augmented and virtual reality, and mobile systems. Professor Höllerer serves on the program committees for IEEE Virtual Reality and the IEEE and ACM International Symposium on Mixed and Augmented Reality. His main research interests lie in augmented reality, 3D interaction, visualization, mobile and wearable computing, and adaptive user interfaces.

Hassan Karimi is assistant professor and director of geoinformatics in the Department of Information Science and Telecommunications at the University of Pittsburgh. He holds a BSc in computer science from the University of New Brunswick and an MSc in computer science and a PhD in geomatics engineering from the University of Calgary. His research interests include mobile computing, mobile GIS, mobile mapping systems, computational geometry, Grid computing, and spatial databases.

Prashant Krishnamurthy is an assistant professor in the Department of Information Science and Telecommunications at the University of Pittsburgh. He has been leading the development of the wireless information systems track for the M.S. in telecommunications curriculum in the telecommunications program there. His research interests are in the areas of wireless data networks, wireless network security and radio propagation modeling.

Jonathan Li is assistant professor in the Department of Civil Engineering at Ryerson University, Toronto, Canada. He received his Ph.D. degree in geomatics engineering from the University of Cape Town, South Africa. He has been working as a research scientist, project manager, and professor in China, Germany, South Africa, and Canada since 1982. Dr. Li heads the CFI-funded Virtual Environment Lab at Ryerson. His teaching and research interests include remote sensing, image analysis, digital mapping, and GIS. He is co-supervising one Ph.D. and six masters students. He is a reviewer for scientific journals including

Geomatica, *Photogrammetric Engineering and Remote Sensing*, *Canadian Journal of Remote Sensing*, and *Environmental Modelling & Software*. He is the author or co-author of over 60 technical articles. Dr. Li is Canada's national correspondent for ISPRS Commission I, member of CIG Photogrammetry & Remote Sensing Committee, and OARS council member. He has also served as conference organizing and scientific committee members. He is a licensed Ontario land surveyor/land information professional and professional Engineer.

Chao Li is a researcher in the Centre for Advanced Spatial Analysis at University College London (UCL) where she is investigating Location-Based Services, particularly issues of urban way-finding using mobile information communication technologies, towards a Ph.D. Previously she worked on testing and compilation of Chinese national standards for electronic measuring instruments and on the implementation of digital Chinese character sets in Beijing, China. She has also worked on the design of graphic user interfaces for GIS cartographic modeling in Hong Kong and as a consultant data analyst in London.

Paul Longley (B.Sc., Ph.D., D.Sc.) is professor of geographic information science (UCL), and Deputy Director of CASA, UCL. He has previously worked in the universities of Bristol, Cardiff, Reading and Karlsruhe. His research is focused around the principles and techniques of geographic information science (GISc), with research applications in geodemographics, urban remote sensing and the analysis of urban morphology. He is the editor of *Computers, Environment and Urban Systems* and has been involved in eight books, including co-authorship of *Geographic Information Systems and Science* (with Goodchild, Maguire and Rhind) and co-editorship (with the same team) of *Geographical Information Systems: Principles, Techniques, Management and Applications* (the second edition of the "Big Book" of GIS).

Xavier Lopez is director of Oracle's location-based services group. Dr. Lopez is responsible for spearheading efforts that incorporate spatial technology across Oracle's database, application server, CRM, and ERP technologies. Prior to joining Oracle, Dr. Lopez was engaged in postdoctoral research on spatial information at U.C. Berkeley. He holds degrees in advanced engineering and planning from University of Maine, Massachusetts Institute of Technology, and the University of California, Davis.

David Maguire is director of products, solutions and international, ESRI and Visiting Professor at the Centre for Advanced Spatial Analysis at University College, London. He is the author of seven books including most recently *Geographical Information Systems and Science* co-authored with Paul Longley, Mike Goodchild and David Rhind (Wiley, New York, 2001). His research interests include very large spatial databases, geographic modeling and analysis and GIS software architectures.

Steve Mann is considered by many to be the inventor of WearComp (wearable computer) and WearCam (eyetap camera and reality mediator). He is currently a faculty member at University of Toronto, Department of Electrical and Computer Engineering. Dr. Mann has been working on his WearComp invention for more than 20 years, dating back to his high school days in the 1970s. He brought his inventions and ideas to the Massachusetts Institute of Technology in 1991, and is considered to have brought the seed that later become the MIT Wearable Computing Project. He also built the world's first covert fully functional WearComp with display and camera concealed in ordinary eyeglasses in 1995, for the creation of his award-winning documentary ShootingBack. He received his Ph.D. degree from MIT in 1997 for work including the introduction of Humanistic Intelligence. He is also inventor of the Chirplet Transform, a new mathematical framework for signal processing, and of Comparametric Equations, a new mathematical framework for computer-mediated reality. Professor Mann has written 139 research publications (39 journal articles, 37 conference articles, 2 books, 10 book chapters, and 51 patents), and has been the keynote speaker at 24 scientific and industry symposia and conferences and has also been an invited speaker at 52 university Distinguished Lecture Series and colloquia.

Kaveh Pahlavan, is a professor of ECE, a professor of CS, and director of the Center for Wireless Information Network Studies, Worcester Polytechnic Institute, Worcester, Massachusetts. He is also a visiting professor of the Telecommunication Laboratory and CWC, University of Oulu, Finland. He has contributed to numerous seminal technical publications in the area of wireless information networks. He is the editor-in-chief of the *International Journal on Wireless Information Networks*. He was the founder, the program chairman and organizer of the IEEE Workshop on Wireless LANs in 1991, 1996, and 2001 and the organizer and technical program chairman of the IEEE International Symposium on Personal, Indoor, and Mobile Radio Communications, Boston, Massachusetts, 1992 and 1998. He was elected as a fellow of the IEEE in 1996 and became a fellow of Nokia in 1999.

Chris Rizos is a graduate of the School of Surveying, The University of New South Wales (UNSW), Sydney, Australia, obtaining a bachelor of surveying in 1975, and a doctor of philosophy in 1980 in satellite geodesy. Chris has been researching the technology and high precision applications of GPS since 1985. He has published over 100 papers, as well as having authored and co-authored several books relating to GPS and positioning technologies. Chris is co-author of the book *Positioning Systems in Intelligent Transportation Systems* with Chris Drane, Artech House, 1998. Chris is currently leader of the Satellite Navigation and Positioning group at UNSW, the premier academic GPS and wireless location technology R&D lab in Australia. He is a fellow of the Australian Institute of Navigation, and a fellow of the International Association of Geodesy.

Narushige Shiode is a research fellow at the Centre for the Advanced Spatial Analysis, University College, London. He is in the final stages of his Ph.D. studies on comparative analyses of urban and cyber spaces. Prior to the current funding from the Joint Information Systems Committee, he has held scholarships from Japanese Federation of Economic Organization and the Graduate School, University College, London. He has degrees from the University of Tokyo (M. Eng. and B. Eng. in urban engineering). His research interests include locational optimization, spatial temporal analysis, 3D visualization, and urban and cyberspace geography.

Jun Tobita is associate professor of earthquake engineering and Architecture at Nagoya University, Japan. His research interests include earthquake response and damage of buildings and urban areas, focusing especially on observation methods and techniques. He has experience in the development of various kinds of observation systems and information networks for disaster prevention and mitigation.

Vladimir Zadorozhny is an assistant professor in the Department of Information Science and Telecommunications, School of Information Sciences, University of Pittsburgh. He received his Ph.D. in 1993 from the Institute for Problems of Informatics, Russian Academy of Sciences in Moscow. Before coming to the US he was a principal research fellow at the Institute of System Programming, Russian Academy of Sciences. Since May 1998 he worked as a research associate and then research scientist at the University of Maryland Institute for Advanced Computer Studies in College Park. He joined the University of Pittsburgh in September 2001. Vladimir's research interests include scalable architectures for wide-area environments with heterogeneous information servers, Web-based information systems, query optimization in distributed databases, semantic interoperability in heterogeneous network environments, distributed object systems and object metamodels. His research has been supported by grants from the National Science Foundation and the Defense Advanced Research Project Agency. He is a member of IEEE and ACM SIGMOD.

Index

P

Q

R

S

T

U

V

Milton Keynes UK
Ingram Content Group UK Ltd.
UKHW021822071024
449327UK00021B/1390

9 780367 578404